The Chemistry of Biosurfaces

(IN TWO VOLUMES)

VOLUME I

THE CHEMISTRY OF BIOSURFACES

(IN TWO VOLUMES)

Volume I

Edited by **MICHAEL L. HAIR**

Xerox Corporation
Rochester, New York

MARCEL DEKKER, INC., New York 1971

Preface

When it was proposed that I edit a book on the chemistry of biosurfaces, the first question that had to be answered was "what do we mean by a biosurface?" In 1968 it was already evident that there was a considerable interest on the part of biochemists in recent developments in surface chemistry and at the same time a realization on the part of industrial chemists that the biological materials area might provide a unique opportunity for a new and potentially profitable field of endeavor. Metallurgists and ceramists were considering bone replacements, engineers were talking membrane processes, and chemists thought of controlling and stabilizing biologically active material. The detergent industry had already added enzymes to its products and the general public had been warned that a biological revolution, comparable to the electronic revolution of the sixties, was at hand. There was one common denominator to these systems: in nearly all cases there would be an interface between an implant and biological material; it was all a problem of surface chemistry.

Surface chemists have struggled for many years with the problems of identifying even the simplest of surfaces and of understanding the processes that occur when one material adsorbs upon another. The frustration of chemists in positively identifying the adsorbed intermediates on catalysts testifies to the difficulty of a problem that has yielded to modern physical techniques only in the most recent decade. Yet the new technology was to combine the unknown complexities of biological systems with the known but unsolved complexities of artificial surfaces. It was clear that at least three disciplines must be combined to provide the final product. The biologist must understand the chemist, the chemist must understand the biologist, and both must work with the engineer and the physician, who would have ultimate responsibility for the final product.

This book represents an attempt to bring these disciplines together, and for the purposes of this book, the biosurface is defined as being any surface, organic or inorganic, which interacts *in vivo* or *in vitro* with an environment of a biological nature. The authors were invited to consider five groups of readers.

1. Newcomers to the subject. People from all branches of science who appreciate that this area is important, but lack specific information.
2. Surface chemists and physicists. This group of people (like myself) might appreciate the chemistry involved, but not the real scope or the biological complexities of the subject.
3. Biologists and related scientists who appreciate the biological complexities, but are perhaps unaware of recent developments in molecular surface chemistry.
4. The medical profession—eventually surgeons will make practical use of the research in this area. A complete understanding of the interactions between man-made materials and natural compounds is of the utmost importance in the design of implant materials, etc.
5. Industrial scientists. The implications of much of this work (for instance, with insolubilized biochemicals) suggest eventual commercial exploitation. All the authors were invited to bear in mind the practical consequences of the work that is now going on and to relate to people who would be interested in following this up from its infancy.

In attempting to discuss these biological interfaces we have taken a very simple tack. In Volume 1 we discuss a series of adsorption phenomena that are analogous to biological situations; Volume 2 is more concerned with actual biological materials.

Before considering any adsorption process it was felt necessary, in the case of biological materials, to introduce the reader to the concepts of hydrophobic bonding and the structure of water, which control to such a large extent the interactions that are actually observed. Lipids are then discussed in some detail as perhaps the most important biological surfactants. We then consider the simplest adsorption process analogous to that occurring in the biological environment—the adsorption of large molecules (simple polymers) on nonbiological surfaces. During the past few years there have been considerable advances in our understanding of the adsorption of organic polymers onto inorganic surfaces. Unfortunately, much of the data generated in this area has been obtained in organic solvents. This is not strictly analogous to the aqueous biological system, but many of the concepts and approaches that have been developed here are of interest, particularly the role of the solvent in controlling the configuration of the adsorbed layer and the effect of small numbers of "active sites" on the process.

Many biological molecules can be considered as complex surfactants. Simple surfactant molecules have been widely utilized by surface chemists in studies of monolayer and micelle formation, and the adsorption of surfactants onto inorganic surfaces is now fairly well understood. The adsorption of these molecules onto nonbiological surfaces is reviewed as a simple analogy of the biological system. A complementary chapter reviews recent advances on studies of the adsorption of lipids and proteins on the same nonbiological surfaces. The adsorption and monolayer studies on lipids and proteins are taken one stage closer to biological reality in the discussion of the preparation and properties of the bilayer membranes. The final chapter in Volume 1 then discusses the thermodynamics of an anesthetic problem—the reversible inhibition of cellular function (narcosis) by some exceedingly simple chemicals, such as nitrogen—and reemphasizes the importance of hydrophobic bonding in the biological environment.

In the second volume the reader is introduced to more biologically oriented problems—the cell surface, the growth of mammalian cells on surfaces, surface effects in hemolysis and hemostasis, stabilized enzymes, antigens and antibodies, surfaces in the extraction of enzymes, and the dental adhesive problem.

Rochester, New York   MICHAEL L. HAIR

Contributors to Volume I

M. B. Abramson, The Saul R. Korey Department of Neurology, Albert Einstein College of Medicine, Bronx, New York

J. L. Brash, Stanford Research Institute, Menlo Park, California

W. B. Dandliker, Department of Biochemistry, Scripps Clinic and Research Foundation, La Jolla, California

V. A. de Saussure, Department of Biochemistry, Scripps Clinic and Research Foundation, La Jolla, California

B. J. Fontana, Chevron Research Company, Richmond, California

D. W. Fuerstenau, Department of Materials Sciences and Engineering, University of California, Berkeley, California

L. S. Hersh, Research and Development Laboratories, Corning Glass Works, Corning, New York

D. J. Lyman, University of Utah, Salt Lake City, Utah

H. T. Tien, Department of Biophysics, Michigan State University, East Lansing, Michigan

Contents

Contents of Volume 2

I

Stabilization of Macromolecules by Hydrophobic Bonding: Role of Water Structure and of Chaotropic Ions*

W. B. DANDLIKER and V. A. de SAUSSURE

Department of Biochemistry
Scripps Clinic and Research Foundation
La Jolla, California

* This work was supported by The John A. Hartford Foundation, Inc., The National Science Foundation (GB 6887), The National Institute of Arthritis and Metabolic Diseases (AM 7508) of The National Institutes of Health, and The American Cancer Society, California Division, Special Grant No. 480.

1

I. Introduction

A. DEFINITION AND NATURE OF "HYDROPHOBIC BONDING"

Hydrocarbons are nearly insoluble in water. The realization that this fact has an important implication for the stability of macromolecules is a milestone in understanding the integrity of biological structures. For this reason hydrophobic interactions, also called hydrophobic bonding, warrant an extended discussion.

From a thermodynamic viewpoint the outstanding feature of the process of dissolving a liquid hydrocarbon in water is a large negative unitary entropy change, which is intimately related to the unique structure of water itself. (For a solution process the unitary entropy change equals the total entropy change minus the ideal entropy of mixing.) To counteract the unfavorable entropy change, apolar groups tend to withdraw from the aqueous phase and to cluster together. This tendency, applied to apolar groups of macromolecules, was named "hydrophobic bonding" by Kauzmann (1).

The term "hydrophobic bond" has frequently been criticized on two grounds recently restated by Hildebrand (2): first, that on the basis of energetics there is an attraction between hydrocarbons and water, rather than a phobia; second, that interactions have none of the characteristics that distinguish chemical bonds from van der Waals' forces. With regard to the first objection, various investigators (3, 4) have pointed out that although the van der Waals' interactions do indeed favor mixing the energy factor is overshadowed by the large, negative, unfavorable unitary entropy change upon mixing. Lumry and Biltonen (5) have remarked that in proteins the tendency of apolar groups to cluster together in compact arrangements is due primarily to the "lipophobicity" of water, and have suggested the term "lipophobic bond." Nevertheless, the term "hydrophobic" is now firmly entrenched in the literature and its usage will undoubtedly continue. On the other hand, there is little argument with Hildebrand's objection to the word "bond," and consequently we shall use the terms "hydrophobic interaction" or "hydrophobic bonding."

A basic difference between the viewpoints of Hildebrand and co-workers (2, 6) and Nemethy et al. (3, 4) hinges upon whether more, or less, hydrogen bonding is present in water containing nonpolar groups as compared with pure water. The answer is very important and is discussed in Sect. II.C and II.D.

B. CONTRIBUTION OF HYDROPHOBIC BONDING TO THE STABILITY OF MACROMOLECULES

Kauzmann, in 1959 (7), was the first to point out that the magnitude of the negative unitary entropy change observed by Frank and Evans (8) for the solution of hydrocarbons in water is sufficient to account in large part for the stability of proteins and other macromolecules in aqueous solutions. Since then, hydrophobic bonding has been shown to play an important role in such diverse phenomena as denaturation, helix-coil transitions, aggregation, solubility, immune reactions, and in the stability of much larger structures such as cellular membranes.

The temperature coefficient of stability of hydrophobic bonding gives considerable insight into the thermodynamics of association or conformational changes of macromolecules. The available evidence (3) on the negative enthalpy changes observed during the dissolution of liquid hydrocarbons in water indicates that the strength of hydrophobic bonding should increase with rising temperature up to about 60°C, at which point $\Delta H \sim 0$ (7, 9). The observation that unfolded conformations are usually favored well below 60° illustrates that hydrophobic interactions are not the sole factor responsible for macromolecular stability (9). Additional stability arises from hydrogen bonding, electrostatic forces between either charges or dipoles, London dispersion forces, and disulfide bonds favorably placed in the native, folded structure. Under the influence of vigorous thermal motion the chance reduction of an S—S bond and its subsequent reformation in a different location may serve to cement the molecule in a new, more random, denatured form that does not readily revert to the original state. This type of alteration may be responsible for the thermal coagulation of proteins (*10*).

C. ASPECTS OF HYDROPHOBIC BONDING TO BE DISCUSSED

As a basis for any discussion on hydrophobic bonding, the structure of liquid water occupies a prime position. Various proposed models for the structure of water will be discussed and, wherever possible, a critical evaluation will be made in the light of available experimental evidence. Next, an account of the thermodynamic and spectroscopic changes that occur when various molecules or ions are added to water will be presented. These include nonpolar molecules, ions (especially chaotropic ions), and urea or other denaturants. Insofar as is possible, models will be given to account for the observed changes in terms of altering the structure of pure, liquid water. The molecular pictures developed for simple systems will then be applied in the interpretation of experimental data on a variety of macromolecular systems. Finally, a number of practical implications arising from the ability to alter the stability of hydrophobic bonding in macromolecules, especially by adding chaotropic agents, will be discussed.

Admittedly, the molecular pictures are qualitative and not susceptible to experimental verification or disproof, and some of the deductions are of an ad hoc nature. This type of conjecture has been criticized (*11*) but offers an intuitive guide toward further experimental and theoretical progress.

II. The Structure of Water and How It Is Affected by the Insertion of Nonpolar Groups or Ions

A. MODELS FOR WATER STRUCTURE

The existence and stability of hydrophobic interactions are intimately and uniquely associated with the kind of immiscibility found in mixtures of nonpolar substances with water. Perhaps the outstanding feature of the microscopic structure of liquid water is the extensive hydrogen bonding between water molecules, and from the following, it seems most reasonable that the existence of hydrophobic bonding is intimately tied up with the hydrogen bonded character of the solvent.

A wide variety of models have been proposed to explain the unique structure of liquid water, and several discussions of these theories have appeared in recent years (12–14). All of these theories assume large amounts of hydrogen bonding in liquid water, but they differ in other important respects and can be classified into two groups: the continuum models and the mixture models. In the continuum models, water is viewed as a hydrogen bonded network with a continuous distribution of bond energies and geometries, including variation in the angle between O—H and O · · · O over a wide range. In the mixture models, water is assumed to consist of a few discrete molecular species in equilibrium with one another and which possess different numbers of H-bonds per molecule.

Most of these theories were developed to explain some particular properties of liquid water. In order to test the models, researchers have contributed additional experimental evidence that either agrees or conflicts with a given theory. Hence, a great deal of information concerning the properties of liquid water is available, but the interpretation of the data is rarely clear cut. As a result, no single model has yet become acceptable to every worker in the field.*

The best known mixture model describes water in terms of "flickering clusters" and was originated by Frank and Wen (15) and further developed by Némethy and Scheraga (3). As may be seen from Fig. 1, this model assumes that water consists of two or more states in equilibrium: molecules not H-bonded and molecules highly H-bonded into short-lived clusters. In order to carry out statistical mechanical calculations with this model, Némethy and Scheraga (3) assumed that a water molecule can exist at any one of several energy levels determined by the number of H-bonds formed by the molecule. Calculations of the thermodynamic parameters of water were in good agreement with experiment except for heat capacity (14).

The gas hydrate model of Pauling (16) pictures water as a network of H-bonded molecules containing large polyhedral cavities that can accommodate non-H-bonded water molecules (Fig. 2). This is one of the few models sufficiently well defined at the molecular level to allow the radial distribution function to be calculated without numerous assumptions. However, Narten and Levy (14) were not able to find a set of parameters for this model which would simultaneously predict both the experimental radial distribution function and density.

Models based upon mixtures of ice structures have been proposed by

* Most recently Dr. H. S. Frank [Science 169, 635 (1970)] has reviewed the problem of the structure of water, and has concluded that the data are still far too meager to specify any exact structure.

Kamb (*17*), Forslind (*18*), and Samoilov (*19*). The first assumes that water consists of a structural mixture of ice I, II, and III (Fig. 3). Appreciable concentrations of sizable clusters would be expected to give rise to much more light scattering than is actually observed.

Forslind's model is an open network which, over short distances, is a slightly expanded ice I structure. The voids in this structure can be

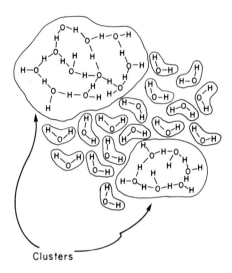

Clusters

Fig. 1. Schematic representation of hydrogen-bonded clusters and unbonded molecules in liquid water according to the flickering-cluster theory of Frank and Wen (*15*). The molecules in the interior of the clusters are quadruply bonded, but this is not shown in the diagram. [Reproduced from G. Némethy and H. A. Scheraga, *J. Chem. Phys.* **36**, 3387 (1962) by permission of the authors and the American Institute of Physics.]

occupied by unbonded water molecules. According to Narten and Levy (*14*), only this model gives agreement with both the small and large angle x-ray scattering, but the success of this model may be partly due to the large number of adjustable parameters.

Much recent evidence conflicts with models involving rigid, structured regions separated by fluid, highly disorganized regions (*20–23*). We have therefore chosen a continuum model as the basis for discussing the effects of apolar groups and ions on water. The actual model used is due to Pople (*24*) and presupposes that liquid water has essentially the same hydrogen bonded structure as that of ice except that the bonds in the

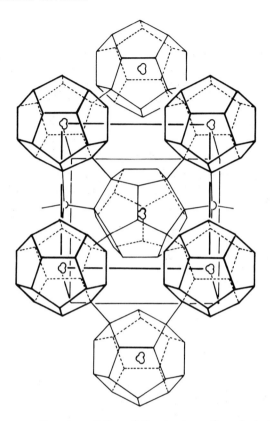

Fig. 2. Diagrammatic representation of the structure of gas hydrates containing a hydrogen-bonded framework of 46 water molecules. Twenty molecules, arranged at the corners of a pentagonal dodecahedron, form a hydrogen-bonded complex about the corners of the unit cube, and another 20 form a similar complex, differently oriented, about the center of the cube. In addition there are 6 hydrogen-bonded water molecules, one of which is shown in the bottom face of the cube. In the proposal for the water–hydrate model additional water molecules, not forming hydrogen bonds, occupy the centers of the dodecahedra and also other positions. [Reproduced from L. Pauling in *Hydrogen Bonding* (D. Hadzi, ed.), 1959, p. 3, by permission of the author and Pergamon Press.]

liquid can bend independently, whereas in the solid they can bend only in ways that maintain the lattice order. According to Pople, the bending of H-bonds in the liquid can occur by rotation and/or translation of water molecules (Fig. 4). This bending leads to a decrease in the H-bond energy, and the variable degree of bending results in the distribution of H-bond energies ranging from zero up to those values for the bonds in

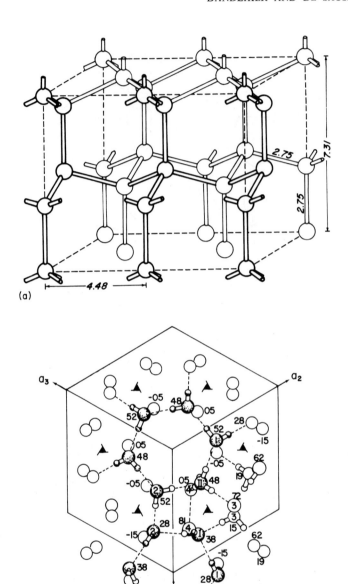

(a)

(b)   0 1 2 3 4 5 Å

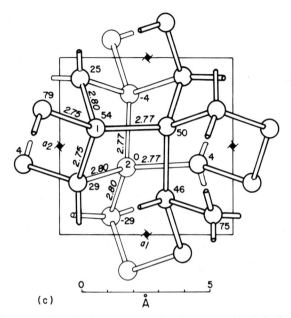

(c)

Fig. 3a. Structure of ice I. Oxygen atom of each water molecule is shown by a ball, and hydrogen bonds by rods. The hydrogen atoms are omitted; they lie essentially on the hydrogen bond lines. The cell outlined is a rectangular (orthohexagonal) cell larger than the true hexagonal unit cell. The hexagonal c axis is labeled 7.31 Å, and one of the hexagonal a axes is labeled 4.48 Å. Bond lengths in Å are indicated. All dimensions are at 100°K. [Reproduced from B. Kamb in *Structural Chemistry and Molecular Biology* (A. Rich and N. Davidson, eds.), 1968, p. 509, by permission of the author and W. H. Freeman and Company.]

Fig. 3b. Structure of ice II viewed in projection along the hexagonal c_H axis. Lower edges of the rhombohedral cell are outlined with solid lines, and threefold screw axes are indicated. Some of the water molecules are here portrayed complete with protons, in the ordered arrangement deduced crystallographically. H-bonds are shown by dashed lines. Nonequivalent water molecules are labeled I and II; other numbering is for reference to neighbors. Bond lengths are: I-1, 2.81 Å; I-3, 2.84 Å; I-4, 2.80 Å; II-1, 2.75 Å. [Reproduced from B. Kamb in *Structural Chemistry and Molecular Biology* (A. Rich and N. Davidson, eds.), 1968, p. 510, by permission of the author and W. H. Freeman and Company.]

Fig. 3c. Structure of ice III shown with the same conventions as in Fig. 3a. The coordinate of each atom along the c axis (height above the plane of projection) is given in hundredths of the axial length. Dimensions are at atmospheric pressure and 100°K. The structure is projected along the c axis (= 6.83 Å); fourfold screw axes (parallel to c) are indicated; nonequivalent water molecules labeled 1 and 2. [Reproduced from B. Kamb in *Structural Chemistry and Molecular Biology* (A. Rich and N. Davidson, eds.), 1968, p. 511, by permission of the author and of W. H. Freeman and Company.]

ice—about 7 kcal/mole (*25*). Those H-bonds with energies approximating zero are those that at that moment are broken or nearly broken and account for the observed fluidity. Upon cooling, the distribution of H-bond energies shifts toward higher values until finally ice is formed. The existence of a quite differently structured, highly ordered "polymeric"

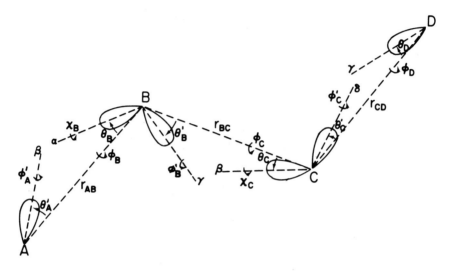

Fig. 4. Four water molecules connected via bent hydrogen bonds, according to Pople. In this figure the bond-forming directions are represented by Greek letters according to the molecule towards which they point. Thus Bα is the bond-forming direction of molecule B which is attached to molecule A. The angle between the bond-forming directions Bα and Bγ of molecule B must have the tetrahedral value $r = \cos^{-1}(-\frac{1}{3})$. This leads to a corresponding tendency for the angle ABC to be tetrahedral. Similar considerations apply to Cβ and Cδ. [Reproduced from J. A. Pople, *Proc. Roy. Soc.*, A**205**, 168 (1951) by permission of the author and The Royal Society.]

form of liquid water induced by adsorption of water molecules onto glass surfaces (*26, 27*) has been the subject of a vigorous and still unsettled controversy during the past two years.

The following discussion on the insertion of an apolar group into water is presented within the framework of the Pople theory. However, it should be noted that the effects would be similar for any somewhat flexible, tetrahedrally bonded structure which could accommodate the apolar group largely into preexisting voids, thus stiffening the entire network.

B. Changes in Properties Observed When Nonpolar Groups Are Inserted into Water

When an apolar group is inserted into water, the following effects have been observed (*3, 6, 7, 28–33*):

(1) a negative unitary entropy change;
(2) a small enthalpy change, usually negative;
(3) an increase in heat capacity;
(4) a decrease in volume;
(5) an NMR shift toward higher fields;
(6) a decrease in the spin lattice relaxation time;
(7) an increase in the intensity of the 3450 cm^{-1} Raman line of water;
(8) a decrease in the frequency of this Raman line.

The thermodynamic changes can be discussed in terms of the process:

Liquid hydrocarbon + water → Solution of hydrocarbon in water

Typical values of $\Delta S_{unitary}$, ΔH, ΔC_p and ΔV are shown in Table 1.

TABLE 1

THERMODYNAMIC CHANGES PER MOLE OF HYDROCARBON FOR THE PROCESS:* LIQUID HYDROCARBON + H_2O → SOLUTION OF HYDROCARBON IN WATER. (STANDARD STATE IS MOLE FRACTION = 1 FOR HYDROCARBON)

Hydrocarbon	$\Delta S_{unitary}$[†] (eu)	$\Delta H°$ (cal/mole)	ΔC_p (cal/mole deg)	ΔV (ml/mole)
Methane	−15.3[a]	−1970[a]	51.5[b]	−1[c]
Ethane	−18.8[a]	−1850[a]	65.1[b]	−4[c]
Propane	−21.2[a]	−1410[a]	69[a]	−21[d]
Butane	−23.0[a]	−1180[a]	80[a]	—
−CH_2−			14–24[e]	—
Benzene	−13.7[a]	570[a]	108[f]	−6.2[d]

* Data at 25° except as noted.
[†] $\Delta S_{unitary} = \Delta S°$ for standard state indicated.
[a] Némethy and Scheraga (*3*).
[b] Claussen and Polglase (*28*).
[c] Miller and Hildebrand (*6*).
[d] Kauzmann (*7*).
[e] Edsall, 20° (*29*).
[f] Bohon and Claussen (*30*).

C. INTERPRETATION OF THE DATA WITHOUT REFERENCE TO ANY
   PARTICULAR MODEL OF WATER STRUCTURE

In this section a summary of the experimental results is given together
with the most direct and simple interpretation of each result.

ΔS. The negative value of the unitary entropy change implies an
overall increase in the degree of order when apolar groups dissolve in
water.

ΔH. The small, negative changes in enthalpy reflect an energetic
component in the interaction between water and apolar groups which
opposes the negative unitary entropy effect and favors mixing. The
relative magnitudes of the heat and entropy effects, expressed as ΔH
and $T\Delta S$, respectively, vary systematically in the series methane, propane
and butane, the entropy contribution becoming relatively more important
as the molecular weight increases.

ΔC_p. The increase in heat capacity accompanying the solution of
hydrocarbons in water can be discussed in terms of the configurational
heat capacity, which has been expressed by Everett (34) as the product of
two terms:

$$C_{p\,\text{config.}} = (\delta H/\delta \xi)_{T,p}(\delta \xi/\delta T)_p \qquad (1)$$

In this equation, ξ is an ordering parameter dependent upon the three-
dimensional structure of the liquid. In this discussion we will use the
convention that an increase in order implies a positive change in ξ. It is
usually a good assumption that an increase in temperature causes a decrease in order and hence decreases ξ. The factor $(\delta \xi/\delta T)_p$ may generally
be expected to be negative. Since $C_{p\,\text{config.}}$ itself must always be positive,
the other factor $(\delta H/\delta \xi)_{T,p}$ must also be negative. The contribution to
the heat capacity from $(\delta H/\delta \xi)_{T,p}$ may be viewed as arising from the
vibrational and rotational motions possible within an existing structure.
The factor $(\delta \xi/\delta T)_p$ represents the contribution arising from the progressive alterations of an existing structure as the temperature is raised.
Hence, the positive ΔC_p observed for the solution of hydrocarbons in
water implies either (a) an increased degree of freedom of the vibrational
and rotational motions within the existing structure, and/or (b) a progressive alteration of an existing structure as the temperature is raised.

ΔV. The decrease in volume noted when hydrocarbons dissolve in
water indicates either that the apolar groups and water molecules have
packed together with a contraction of some structure or that the apolar
groups are accommodated into open spaces or voids preexisting within

the water structure itself. The former seems unlikely in view of the small ΔH's involved.

NMR. When hydrocarbons dissolve in water, or when molecules with large apolar groups are inserted into water, there is an observed NMR shift toward higher applied fields for proton resonance.

While the theoretical intepretation of the field shift observed when H-bonds are formed is complex (*25, 35*), there is an overwhelming amount of experimental evidence indicating that the formation of H-bonds produces a shift toward lower fields. Early work on the interpretation of this field shift has been reviewed by Kavanau (*12*). One of the earliest findings of NMR work was that the proton resonance signal of ethyl alcohol was shifted toward higher fields either when the temperature was raised or when the alcohol was diluted in an inert solvent. The shift was interpreted in terms of decreased hydrogen bonding between the ethyl alcohol molecules. Both Muller and Reiter (*36*) and Hindman (*37*) concluded that the upfield change in the chemical shift in going from ice to water could be accounted for by the increase in H-bond bending (decrease in H-bond strength) when melting occurs. For both gaseous and liquid water the chemical shift is downfield (with respect to a bare proton), but this shift is greater for liquid water than for the gas. If it is accepted that the degree of hydrogen bonding is greater in liquid water than in water vapor, an NMR shift toward higher fields implies a decrease in the number and/or strength of hydrogen bonds.

By analogy, the upfield shifts that occur when hydrocarbons dissolve in water should imply a decrease in the number and/or strength of hydrogen bonds in water. However, this interpretation is not accepted by workers in this field because it is contradicted by evidence (following) on spin lattice relaxation times and Raman spectra.

NMR data have shown that the spin lattice relaxation time of water protons decreases when apolar groups are present (*31*). This decrease is equivalent to an increase in the correlation time and implies an increase in the number and/or strength of H-bonds present. While this conclusion, as pointed out above, is diametrically opposed to that inferred from the field shifts, workers in this area have almost universally given greater weight to the conclusion from the spin lattice relaxation time and generally agree that the presence of the apolar groups results in an increase in the number and/or strength of hydrogen bonds in water; in other words, the hydrogen bonds in the presence of apolar groups have a greater degree of covalent character (*32*). In addition, data on the activation energy for spin lattice relaxation led to the conclusion that "in the presence of a hydrocarbon chain the forces restricting the mobility of the water protons are increased" (*32*).

Raman Measurements. Performed on water and on solutions of tetraalkyl-ammonium salts (*33*) these showed that in the presence of large tetraalkyl groups there is both an increase in intensity and a decrease in frequency of the 3450 cm^{-1} Raman line. These shifts have been widely interpreted as indicating an increase in the number and/or strength of hydrogen bonds in the presence of apolar groups.

In summary, the overall conclusion from the spectroscopic data is that the insertion of an apolar group into water produces an increase in the number and/or strength of the H-bonds.

D. Models That Account for the Changes Observed When Apolar Groups Are Inserted into Water

Of all the thermodynamic changes observed upon inserting an apolar group into water, the negative unitary entropy effect has historically attracted the greatest attention and has thus received the greatest effort to explain it. The earliest explanation originated as the "iceberg" concept of Frank and Evans (*8*), in which water molecules in the vicinity of an apolar group become more ordered. An extension of this concept (*3, 4, 15*) postulates "flickering clusters" in which, by a cooperative effect, existing hydrogen bonds facilitate the formation of many more, resulting in a temporary cluster of water molecules in the vicinity. In a somewhat different picture stemming from Pauling's pentagonal dodecahedron structure for liquid water (*16, 38*), the apolar group is situated inside the dodecahedron as in a clathrate structure (*39*). The presence of the apolar molecule inside the water tends to stabilize it, making it more "crystalline." Powell and Latimer (*40*) have stated their conclusions in more general terms by saying that in this process "there occurs an increased degree of hydrogen bonding or a greater restriction of vibrational motion of the neighboring water molecules."

A different approach has been taken by Miller and Hildebrand (*6*). These authors discuss the solubility of inert gases in water in terms of the Pople structure and show that the entropy of solution of inert gases in water is proportional to the total surface area of inert gas introduced. (The surface area per mole was calculated by assuming it to be proportional to the two-thirds power of the molal volume of the liquid at its boiling point.) Miller and Hildebrand propose an explanation for the negative unitary entropy change upon insertion of an inert gas molecule into the water structure on the basis that around the surface of the inserted molecule a number of H-bonds are "deactivated or destroyed," resulting in a loss of heat capacity and entropy of the water proportional to the surface area of the dissolved gas. This loss of heat capacity postulated by Miller and Hildebrand is at a variance with the experimental results cited above. Moreover, the suggestion that H-bonds are broken disagrees with current interpretations of the thermodynamic and spectroscopic data.

The model we propose for hydrophobic bonding interprets the above experimental evidence in terms of the decrease in the freedom of H-bond bending in the water molecules adjacent to apolar groups. The model is discussed from the viewpoint of the dissolution of apolar groups in water, which (it should be carefully noted) is the reverse of hydrophobic "bond" formation.

The insertion of an apolar molecule into water having the structure visualized by Pople sterically impedes the motions of nearby water molecules, restricting H-bond bending (produced by rotation and/or translation of water molecules) and increasing the average H-bond energy. Figure 5 shows by models that a hydrocarbon—in this case nonane—can indeed be inserted into a tetrahedrally bonded water structure without noticeable distortion of the preexisting network. The presence of the hydrocarbon molecule (in the model) greatly stiffens the structure and restricts the bending of H-bonds. In the context of the Pople model the insertion of the hydrocarbon will tend to keep the structure more regularly tetrahedral and will lead to the negative entropy effect by interfering with translation and rotation of water molecules. The stiffening of the structure should be accompanied by a strengthening and shortening of the H-bonds in accordance with the angular dependence of H-bond energy assumed by Pople, i.e., greatest energy when the O—H – – – O forms a straight line. This strengthening of H-bonds evolves heat and reduces the entropy as is experimentally observed when an apolar group is inserted.

As one would expect, these changes are of the same sign as the ΔH and ΔS of formation of H-bonds, viz., -7 kcal and -18 eu per mole of H-bond, respectively (25). Sizable differences in entropy can be obtained by changes in arrangement alone without either rupture or formation of H-bonds, as can be seen from data assembled by Kamb (17) on ice II and III. Both of these have tetrahedral frameworks, exist at the same pressures, and have nearly the same densities, yet they differ greatly in molar entropies, viz., -0.77 and 0.26 eu per mole of water, respectively (hypothetically decompressed to 1 atm pressure and compared to ice I whose entropy is taken as zero).

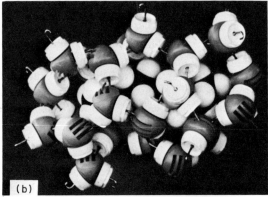

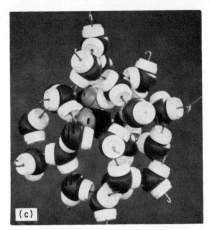

Fig. 5. Corey–Pauling–Koltun models showing a tetrahedrally bonded structure of (a) water, (b) water into which a nonane molecule has been inserted, and (c) water in which one molecule has been replaced by urea (see Sect. II.H).

If the assumption is made that the entropy of fusion is accounted for by the entropy of H-bond bending (Pople theory), then an estimate may be made of the number of water molecules that must be involved about the apolar group to account for the observed entropy change. The entropy of fusion of water is -5 eu per mole. The unitary entropy of solution of butane is -23 eu per mole of hydrocarbon, implying that the insertion of a butane molecule into water impedes bending (to the degree present in ice) of H-bonds associated with about 5 water molecules. This is an altogether reasonable number in terms of the model, taking into consideration that the impedance is probably not to the degree present in ice and hence that a somewhat larger number of water molecules is involved. This picture is in harmony with the findings of Miller and Hildebrand (6) discussed on p. 14. These workers found that the entropy of solution of inert gases in water is proportional to the total surface area of inert gas introduced. It seems reasonable in the context of our picture to assume that the number of H-bonds affected, so as to stiffen them, should also be proportional to the "surface" of contact between the apolar groups and water molecules.

The strengthening of H-bonds noted above also manifests itself experimentally in a decrease in NMR spin lattice relaxation time, and an increase in intensity and decrease in frequency of the $3450\ \mathrm{cm^{-1}}$ Raman line of water when apolar groups are inserted into water.

It is tempting to infer similarities between the freezing of water and the insertion of apolar groups inasmuch as the strengthening of H-bonds plays an important part in both. However, the two processes differ in important respects since the freezing of water results in an increase in volume and a decrease in heat capacity, whereas the opposite effects are observed upon inserting an apolar group into water. The strengthening of H-bonds in the latter process presumably is accompanied by some resultant expansion. The fact that the experimentally determined partial molal volumes of hydrocarbons in water are less than the molar volumes indicates that this expansion has been more than compensated for by the accommodation of the apolar molecules into void space in the water structure—estimated to be 57% for spherical water molecules of 2.9 Å diameter (13).

The increase in heat capacity accompanying the decrease in entropy can be discussed in terms of the configurational heat capacity [Eq. (1)], but no definitive conclusions are apparent. According to our model, the addition of an apolar group would be expected to increase the absolute magnitudes of both $(\delta H/\delta \xi)_{T,p}$ and $(\delta \xi/\delta T)_p$, the former by virtue of the stiffening of the structure, hence leading to a greater energy input required to produce a given distortion, and the latter by the progressive reversal of the stiffening as the temperature is raised, leading to reappearance and gradual enhancement of molecular vibrational and rotational motion.

In both this picture and that of Miller and Hildebrand (6), inert molecules are inserted into water having a Pople-type structure. However, our conclusions with regard to how H-bonds are affected differ from theirs. A consideration of all the experimental evidence suggests that the H-bonds are strengthened, as we have pointed out above, rather than that they have been "deactivated or destroyed" as claimed by Miller and Hildebrand.

The thermodynamic effects of inserting an inert molecule into water may also be discussed in terms of a simple statistical picture. Because certain complexions are ruled out by the restriction of molecular motion adjacent to the inert molecule, the number of distinguishable micro-molecular states (complexions) in a given amount of this highly structured solvent is decreased by the presence of the inert molecule (after due allowance for those complexions corresponding to the ideal entropy of mixing).

As indicated, the molecular picture for hydrophobic "bond" formation is the reverse of that for introducing apolar groups into water. Hydrophobic bonding may therefore be defined as the tendency of apolar groups to withdraw from the aqueous phase, cluster together, remove restrictions on H-bond bending, and to thus achieve a positive unitary entropy change. This tendency is responsible for the low solubility of hydrocarbons in water and for the preferential location of apolar groups of macromolecules in the interior of the molecules in aqueous solutions.

E. Effects of Chaotropic Ions on Water Structure without
    Reference to Any Model

The existence of specific ion effects on macromolecular systems has been recognized since Hofmeister (41) noted that salts differ greatly in their ability to salt out globulins. A wide variety of salt effects on biological systems has since been found to give rise to the same sequence observed by Hofmeister. Specifically, those ions that are the most effective precipitants lead to folding, coiling, and association, while those least effective in precipitation promote unfolding, extension, and dissociation. The latter ions, particularly perchlorate and thiocyanate, have been termed "chaotropic" (tending to disorder) by Hamaguchi and Geiduschek (42). The breadth and generality of anionic effects will be discussed in Sect. III.

A statement by Hamaguchi and Geiduschek suggests a common thread connecting all these areas: "... this class of denaturing agents are 'hydrophobic bond' breakers and ... they act in this manner, at least in part, by virtue of their effect on the structure of water." This thought has been reiterated with additional examples by numerous investigators, including von Hippel and Wong (43), Dandliker et al. (44), Schrier and Schrier (45), Hatefi and Hanstein (46), and von Hippel and Schleich (47).

Using new data on the structure of water and salt solutions, it is now possible to develop this idea into a qualitative theory which provides a unified explanation for the manifold effects of chaotropic ions. (The comprehensive monographs of Kavanau (12, 48) summarize earlier thinking in this area.)

1. Entropies of Hydration

Interpretation of the thermodynamic changes for the solution of ions in water is much more complicated than for the solution of hydrocarbons because of the strong interactions between the ions and water. These strong interactions must certainly disrupt the regular tetrahedrally H-bonded structure of water while imposing a new kind of order around each ion. These changes are reflected in the trends of the partial molal entropies of hydration, which are the least negative for highly chaotropic ions. The partial molal entropy of hydration $\bar{S}_h^o$ of an ion is equal to the partial molal entropy of solution of the ion minus the sum of the translational and rotational entropies of the gaseous ion as calculated from the Sackur–Tetrode equation and from the rotational partition function, respectively.

It should be pointed out that the absolute magnitude of these entropies depends upon the zero point taken. Latimer (49) based his scale upon an earlier determination of the absolute partial molal entropy of the chloride ion as $+18.1 \pm 0.5$ eu. From this value the partial molal entropy of hydration of the iodide ion was found to be -10.4 eu. Gurney (50) based his reference point on the behavior of the "B" coefficient of viscosity. By assigning an arbitrary value of -5.5 eu for the partial molal ionic entropy of the hydrogen ion, he found that the B-coefficients for all univalent ions (positive and negative) fell on a single straight line. Gurney's reference point sets the partial molal ionic entropy of the iodide ion at $+30.8$ eu, compared to $+30.1$ for the same quantity on Latimer's scale; hence, the two scales are essentially identical.

Nightingale (51) improved on Gurney's treatment by calculating the partial molal entropies of hydration $\bar{S}_h^o$ of ions relative to that for the ions in the hypothetical ideal gas state. These values were calculated by

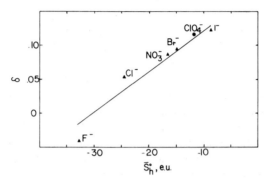

Fig. 6. Correlation of NMR shifts for solutions of sodium salts with partial molal entropies of hydration for the various anions. Ordinate: δ—ppm of NMR diamagnetic shift (toward higher fields) per mole of salt added per liter. Abscissa: $\bar{S}^\circ_h$—partial molal entropy of hydration of the ion. NMR data: ▲ Franconi and Conti (*56*); ● Shoolery and Alder (*53*). Entropy data: Nightingale (*51*).

subtracting from the partial molal entropies of solution the sum of the translational and rotational entropies of the gaseous ions as calculated from the Sackur–Tetrode equation and from the rotational partition function, respectively. An additional shift in the entropies occurs if the partial molal unitary entropies are considered; the latter as tabulated by von Hippel and Schleich (*35*) are 8 eu smaller than the partial molal ionic

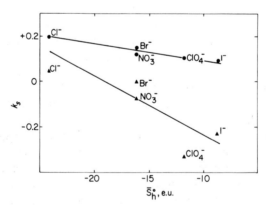

Fig. 7. Correlation of the solubility constant k_s of two nonelectrolytes in various sodium salt solutions with the partial molal entropies of hydration of the anions. Ordinate: k_s—equal to $1/C$ [$\log_{10}(m_0/m)$], where C is the molar concentration of salt and m_0 and m are the molar solubilities of the nonelectrolyte in water and in salt solution, respectively. Abscissa: $\bar{S}^\circ_h$—partial molal entropy of hydration of the ion. Solubility constant data: ● benzene, McDevit and Long (*57*); ▲ acetyltetraglycine ethyl ester, Robinson and Jencks (*58*). Entropy data: Nightingale (*51*).

entropies (taking the value for H^+ to be zero) as given by Gurney (50). In the graphical data of Figs. 6 and 7 we have used the same scale as that of Nightingale. Regardless of the conventions adopted it is important to note that the trends, and the resulting relative ordering of the ions, are the same. What is especially significant for the present discussion is that the entropies are *least* negative for those ions that are most highly chaotropic. These same ions are also implicated by several types of evidence—e.g. viscosity and NMR spectra—as being the most effective as "water structure-breakers."

2. Viscosity B-Coefficients

In the Jones and Doyle equation (52)

$$\eta/\eta_0 = 1 + A\sqrt{c} + Bc \qquad (2)$$

the concentration dependence of the relative viscosity (η/η_0) of ionic solutions is described in terms of A and B, empirical parameters supposedly representative of ion–ion interactions and ion–water interactions, respectively. A linear correlation exists between the ionic B coefficients of viscosity and the partial molal ionic entropies of hydration (51). The greater the structure-breaking effect (larger negative B values), the less negative are the entropies of hydration, in keeping with the idea that greater disorder means greater entropy. Since the anomalously high viscosity of water (compared with H_2S for example) is due largely to extensive H-bonding, it is reasonable to suppose that the structure-breaking ions produce a net decrease in the number of H-bonds when introduced into water.

3. NMR Data

Data on the proton magnetic resonance of water in the presence of various salts have also been interpreted in terms of the breakdown of the H-bonded structure of water (53–56). Large diamagnetic shifts (toward higher fields) were observed for sodium salts containing ClO_4^-, I^-, NO_3^-, and Cl^- and indicate H-bond breaking, whereas paramagnetic shifts (toward lower fields) were observed for sodium compounds containing OH^- and PO_4^{3-}. All of these data are consistent with the view that chaotropic ions lead to a net rupture of the H-bonds of water.

A relationship connecting the disordering of water by chaotropic ions and the breaking of H-bonds is shown in Fig. 6. The correlation between the partial molal ionic entropies of hydration and the NMR shifts for solutions of various sodium salts indicates that there is an intimate relationship between NMR shifts as an index of H-bond rupture, the entropies of hydration as an index of the degree of disorder, and chaotropic effects.

F. Molecular Picture of the Action of Chaotropic Ions on Water

The effects of chaotropic ions on water have been interpreted by many workers as a structure-breaking action. In terms of the Pople model of water, the rupture of H-bonds caused by the addition of chaotropic ions may be viewed as resulting in a breakdown of the tetrahedral framework of water and a disordering of the native water structure to a degree governed by the nature of the chaotropic ion and its concentration. This breakdown of the water structure may be regarded as an inevitable consequence of the new order imposed by the ion in orienting water molecules about itself. The nature and extent of this order would be expected to depend mostly upon ionic net charge and crystallographic radius. The water dipoles near a negative ion must be preferentially oriented with the hydrogens toward the ion, while the reverse should be true for water in the region of a positive ion. The innermost water shells are held very tightly, probably with a high degree of order, which falls off gradually as the distance from the ion increases. For a given net charge, the electric field about an ion increases as the crystallographic radius decreases, and small ions may hence be expected to be more highly hydrated than large ions of the same charge and symmetry. These ideas are borne out, for example, by electrolytic conductance measurements. For a recent discussion see von Hippel and Schleich (35).

The important point for the present discussion is that although the insertion of ions into water breaks down the native water structure, the ion in turn creates a new kind of structure which should be the most ordered for ions of high charge density (relatively nonchaotropic ions). Hence, it follows that the partial molal entropies of hydration for the most highly chaotropic ions are the least negative.

G. The Interrelationships between Water Structure, Chaotropic Ions, and Hydrophobic Bonds

The disruption of water structure by chaotropic ions is intimately related to the stability of hydrophobic bonds. As a model system we have considered the solubility of benzene in water and salt solutions (57).

Whereas both aliphatic and aromatic groups form hydrophobic bonds, we have chosen benzene as a model compound for hydrophobic bonding because of the greater amount of data available. Figures 7 and 8 show the values of a solubility constant (k_s) of benzene and of a more polar solute, acetyltetraglycine ethyl ester, as a function of the partial molal entropies of hydration of the anion, and of the ionic B coefficient of viscosity, respectively. The solubilities follow the usual chaotropic order for macromolecular phenomena. Insofar as benzene solubility is an

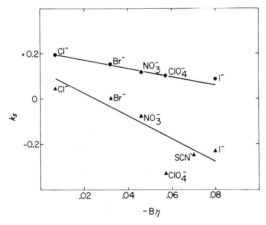

Fig. 8. Correlation of the solubility constant k_s of two nonelectrolytes in various sodium salt solutions with the ionic coefficients of viscosity for the anions. Ordinate: k_s—as defined in Fig. 7. Abscissa: B_η—defined by $\eta/\eta_0 = 1 + A\sqrt{C} + B_\eta C$. Solubility constant data: ● benzene, from McDevit and Long (57); ▲ acetyltetraglycine ethyl ester, Robinson and Jencks (58). Viscosity data: Nightingale (51), except for SCN which was calculated from viscosity data of 0.1–0.5 M NaSCN (International Critical Tables, Vol. V).

acceptable model for hydrophobic bonding, Figs. 7 and 8 show that the tendency toward hydrophobic bonding (i.e. increased values of k_s, or lower benzene solubility) is inhibited by highly chaotropic ions and that the extent of this inhibition is dependent upon the degree of disorder of the solution as measured either by ionic entropies (Fig. 7) or viscosity (Fig. 8).

The absolute solubility of benzene is greater in water than in any salt shown, whereas for the ester, mostly the reverse is observed. Such differences are manifestations of other important effects governing solubility generally, such as polarity, the role of cations (8), the direct interaction of the ions with amide and peptide dipoles (58), electrostatic shielding, ion binding and salting-out (*44*). Despite the contributions of these and possibly other factors, the influence of the chaotropic effect is clearly seen.

The correlation of benzene solubility with the partial molal entropies of hydration of the ions and with the chaotropic order makes it attractive to suppose that the action of chaotropic ions is on the water and not primarily on the macromolecule itself. However, it is not justifiable to draw such a conclusion without having a reasonable explanation of why it might be so. We are thus brought to a crucial question: why is it that water that has been disrupted by the addition of strongly chaotropic ions is an unfavorable medium for the stability of hydrophobic bonds? The answer lies in increased randomness and entropy of the disrupted solvent and may be stated in a simple, intuitive, thermodynamic way: once the water structure has been broken up, it cannot now undergo the further positive entropy change required for the formation of hydrophobic bonds. The rupture of H-bonds of water by the addition of chaotropic ions introduces additional vibrational, translational, and rotational motion over that of water alone. The introduction of apolar molecules now has negligible effects on the heat capacity and the entropy since the tetrahedral structure of water has been disrupted. Hence, there is no mechanism by which the withdrawal of the apolar group can appreciably affect the degree of H-bond bending.

A slightly different view focuses attention upon the new order imposed when ions are dissolved in water. The apolar group can be inserted into pure water or into solutions of nonchaotropic ions only with difficulty because of the existing structures—in one case the tetrahedral structure of water, in the other the organized hydration shells around the ions. In solutions of chaotropic ions the water structure is largely broken down but little new order has been reestablished, and the apolar groups are accommodated with few nonideal contributions to the free energy. For pure water these nonideal contributions are mainly entropic. Data on the thermodynamics of solution of hydrocarbons (especially unbranched, aliphatic hydrocarbons) in ionic solutions should reveal whether or not the barriers to solubility are entropic, as they are in water, or whether large enthalpy changes are also involved.

H. THE SPECIAL CASE OF UREA

The action of urea on macromolecular structures closely resembles that of chaotropic ions in that both strongly favor unfolding and denaturation. Whether or not the action of urea is similarly mediated by an effect on the structure of water is not clear. However, there is widespread belief that in the presence of urea, hydrophobic bonding is decreased (59). In any case, the water–urea interaction itself presents a baffling problem. Various data have been invoked to elucidate the nature of the water–urea interaction and the structure of urea solutions (11, 59, 60). These investigators have pointed out the many, sometimes conflicting, interpretations that can be made of the existing information, including:

(1) a small, positive B coefficient of viscosity;

(2) a small, positive entropy of dilution ($\sim$0.034 eu/mole at 4 M);

(3) NMR measurements showing no alteration in the chemical shift of water protons until the urea concentration is 7 M, at which point a very small upfield shift was observed;

(4) the observation that the heat capacity of urea at room temperature is the same as the partial molal heat capacity of urea in water;

(5) the enhancement of the dielectric constant of water by the addition of urea.

Subramanian et al. (59) have suggested that urea "behaves neither as a structure breaker nor as a structure maker, but participates with great facility in the mobile hydrogen bonding amidst the 'flickering clusters' of water." In terms of the Pople model of water, our indications with models (Fig. 5) are that urea molecules can fit reasonably well into the tetrahedrally-bonded water structure, taking the place of a water molecule by forming four H-bonds, one with each amino group and two to the oxygen. Whether or not such a hybrid structure has a relationship to the action of urea on macromolecules is not known.

III. Hydrophobic Bonding in Macromolecular Systems

A. The Conformation of Macromolecules

The physicochemical properties and biological functions of macromolecules are determined in large part by their conformation. The detailed conformation is a function of both the primary sequence (of amino acid residues in the case of proteins) and of the environment.

The sensitivity of conformation to the environment is directly related to the basic question of whether the native conformation of a protein is the thermodynamically stable form of lowest free energy. There is no doubt that after equilibration the conformation of a protein in solution is in the state of lowest free energy. There is, however, a good possibility that the conformation normally obtained in solution is not that which preexisted in the cell where the environment was quite different and which in turn could be quite different from that on the ribosome where the protein was originally synthesized. The observation that enzymes extracted from tissues often undergo large increases in activity in the presence of salts or urea is possibly a manifestation of these differences.

The types of interactions maintaining a given conformation include H-bonds, electrostatic forces, dispersion forces, S-S bonds, and hydrophobic bonding. The balance between all these influences for a protein molecule in aqueous solution is very delicate, so that small changes in environment can induce large changes in arrangement. Thermodynamically this delicate balance means that $\Delta F \sim 0$ for a transition between one conformation and another. For this to be true, there must be "compensation" for the process, i.e. $\Delta H \sim T \Delta S$. This condition can hold even though drastic changes in the arrangement of the macromolecule may be proceeding. In order to arrive at a more intimate understanding of conformational changes, the net reaction may be conceptually split into two steps: (a) the spatial rearrangement of the macromolecule itself, and (b) altered interactions between the solvent and groups on the protein. Our discussion is centered mainly on the second step—specifically, upon the interaction between the water and the apolar groups of the protein. It is this interaction that is responsible for hydrophobic bonding.

B. Relationship between Solution of Apolar Molecules in Water and Hydrophobic Bonding in Macromolecules

The dissolution of small, apolar molecules in water would seem to be a reasonable model for hydrophobic bonding in macromolecules since in the latter case, interactions between water and apolar side chains—such as those of leucine, valine and phenylalanine—must play an important role; hence, the two phenomena may be expected to show parallel thermodynamic behavior. (It should be kept in mind that hydrophobic bonding is just the reverse of the dissolution of a hydrocarbon.) Because additional factors are operative in macromolecular changes, objections can obviously be raised to procedures in which an extrapolation is made from the solubilities of small molecules in water to hydrophobic bonding in macromolecules. Nevertheless, beginning with Kauzmann's initial calculations, the idea of relating thermodynamic properties of solutions of apolar groups to those of protein solutions has been used by numerous investigators to conjecture about the much more complicated macromolecular processes of unfolding, denaturation, transition from helix to coil, solubilization, dissociation, and disaggregation.*

Various approaches have been used to bridge the gap between small and large molecules as discussed immediately below.

C. APPROACHES TO THE STUDY OF HYDROPHOBIC BONDING IN MACROMOLECULAR SYSTEMS

1. Models Involving Small Molecules

In this approach, thermodynamic functions ΔF, ΔH, ΔS, or ΔC_p are calculated for the transfer of macromolecules from one environment to another, using corresponding data on model systems such as hydrocarbons or amino acids or low molecular weight peptides containing nonpolar side chains. Brandts (61) has chosen the quantity ΔC_p as such a criterion. As shown in Sect. II.B, ΔC_p is positive and large for the process of inserting apolar groups into water. Similarly, Fasman et al. (62) have found that large values of ΔC_p exist for the helix-coil transition in copolymers containing nonpolar side chains, such as copolymers of L-glutamic acid and L-leucine, but not in poly-L-glutamic acid where non-hydrophobic effects predominate. From evidence of this type, Brandts (61) concluded that the magnitude of ΔC_p obtained, for example, by calorimetric measurements of the temperature coefficient of the heat of transition, can be a useful criterion for detecting whether the transition has been accompanied by a net increase in exposure of apolar groups to water.

The free energy of transfer of small molecules from one solvent to another has been used to measure the hydrophobic character of proteins. Thus, Tanford (63) suggested that the free energy of transfer of amino acids from ethanol to water is a reasonable measure, especially since many data are available on solubilities in both solvents. These free energies of transfer become more positive as the hydrophobic character of a compound becomes more pronounced, and Dunhill (64) has termed

---

* One factor in the thermodynamics of large molecules that cannot be directly deduced from small molecule behavior is the configurational entropy. This must be estimated on the basis of statistical mechanical arguments. An approximate value of this parameter for the process of unfolding is between 2 and 6 eu per mole of residue unfolded (61).

them "hydrophobicities." Bigelow (65) has calculated the average hydrophobicities for several proteins from their amino acid composition.

Still another measure of hydrophobic character is that of Waugh (66), who found that the frequency of nonpolar side chains in proteins varied from about 0.21 to 0.47. Waugh recognized that this large number of hydrophobic groups could be a major structural determinant. Fisher (67) employed the ratio of the volumes of the polar groups to those of the nonpolar groups as a measure of hydrophobic character. A possible fault with this particular parameter is that a group is either termed polar or nonpolar and no allowance is made for a gradation of properties between these two extremes.

Studies of micellar systems have afforded some insight into the role of hydrophobic interactions in proteins. Recent studies include those of Heitmann (68) on the hydrophobic interactions of cysteine side chains in micelles, and of Gratzer and Beaven (69) on the effect of protein denaturants on micelle stability. Heitmann's measurements of critical micelle concentrations (CMC) for long chain N-acylcysteines, N-acylglycines, and N-acylserines point up a possible important role for cysteine in hydrophobic bonding. The CMC of the cysteine series was found to be only about one-half that of the glycine or serine series. Heitmann attributed this increased stability for the micelles of the cysteine series to a participation of cysteine side chains in hydrophobic bonding. (The possibility that sulfhydryl groups were being oxidized to form disulfide bonds appears to be precluded by analyses that showed that the sulfhydryl content remained constant throughout the experiments.)

These results support contentions that sulfhydryl groups in proteins tend to be buried in the interior of the molecule along with other hydrophobic groups. This would adequately account for the frequently observed low reactivity of many cysteine groups which have to be "unmasked" before reactivity can be detected, as well as for the stability of proteins, such as serum albumin and ovalbumin, which simultaneously contain both sulfhydryl and disulfide bonds. If sulfhydryl, and perhaps disulfide, bonds are buried within apolar groups, this would serve as protection against disulfide exchange which might otherwise lead to cross-linking in locations other than the native ones, yet would allow rearrangement or formation of disulfide bonds where environmental changes exposed the protected sulfur-containing groups. It is reasonable to conclude that both cysteine and cystine could bond very well hydrophobically with themselves or with other apolar groups, as the pK of cystine groups is well above physiological pH.

Increases in the CMC for a nonionic detergent, Triton X-100, have been induced by protein denaturants such as urea and guanidine hydrochloride (69). This parallels the effect of these denaturants on the solubility of hydrocarbons as well as on the stability of the native conformation of proteins such as ribonuclease. In all cases these denaturants appear to weaken the hydrophobic interactions.

A novel experimental approach for measuring hydrophobic interactions separately from hydrogen bonding effects has been devised by Susi and Ard (70). Using the dimerization of ϵ-caprolactam as a model system involving both types of interactions, these workers have combined the results of vapor pressure measurements, which reflect all types of interaction, and IR spectra, which measure the contribution from H-bonding. After deducting the calculated contribution from hydrogen bonding, the residual stability was attributed to hydrophobic bonding. The results were expressed in terms of deviation from Raoult's law and showed that in dilute solution the major part of the interaction between caprolactam molecules is due to nonpolar interactions. In intermediate concentrations the contribution from hydrogen bonding and hydrophobic bonding may be comparable.

Complex formation between small molecules shows definite evidence of hydrophobic contributions to the free energy of binding. Crothers and Ratner (71) have investigated the formation of complexes between actinomycin and deoxyguanosine in a series of solvent mixtures containing various relative amounts of methanol and water. They found that the complexes were destabilized at higher methanol concentrations due to an increasingly negative ΔS of complex formation. These results are compatible with the conventional view that hydrophobic bonding is due to ordering of solvent molecules around the solute, rather than the view held by Sinanoğlu and Abdulnur (72) that surface tension effects predominated. Crothers and Ratner argue that in the latter model the effects of surface entropy would lead to a *less* negative ΔS of association in going from water to a nonaqueous solvent, which is contrary to the experimental findings for this system.

Shorenstein et al. (73) observed hydrophobic bonding in an interacting system which mimics the function of enzymes. The hydrolysis of p-nitrophenyl-N-dodecyl-N,N-dimethylammonioethyl carbonate bromide is markedly accelerated in the presence of long-chain N-acylhistidine derivatives. This enhancement was accounted for in terms of a prereaction complex formed by hydrophobic bonding between the hydrocarbon components of the two compounds. The equilibrium concentration of the complex was assumed to be proportional to the observed initial rate of hydrolysis. With this assumption, it was found that the complex formation was characterized by a large negative, excess free energy change which was attributed to hydrophobic bonding, inasmuch as the entropy change was large and positive, and the enthalpy change was close to zero. This model of hydrophobic bonding agrees with the Kauzmann model of a partial reversal of solution of apolar groups. The interaction between the hydrocarbon groups during complex formation is thought to cause water that was initially in contact with the hydrocarbons to be released into the bulk water phase where it "can interact more strongly with other water molecules."

2. Studies Involving Interactions between Small Molecules and Large Molecules

Since 1962, Wishnia and co-workers (74–76) have presented an intriguing series of studies on the binding of alkanes such as butane and pentane to proteins. The studies were based on the premise that the transfer of alkanes from water to protein closely resembles the transfer of an alkyl group of a protein from the aqueous environment to the interior of the protein. Early work (74) revealed that the solubility of alkanes in protein solutions was greatly enhanced over that in water, and that the ΔH of transfer from water to protein was near zero. Both observations are in keeping with the picture of hydrophobic bonding. Binding of the alkanes to protein was found to be highly sensitive to the conformation of the protein (75). At pH 4.1, bovine serum albumin was converted to a form that bound only one-quarter (or one-fifth) as much butane (or pentane) as did the native protein. This lowered solubility, under conditions that would be expected to increase the exposure of individual hydrophobic groups to the solvent, suggests that the site for hydrophobic binding is large and that its structure is disrupted at low pH.

Studies on the binding of alkanes to β-lactoglobulin monomers, dimers, and octamers indicated that the degree of aggregation of the globulin did not affect the amount of binding (76). Wishnia and Pinder concluded that the sites associated with aggregation must be at least relatively nonhydrophobic and that any conformational changes resulting from the polymerization did not exert an effect at the hydrophobic binding site for the alkanes.

A different approach to locating hydrophobic sites by means of interactions between small and large molecules involves the use of fluorescent probes. Upon interaction with macromolecules, certain compounds, known as fluorescent probes, undergo changes in their fluorescent properties. Some of these dyes, such as 2-p-toluidinyl-naphthalene-6-sulfonate (TNS), are probes for hydrophobic regions and can be used to detect many kinds of conformational changes (77).

3. Studies Involving Conformational Changes in Large Molecules

The reversible process of folding ⟷ unfolding is an excellent example of the marginal stability of protein structures in aqueous solutions, where

TABLE 2

THERMODYNAMIC STABILITY OF CHYMOTRYPSIN

$\Delta F°$ unfolding at pH 3.0 and 27°C is 7 kcal/mole
at pH 7.0 and 27°C is 14 kcal/mole
of which the total is distributed as follows at pH 3:

$\Delta H°$ (hydrogen bonds + hindered rotation)	=	184 kcal
$\Delta F°$ (abnormal ionization + repulsion)	=	−9 kcal
$-T\,\Delta S°$ (conformational entropy)	=	−364 kcal
$\Delta H°$ (hydrophobic bonds)	=	−123 kcal
$-T\,\Delta S°$ (hydrophobic bonds)	=	319 kcal

($\Delta H°_{\text{NET}}$ = 61 kcal; $\Delta S°_{\text{NET}}$ = 177 eu) NET = 7 kcal
$\Delta h°_{\text{H.B.}} + \Delta h°_{\text{rot}}$ = 800 cal per bond
$\Delta S°_{\text{conform}}$ = 5.2 eu per residue

[Reproduced from R. Lumry and R. Biltonen, in *Structure and Stability of Biological Macromolecules* (S. N. Timasheff and G. E. Fasman, eds.), 1969, vol. 2, p. 111, by permission of the authors and Marcel Dekker, Inc.]

slight changes in environment can shift the equilibrium from one state to the other. The concept of "compensation" is illustrated for the unfolding of chymotrypsinogen in Table 2. Large changes occur in enthalpies and entropies, whereas the overall free energy change for unfolding is only 7 kcal. As Lumry and Biltonen (5) clearly pointed out, the contributions to ΔH and ΔS can be considered conceptually to arise from changes both in the macromolecule itself and in the solvent. A major factor in "compensation" is the negative entropy change accompanying penetration of apolar groups into the water structure. The large positive heat capacity changes, suggested by Brandts (61) as a criterion for unfolding, also are present for this system and amount to 5 ± 2 kcal/mole degree. NMR data further confirm that the transition of chymotrypsinogen to chymotrypsin is an unfolding process (78). A dramatic sharpening of various proton resonance bands was observed for unfolding and this has been interpreted as indicating an increase in the motional freedom of the side chains and of their protons.

The change in the degree of exposure of hydrophobic groups to the solvent mentioned above would be expected to change the amount of ordering produced in the solvent by the presence of the macromolecule. A possibly more important cause of solvent ordering produced by the presence of highly folded macromolecules in water is that arising from the interaction of water with the ionic and polar groups on the surface. A wealth of experimental evidence indicates that such order exists. The types of evidence include NMR measurements on nucleic acids and proteins (79), myelinated nerves (80), actomyosin (81), muscle fibers (82), collagen, DNA and keratin (83), as well as x-ray diffraction (84, 85), dielectric dispersion (86), and density measurements (87).

An interesting example of structured water was revealed by the work of Schultz and Asunmaa (88) in which a kind of organized water, which excludes ions and small organic molecules, was found to exist in the pores of glass membranes. The possible implications of these findings to the structure and action of biological membranes have been discussed by Schultz (89).

4. Studies of Interactions between Macromolecules

An interesting confirmation of the role of water in interactions between macromolecules comes from measurements on the polymerization of the protein of tobacco mosaic virus (90, 91). This endothermic process actually squeezes out water, which can be measured directly by equilibrium dialysis in a quartz spring balance. It was found that about 96 moles of water per mole of protein trimer were lost by the protein upon polymerization. This association was found to exhibit a positive entropy change, indicating increased hydrophobic bonding.

D. Effects of Chaotropic Ions on the Stability of Biological Systems

For most studies on macromolecular interactions and conformational changes, a basically different approach has been used to assess the importance of hydrophobic bonding. This approach consists of examining the effects of neutral salts, particularly anions, on macromolecular systems.

The breadth and generality of anionic effects on macromolecular systems are illustrated in Table 3, adapted and updated from von Hippel and Wong (43). Additional examples have been tabulated by Hatefi and Hanstein (46). For a given cation, the effectiveness of anions in expanding and disrupting macromolecular structures generally follows the order from left to right in the table. As mentioned in Sect. II.E, those ions that promote unfolding, extension, and dissociation have been termed "chaotropic." The unifying principle underlying the effects of salts on macromolecules, on simple organic systems, and on water itself is that the order of effectiveness of anions is nearly the same for all. The conclusion is that chaotropic ions exert their effects by altering the structure of water, thus weakening the tendency toward hydrophobic bonding.

These changes in hydrophobic bonding are clearly evident despite the fact that other effects such as electrostatic forces and the direct action of ions on H-bonds within the macromolecules are obviously superimposed. For example, Robinson and Jencks (58) provide strong evidence for direct interaction between certain large anions and amide dipoles. Schrier and Schrier (45) propose a model that describes the effects of salts on macromolecular transitions as consisting of (1) a salting-in of amide groups exposed as the molecule unfolds and (2) a salting-out of the nonpolar groups.

A comprehensive review of salt effects on macromolecules has been written by von Hippel and Schleich (47). A typical salt effect is the thermal transition of ribonuclease, which is plotted in Fig. 9 as a function of salt concentration. The usual order of the Hofmeister series is seen, the most chaotropic ions being most effective in lowering the transition temperature. von Hippel and Schleich suggest that the curvature of the lines may possibly be attributed to an electrostatic component of the transition, which would tend to destabilize the native form of the molecule. Other examples of the powerful effects of chaotropic ions follow.

1. Immune Reactions

The discovery by Dandliker et al. (44) that antigen–antibody complexes can be dissociated by chaotropic ions resulted from the idea that these complexes owe their stability to the same kinds of forces that maintain the tertiary structure of macromolecules. Dissociation was achieved at neutral pH, and the order of decreasing effectiveness for the ions was thiocyanate > perchlorate > iodide. The dissociation process was followed by observing either the inhibition of precipitation or the dissolution

TABLE 3

EFFECTS PRODUCED BY CHAOTROPIC IONS

$$\left.\begin{array}{l}\text{Helix}\\\text{Stabilization}\\\text{Precipitation}\\\text{Association}\\\text{Folding}\end{array}\right\} \longleftrightarrow \left.\begin{array}{l}\text{Coil}\\\text{Denaturation}\\\text{Solubilization}\\\text{Dissociation}\\\text{Unfolding}\end{array}\right\} \longrightarrow$$

Nature of effect	Order of effectiveness	Reference
Solubilization of euglobulins	$SO_4^{2-} < HPO_4^{2-} < CH_3COO^- < Cl^- < NO_3^- < ClO_3^-$	Hofmeister (*41*)
Solubility of benzene derivatives: leucine, benzamide, *p*-toluidine, caffeine, phenylalanine	$SO_4^{2-} < CH_3COO^- < Cl^- < Br^- < NO_3^- < I^- < CCl_3COO^- < SCN^-$	Schryver (*92*)
Inhibition of immune reaction between ovalbumin and antiovalbumin	$F^- < Cl^- < Br^- < I^- < SCN^-$	Kleinschmidt and Boyer (*93*)
Solubilization of benzoic acid	$Cl^- < Br^- < I^- < NO_3^- < ClO_4^- < SCN^-$	Long and McDevit (*94*)
Lowering shrinking temperature of collagen; lowering melting point of gelatin	$SO_4^{2-} < CH_3COO^- < Cl^- < Br^- < NO_3^- < ClO_4^- < I^- < SCN^-$	Bello et al. (*95*) von Hippel and Wong (*43, 96*)
Lowering transition temperature of DNA	$Cl^-, Br^- < CH_3COO^- < I^- < ClO_4^- < SCN^- < CCl_3COO^-$	Hamaguchi and Geiduschek (*42*)
Dissociation of hapten–antibody complexes	Phosphate $< MoO_4^{2-} < CrO_4^{2-} < SO_4^{2-} < CH_3COO^- < NO_3^- < SCN^-$	Pressman et al. (*97*)

Description	Series	Reference
Raising cloud point of polyvinyl-methyloxazolidinone	$F^- < Cl^- < Br^- < SCN^-$	Klotz (98)
Solubilization of acetyltetraglycine ethyl ester	$SO_4^{2-} < H_2PO_4^- < CH_3COO^- < Cl^- < Br^- < NO_3^- < I^- < SCN^- < ClO_4^-$	Robinson and Jencks (58)
Solubilization of purine and pyrimidine bases	$SO_4^{2-} < CH_3COO^- < Cl^- < Br^- < I^- < ClO_4^-, SCN^- < CCl_3COO^-$	Robinson and Grant (99)
Inhibition of enzyme activity (trypsin, γ-chymotrypsin, etc.)	$CH_3COO^- < Cl^- < Br^-, NO_3^- < I^- < ClO_4^-, SCN^-$	Warren and Cheatum (100)
Dissociation of antigen–antibody complexes	$Cl^- < I^- < ClO_4^- < SCN^-$	Warren et al.(101) Dandliker et al. (44); de Saussure and Dandliker (102)
Solubilization of particulate proteins	$F^- < Cl^- < ClO_4^- < SCN^-$	Hatefi and Hanstein (46)
Shift from fractional to first order kinetics for antigen–antibody reactions; decrease in the forward rate constant for antigen–antibody association	Phosphate, $SO_4^{2-} < F^- < Cl^- < ClO_4^- < SCN^-$	Levison and Dandliker (103, 106, 107)
Lowering of association constant for penicillin–antipenicillin antibody reaction	$F^- <$ phosphate $< Cl^- < ClO_4^- < SCN^-$	de Saussure and Dandliker (110)

of preformed antigen-antibody precipitates; the latter effect is shown in Fig. 10 for an ovalbumin–antiovalbumin precipitate. As the ionic concentrations were increased, or in going from iodide to perchlorate to thiocyanate, the equilibrium constant for association was decreased. The effects observed are in agreement with the idea that hydrophobic bonding

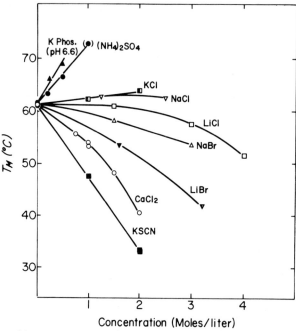

Fig. 9. Transition temperatures of ribonuclease as a function of concentration of various added salts. All the solutions were adjusted to pH 7.0 and also contained 0.15 M KCl and 0.013 M sodium cacodylate; ribonuclease concentration, ≈ 5 mg/ml. [Reproduced from P. H. von Hippel and K.-Y. Wong, *J. Biol. Chem.* **240**, 3909 (1965), by permission of the authors and the American Society of Biological Chemists.]

plays an important part in the stability of antigen–antibody complexes. Chaotropic ions lead to the dissociation of these complexes by destabilizing hydrophobic bonding via a breakdown of the structure of water as discussed above.

Ultracentrifuge studies by de Saussure and Dandliker (*102*) confirmed that, for the systems studied the dissociation was complete in 3 M SCN$^-$, and that antibody purified by elution with 3 M SCN$^-$ from immunospecific columns retained its ability to combine with antigen to form both soluble and insoluble complexes. These findings have been developed into

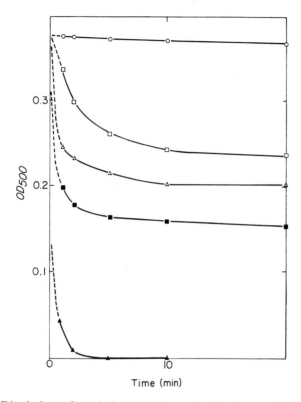

Fig. 10 Dissolution of washed ovalbumin–antiovalbumin precipitates (optimal precipitation zone) in various ions added to diluent buffer. A small measured portion of the specific precipitate finely suspended in diluent buffer was added to the ion and the light transmission measured as a function of time (pH 6.2). Corrections for the effect of refractive index on the scattering have been applied. ○ control; □ 2 M potassium iodide; △ 2 M sodium perchlorate; ▽ 2 M sodium thiocyanate; ▼ 4 M thiocyanate, pH 6.0. [Reproduced from W. B. Dandliker, R. Alonso, V. A. de Saussure, F. Kierszenbaum, S. A. Levison and H. C. Schapiro, *Biochemistry* **6**, 1460 (1967), by permission of the authors and the American Chemical Society.]

a general method for the immunospecific purification, at neutral pH, of antibodies to various antigens, including viruses (*104, 105*).

Long before the present picture of the action of chaotropic ions was developed, Pressman et al. (*97*) observed specific ion effects on the stability of hapten–antibody complexes (Table 3). The effects were explained on the basis of electrostatic and steric factors which prevented large ions from entering the antibody active site but which allowed small ions to decrease electrostatic repulsive effects between hapten and antibody. In light of the present discussion it appears that the effects observed by Pressman are explicable on the basis of hydrophobic bonding.

Evidence on the influence of chaotropic ions on the kinetics of macromolecular reaction has recently been provided by antigen–antibody studies (*103, 106, 107*). These investigations showed that the antigen–antibody reaction is generally first order with respect to both reactants in chaotropic media, but fractional order with respect to antibody in relatively nonchaotropic media. An additional effect was found in that the magnitudes of the second order rate constants (for the forward reaction) progressively decrease as the medium becomes more chaotropic. The reaction mechanism proposed consisted of two steps: the formation of a loose encounter pair between antigen and antibody followed by a rate determining step involving the rearrangement of the encounter pair to give a stable complex.

As pointed out by Dandliker et al. (*108*), a logical consequence of the assumed mechanism is that effects on the order of reaction are assignable to encounter pair formation, while effects on overall rate are assignable to the rearrangement. Where the reaction is second order the effect of chaotropic ions manifests itself as a decrease in the second order rate constant, suggesting that the rearrangement of the encounter pair involves hydrophobic bonding which is inhibited by the presence of the chaotropic ions. In light of the proposed mechanism, the fractional order observed in less chaotropic media must be attributed to a nonuniformity of the binding sites involved in encounter pair formation. This nonuniformity may be related to the folded conformation of the bivalent antibody molecule in nonchaotropic media since the effect disappears either by placing it in chaotropic media or by cleaving it with papain to give univalent antibody.

This picture is in harmony with the antibody model of Feinstein and Rowe (*109*) in which the molecule clicks open about a central hinge prior to reaction with hapten or antigen. In chaotropic media perhaps the molecule is already in the open conformation. After the formation of the encounter pair between the open conformation and the antigen molecule, the subsequent rearrangement to give a stable complex is impeded by the decreased tendency for hydrophobic bonding in the presence of chaotropic ions.

The relative insensitivity of the rates of the back reaction ties in closely with the concepts presented above. Once the complex has been formed, solvent is largely excluded in the region of contact of the active sites.

Hence, the presence of chaotropic ions in the solvent has relatively negligible effects on the back reaction as compared to the forward reaction. By virtue of the dynamic equilibrium between free antigen, antibody and complex, chaotropic ions lead to a lowering of the association constant by slowing the forward reaction. It is tempting to generalize that a similar relative insensitivity of the back reaction to the nature of the ionic environment may be found for a variety of associations involving macromolecules.

2. Solubilization

Solubilization of large biological structures may also be achieved by chaotropic ions. We have found that a combination of ether extraction and treatment with chaotropic ions disrupted the structure of influenza virus to yield small soluble fragments (fluorescence polarization with fluorescein labeling is about 0.15 at room temperature in aqueous solution). Erythrocyte membranes prepared by lysis in distilled water and by ether extraction showed the same degree of solubilization at pH 7 in the presence of 2.5 M thiocyanate (but not with chloride) (110) as was achieved at pH 9 at low ionic strengths (111, 112). Hatefi and Hanstein (46) have reported that chaotropic ions induce a rapid oxidation of lipids in mitochondria and other subcellular particles.

E. Practical Implications of Being Able to Affect the Stability of Macromolecules in Aqueous Solution via Hydrophobic Bonding

Because of the importance of hydrophobic bonding in stabilizing macromolecular structures such as proteins, nucleic acids, viruses and cellular membranes, an understanding of the action of chaotropic ions immediately suggests important new methods for the dissociation of these structures into simpler subunits under mild conditions without breaking covalent bonds. By variation of both the nature of the ion and its concentration, a virtually unlimited range of experimental conditions can be attained and empirically optimized for a particular system under study. Because the action of chaotropic ions is so powerful, the exact conditions must be carefully determined so as to accomplish the desired degree of dissociation or unfolding without causing undesirable, apparently irreversible changes such as have been observed at 6 M thiocyanate for ovalbumin (102).

Chemically, the most gentle approach would be to add chaotropic ions to weaken or disrupt the hydrophobic bonding and at the same time to affect hydrogen bonding and electrostatic attractions. If under these circumstances a structure does not readily dissociate or break down into subunits, it may well be that it is cross-linked by disulfide bonds. This possibility can be tested by examining the combined effects of chaotropic ions and disulfide reducing agents such as cysteine or mercaptoethanol.

A structure still resistant to this combination may possess an extraordinarily high concentration of lipids that simply prevent aqueous solutions from penetrating into the core of the structure. In these cases it may be found that very gentle lipid extraction, beginning with nonpolar solvents such as aliphatic hydrocarbons and gradually progressing toward more polar materials such as diethylether, will serve to remove lipid materials, which are bound relatively loosely. The simultaneous presence of diethylether and water sometimes induces irreversible changes in macromolecular structures, so that dry conditions may be required. The further possibilities of utilizing combinations of water soluble solvents together with water, chaotropic ions, and perhaps reducing agents are intriguing.

ACKNOWLEDGMENT

The authors gratefully acknowledge the helpful suggestions and criticisms of Professor John T. Edsall during the preparation of the manuscript.

REFERENCES

1. Kauzmann, W., in *The Mechanism of Enzyme Action* (W. D. McElroy and B. Glass, eds.), Johns Hopkins, Baltimore, 1954.
2. Hildebrand, J. H., *J. Phys. Chem.* **72**, 1841 (1968).
3. Némethy, G. and Scheraga, H. A., *J. Chem. Phys.* **36**, 3382, 3401 (1962).
4. Némethy, G., Scheraga, H. A. and Kauzmann, W., *J. Phys. Chem.* **72**, 1842 (1968).
5. Lumry, R. and Biltonen, R., in *Structure and Stability of Biological Macromolecules* (S. N. Timasheff and G. D. Fasman, eds.), Vol. 2, Dekker, New York, 1969.
6. Miller, K. W. and Hildebrand, J. H., *J. Amer. Chem. Soc.* **90**, 3001 (1968).
7. Kauzmann, W., *Advan. Prot. Chem.* **14**, 1 (1959).
8. Frank, H. S. and Evans, M. W., *J. Chem. Phys.* **13**, 507 (1945).
9. Scheraga, H. A., Némethy, G. and Steinberg, I. Z., *J. Biol. Chem.* **237**, 2506 (1962).
10. Steinrauf, L. K. and Dandliker, W. B., *J. Amer. Chem. Soc.* **80**, 3833 (1958).
11. Holtzer, A. and Emerson, M. F., *J. Phys. Chem.* **73**, 26 (1969).
12. Kavanau, J. L., *Water and Solute–Water Interactions*, Holden-Day, San Francisco, 1964.
13. Berendsen, H. J. C., in *Theoretical and Experimental Biophysics* (A. Cole, ed.), Dekker, New York, 1967, p. 26.
14. Narten, A. H. and Levy, H. A., *Science* **165**, 447 (1969).
15. Frank, H. S. and Wen, W-Y, *Discuss. Faraday Soc.* **24**, 133 (1957).
16. Pauling, L., in *Hydrogen Bonding* (D. Hadzi, ed.), Pergamon, New York, 1959, p. 1.

17. Kamb, B. in *Structural Chemistry and Molecular Biology* (A. Rich and N. Davidson, eds.), Freeman, San Francisco, 1968.
18. Forslind, E., *Acta Polytech. Scand.* **115**, 9 (1952).
19. Samoilov, O. Y. and Nosova, T. A., *J. Struct. Chem.* **6**, 767 (1965).
20. Mysels, K. J., *J. Amer. Chem. Soc.* **86**, 3503 (1964).
21. Wall, T. T. and Hornig, D. F., *J. Chem. Phys.* **43**, 2079 (1965).
22. Scatchard, G., *Fed. Proc.* **25**, 954 (1966).
23. Stevenson, D. P., in *Structural Chemistry and Molecular Biology* (A. Rich and N. Davidson, eds.), Freeman, San Francisco, 1968, p. 490.
24. Pople, J. A., *Proc. Roy. Soc. (London)*, **A205**, 163 (1951).
25. Pimentel, G. C. and McClellan, A. L., *The Hydrogen Bond*, Freeman, San Francisco, 1960.
26. Willis, E., Rennie, G. K., Smart, C., and Pethica, B. A., *Nature* **222**, 159 (1969).
27. Lippincott, E. R., Stromberg, R. R., Grant, W. H., and Cessac, G. L., *Science* **164**, 1482 (1969).
28. Claussen, W. F. and Polglase, M. F., *J. Amer. Chem. Soc.* **74**, 4817 (1952).
29. Edsall, J. T., *J. Amer. Chem. Soc.* **57**, 1506 (1935).
30. Bohon, R. L. and Claussen, W. F., *J. Amer. Chem. Soc.* **73**, 1571 (1951).
31. Hertz, H. G. and Zeidler, M. D., *Ber. Bunsenges. Physik. Chem.* **67**, 774 (1963).
32. Clifford, J., Pethica, B. A. and Senior, W. A., *Ann. N.Y. Acad. Sci.* **125**, 458 (1965).
33. Hertz, H. G. and Zeidler, M. D., *Z. Elektrochem.* **68**, 821 (1964).
34. Everett, D. H., *Discuss. Faraday Soc.* **24**, 216 (1958).
35. von Hippel, P. H. and Schleich, T., in *Structure and Stability of Biological Macromolecules* (S. N. Timasheff and G. D. Fasman, eds.), Vol. 2, Dekker, New York, 1969.
36. Muller, N. and Reiter, R. C., *J. Chem. Phys.* **42**, 3265 (1965).
37. Hindman, J. C., *J. Chem. Phys.* **44**, 4582 (1966).
38. Frank, H. S. and Quist, A. S., *J. Chem. Phys.* **34**, 604 (1961).
39. Klotz, I. M., *Brookhaven Symp. Biol.* **13**, 25 (1960).
40. Powell, R. E. and Latimer, W. M., *J. Chem. Phys.* **19**, 1139 (1951).
41. Hofmeister, F., *Arch. Exp. Pathol. Pharmakol.* **24**, 247 (1888).
42. Hamaguchi, K. and Geiduschek, E. P., *J. Amer. Chem. Soc.* **84**, 1329 (1962).
43. von Hippel, P. H. and Wong, K.-Y. *Science*, **145**, 577 (1964).
44. Dandliker, W. B., Alonso, R., de Saussure, V. A., Kierszenbaum, F., Levison, S. A. and Schapiro, H. C., *Biochemistry* **6**, 1460 (1967).
45. Schrier, E. E. and Schrier, E. B., *J. Phys. Chem.* **71**, 1851 (1967).
46. Hatefi, Y. and Hanstein, W. G., *Proc. Nat. Acad. Sci. U.S.* **62**, 1129 (1969).
47. von Hippel, P. H. and Schleich, T., *Accts. Chem. Res.* **2**, 257 (1969).
48. Kavanau, J. L., *Structure and Function in Biological Membranes*, Vols. 1 and 2, Holden-Day, San Francisco, 1965.
49. Latimer, W. M., *Chem. Rev.* **18**, 349 (1936).
50. Gurney, R. W., *Ionic Processes in Solution*, McGraw-Hill, New York, 1953.
51. Nightingale, E. R., Jr., *J. Phys. Chem.* **63**, 1385 (1959).
52. Jones, G. and Doyle, M., *J. Amer. Chem. Soc.* **51**, 2950 (1929).
53. Schoolery, J. N. and Alder, B. J., *J. Chem. Phys.* **23**, 805 (1955).
54. Hindman, J. C., *J. Chem. Phys.* **36**, 1000 (1962).
55. Franconi, C. and Conti, F., in *Nuclear Magnetic Resonance in Chemistry* (B. Pesce, ed.), Academic, New York, 1965.

56. Franconi, C., Dejak, C. and Conti, F., in *Nuclear Magnetic Resonance in Chemistry* (B. Pesce, ed.), Academic, New York, 1965.

57. McDevit, W. F. and Long, F. A., *J. Amer. Chem. Soc.* **74**, 1773 (1952).

58. Robinson, D. R. and Jencks, W. P., *J. Amer. Chem. Soc.* **87**, 2470 (1965).

59. Subramanian, S., Balasubramanian, D. and Ahluwalia, J. C., *J. Phys. Chem.* **73**, 266 (1969).

60. Vidulich, G. A., Andrade, J. R., Blanchette, P. P. and Gilligan, T. J., III., *J. Phys. Chem.* **73**, 1621 (1969).

61. Brandts, J. F., in *Structure and Stability of Biological Macromolecules* (S. N. Timasheff and G. D. Fasman, eds.), Vol. 2, Dekker, New York, 1969.

62. Fasman, G. D., Lindblow, C. and Bodenheimer, E., *Biochemistry* **3**, 155 (1964).

63. Tanford, C., *J. Amer. Chem. Soc.* **84**, 4240 (1962).

64. Dunnill, P., *Sci. Progr.* **53**, 609 (1965).

65. Bigelow, C. C., *J. Theor. Biol.* **16**, 187 (1967).

66. Waugh, D. F., *Advan. Protein Chem.* **9**, 243 (1954).

67. Fisher, H. F., *Proc. Nat. Acad. Sci. U.S.* **51**, 1285 (1964).

68. Heitmann, P., *Eur. J. Biochem.* **3**, 346 (1968).

69. Gratzer, W. B. and Beaven, G. H., *J. Phys. Chem.* **73**, 2270 (1969).

70. Susi, H. and Ard, J. S., *J. Phys. Chem.* **73**, 2440 (1969).

71. Crothers, D. M. and Ratner, D. I., *Biochemistry* **7**, 1823 (1968).

72. Sinanoğlu, O. and Abdulnur, S., *Fed. Proc.* **24**, Suppl. 15, S-12 (1965).

73. Shorenstein, R. G., Pratt, C. S., Hsu, Chen-Jung, and Wagner, T. E., *J. Amer. Chem. Soc.* **90**, 6199 (1968).

74. Wishnia, A., *Proc. Nat. Acad. Sci. U.S.* **48**, 2200 (1962).

75. Wishnia, A. and Pinder, T. W., Jr., *Biochemistry* **3**, 1377 (1964)

76. Wishnia, A. and Pinder, T. W., Jr., *Biochemistry* **5**, 1534 (1966).

77. Edelman, G. M. and McClure, W. O., *Accts. Chem. Res.* **1**, 65 (1968).

78. Hollis, D. P., McDonald, G., and Biltonen, R. L., *Proc. Nat. Acad. Sci. U.S.* **58**, 758 (1967).

79. Kuntz, I. D., Brassfield, T. S., Law, G. D. and Purcell, G. V., *Science* **163**, 1329 (1969).

80. Chapman, G., and McLauchlan, K. A., *Nature* **215**, 391 (1967).

81. Cope, F. W., *J. Gen. Physiol.* **50**, 1353 (1967).

82. Hazelwood, C. F. and Nichols, B. L., *Nature* **222**, 747 (1969).

83. Berendsen, H. J. C. and Migchelsen, C., *Ann. N.Y. Acad. Sci.* **125**, 365 (1965).

84. Jacobsen, B., *Kem. Tidskr.* **67**, 1 (1955).

85. Beeman, W. W., Geil, P., Shurman, M., and Malmon, A. G., *Acta Crystallogr.* **10**, 818 (1957).

86. Grant, E. H., *Ann. N.Y. Acad. Sci.* **125**, 418 (1965).

87. Hearst, J. E. and Vinograd, J., *Proc. Nat. Acad. Sci. U.S.* **47**, 1005 (1961).

88. Schultz, R. D. and Asunmaa, S. K., *Recent Progress in Surface Science* (J. F. Danielli et al., eds.) Vol. 3, Academic, New York, 1969.

89. Schultz, R. D., Presentation before the Intra-Science Research Foundation, Santa Monica, California, January 23, 1970.

90. Stevens, C. L. and Lauffer, M. A., *Biochemistry* **4**, 31 (1965).

91. Jaenicke, R. and Lauffer, M. A., *Biochemistry* **8**, 3083 (1969).

92. Schryver, S. B., *Proc. Roy. Soc. (London)*, **B83**, 96 (1910).

93. Kleinschmidt, W. J. and Boyer, P. D., *J. Immunol.* **69**, 247 (1952).

94. Long, F. A. and McDevit, W. F., *Chem. Rev.* **51**, 119 (1952).
95. Bello, J., Riese, H. C. A. and Vinograd, J. R., *J. Phys. Chem.* **60**, 1299 (1956).
96. von Hippel, P. H. and Wong, K.-Y., *Biochemistry* **2**, 1387 (1963).
97. Pressman, D., Nisonoff, A. and Radzimski, G., *J. Immunol.* **86**, 35 (1961).
98. Klotz, I. M., *Fed. Proc.* **24**, Suppl. 15, S-24 (1965).
99. Robinson, D. R. and Grant, M. E., *J. Biol. Chem.* **241**, 4030 (1966).
100. Warren, J. C. and Cheatum, S. G., *Biochemistry* **5**, 1702 (1966).
101. Warren, J. C., Stowring, L. and Morales, M., *J. Biol. Chem.* **241**, 309 (1966).
102. de Saussure, V. A. and Dandliker, W. B., *Immunochemistry* **6**, 77 (1969).
103. Levison, S. A. and Dandliker, W. B., *Immunochemistry* **6**, 253 (1969).
104. Dandliker, W. B., de Saussure, V. A. and Levandoski, N., *Immunochemistry* **5**, 357 (1968).
105. Dandliker, W. B., de Saussure, V. A. and Grow, T. E., *J. Virol.* **3**, 283 (1969).
106. Levison, S. A., Jancsi, A. N. and Dandliker, W. B., *Biochem. Biophys. Res. Commun.* **33**, 942 (1968).
107. Levison, S. A., Kierszenbaum, F. and Dandliker, W. B., *Biochemistry* **9**, 322 (1970).
108. Dandliker, W. B., Portmann, A. J. and Levison, S. A., Eighth International Congress of Biochemistry, Switzerland, September, 1970.
109. Feinstein, A. and Rowe, A. J., *Nature* **205**, 147 (1965).
110. de Saussure, V. A. and Dandliker, W. B., unpublished work.
111. Moskowitz, M., Dandliker, W. B., Calvin, M. and Evans, R. S., *J. Immunol.* **65**, 383 (1950).
112. Dandliker, W. B., Moskowitz, M., Zimm, B. H. and Calvin, M., *J. Amer. Chem. Soc.* **72**, 5587 (1950).

2

Lipids in Water

MORRIS B. ABRAMSON

The Saul R. Korey Department of Neurology
Albert Einstein College of Medicine
Bronx, New York

Before beginning a description of aqueous systems of lipids it is necessary to define the scope of the presentation. The term "lipid" has been used to include a large number of compounds of biological origin representing many types with a wide variety of chemical structures. The classification of these compounds is based upon their common solubility in solvents possessing nonpolar groups: chloroform, carbon tetrachloride, hexane, petroleum ether, and in more limited instances, ethanol or ether. The diversity of these compounds can be seen when we find that they include fatty acids, neutral fats (mono-, di-, and triglycerides), phospholipids, sphingolipids, long-chain aliphatic alcohols, terpenes, steroids as well as complex compounds of lipids with proteins, and polysaccharides. This survey, however, will be limited to those lipids that have been shown to be present in membrane systems or that have properties leading to an involvement at an aqueous interface. All such lipids possess one or more relatively large hydrocarbon groups that are insoluble in water (hydrophobic) but soluble in nonpolar solvents (lipophilic), and at least one terminal polar or ionic group that is water-soluble (hydrophilic). Eliminated from this discussion are such lipids as the neutral fats and waxes.

I. Lipid Structures

The great variety of lipids that have already been identified in plants and microorganisms as well as in animals indicate that an almost endless number of these compounds may exist. However, the relatively small number of lipids that have been thoroughly studied have been obtained from mammalian tissue or from some common source such as egg yolk, yeast, or soya. The characteristics shown by these representative lipids help establish principles that are applicable to a larger number of other lipids with similar chemical structures. In elucidating these principles we will refer to a few classes of lipids. These will include long-chain acids, soaps, phospholipids, sphingolipids, and cholesterol and cholesterol esters.

A. Phospholipids

The phospholipids, which constitute the major lipid components of mammalian membrane systems, all contain a phosphoric acid ester. Many are related to fats in that they are diacylglycerides with two fatty acid chains similar to those found in fats (triacylglycerides). A simple member of this group is phosphatidic acid (1).

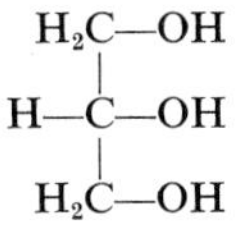

$$H_2COOCR$$
$$R'COOCH \quad O$$
$$H_2C-OP-OH$$
$$O$$
$$H$$

H_2C-OH	$H_2C-OOCR$	
$H-C-OH$	$HC-OOCR'$	
H_2C-OH	H_2C-OH	
Glycerol	Diacylglyceride	Phosphatidic acid
		(1)

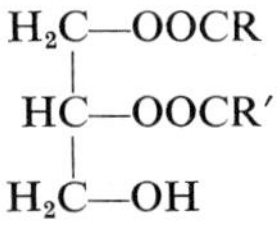

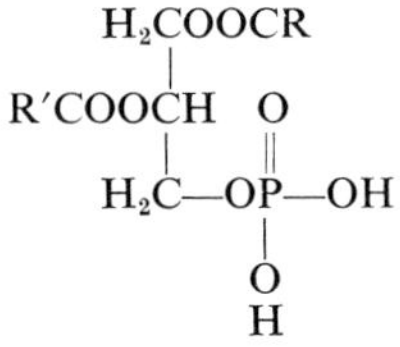

Related phospholipids can be derived from phosphatidic acid by esterification of the phosphate groups. The most important derivatives contain choline, ethanolamine, L-serine, or inositol. The lipids with these groups are, respectively, 1,2-diacyl-L-phosphatidyl choline (lecithin) **(2)**, 1,2-diacyl-L-phosphatidyl ethanolamine **(3)**, 1,2-diacyl-L-phosphatidyl serine **(4)**, and 1,2-diacyl-L-phosphatidyl inositol **(5)**. A somewhat more complex compound is diphosphatidylglycerol (cardiolipin), which appears to consist of two phosphatidic acid units esterified at the 1 and 3 position of glycerol.

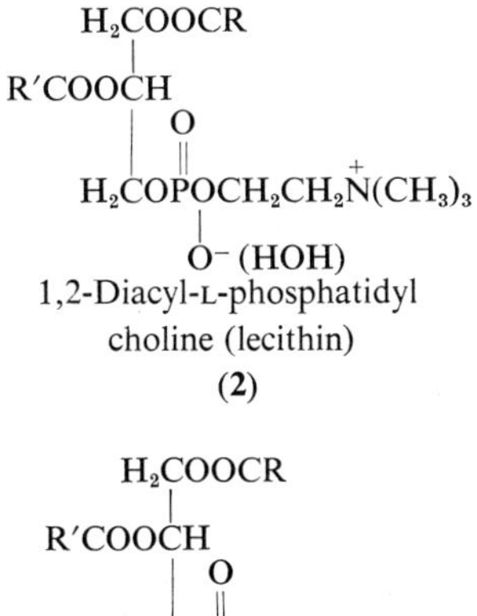

$$H_2COOCR$$
$$R'COOCH$$
$$O$$
$$H_2COPOCH_2CH_2\overset{+}{N}(CH_3)_3$$
$$O^- \text{(HOH)}$$
1,2-Diacyl-L-phosphatidyl
choline (lecithin)
(2)

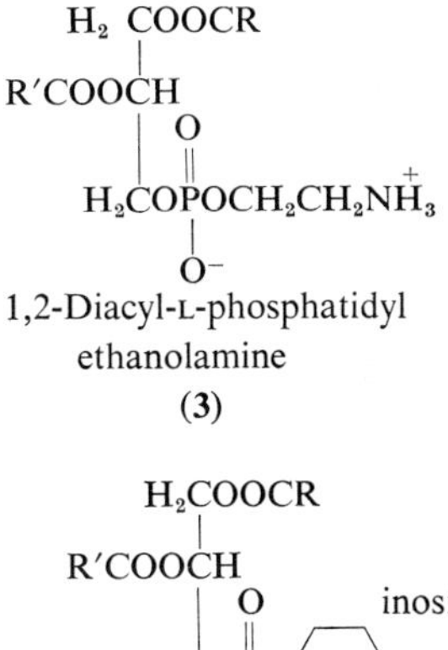

$$H_2 \quad COOCR$$
$$R'COOCH$$
$$O$$
$$H_2COPOCH_2CH_2\overset{+}{N}H_3$$
$$O^-$$
1,2-Diacyl-L-phosphatidyl
ethanolamine
(3)

$$H_2COOCR$$
$$R'COOCH$$
$$O$$
$$H_2COPOCH_2CHCOO^-Na^+$$
$$O^- \quad \overset{+}{N}H_3$$ 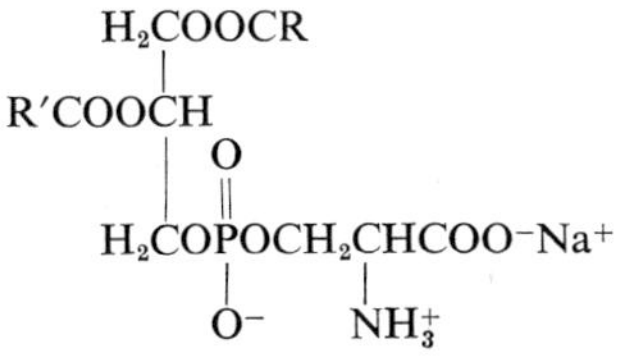
1,2-Diacyl-L-phosphatidyl serine
(4)

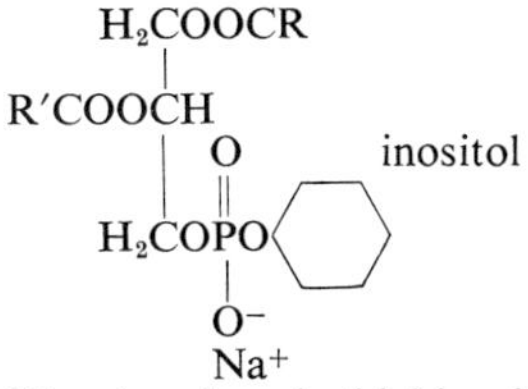

$$H_2COOCR$$
$$R'COOCH$$
$$O \qquad \text{inositol}$$
$$H_2COPO$$
$$O^-$$
$$Na^+$$
1,2-Diacyl-L-phosphatidyl inositol
(5)

B. SPHINGOLIPIDS

The sphingolipids are not glycerides. Instead, they contain a long-chain amino alcohol, sphingosine ($CH_3(CH_2)_{12}CH:CHCHOHCHNH_2CH_2OH$),

as well as a fatty acid. Sphingomyelin (6) is related to lecithin (2) as
shown. Another important sphingolipid is cerebroside (monoglycosylceramide) (7). This does not contain any acid group but has a galactose
molecule in ether linkage to the sphingosine. Sulfatide (8) is somewhat
more complex, with a sulfate joined to the galactose.

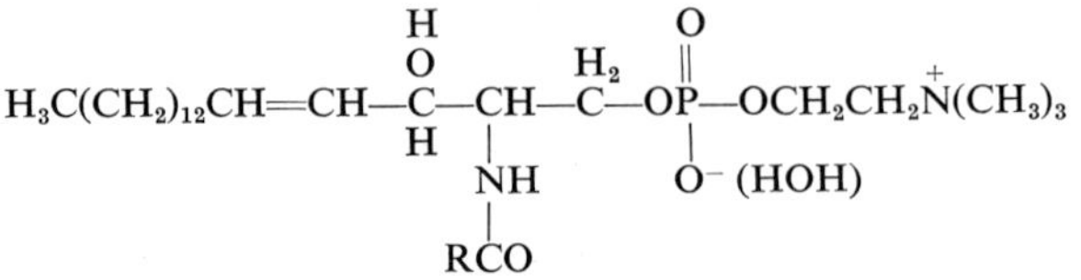

$$H_3C(CH_2)_{12}CH=CH\underset{\underset{RCO}{\overset{|}{NH}}}{\overset{\overset{H}{\overset{O}{|}}}{\underset{|}{C}}}-CH-\underset{\underset{\underset{(HOH)}{O^-}}{\overset{H_2}{C}}}{\overset{O}{\overset{||}{OP}}}-OCH_2CH_2\overset{+}{N}(CH_3)_3$$

Sphingomyelin

(6)

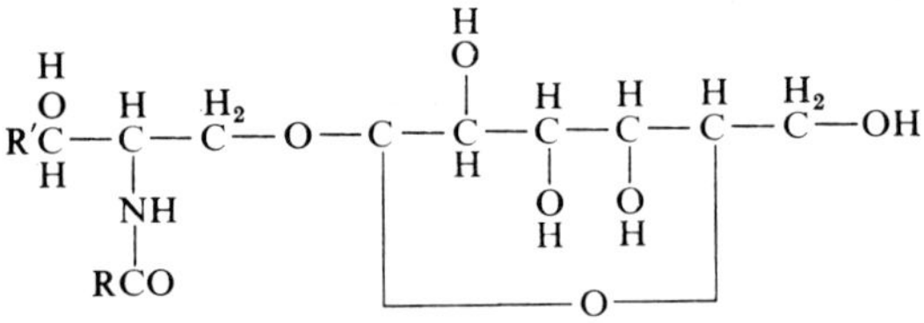

monoglycosylceramide

(cerebroside)

(7)

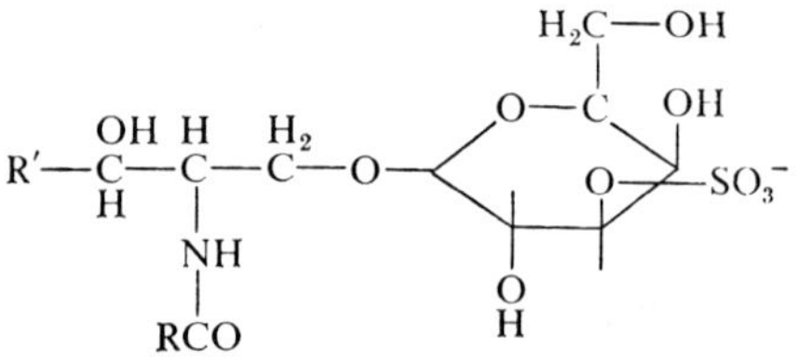

sulfatide

(8)

C. CHOLESTEROL AND CHOLESTERYL ESTERS

The sterols comprise a group of lipids that are alcohols or their esters.
Cholesterol (9) is probably the one of greatest interest. It consists of a
phenanthrene ring to which is attached a five-carbon ring carrying a
branched side chain. An alcohol on the first ring provides the group for
the formation of esters. The cholesteryl ester may be formed from

relatively small organic acids such as in cholesteryl formate or from more complex ones as in cholesteryl linoleate.

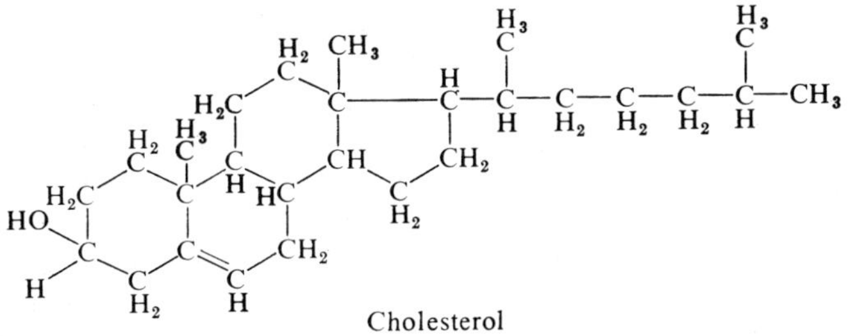

Cholesterol

(9)

II. Membrane Lipids

The lipids that are present in membranes or are active at an aqueous interface have unique structural characteristics, with distinctly lipophilic and hydrophilic regions. Such compounds are called *amphipathic*. Structurally the two regions are so far apart that the unique properties of the two regions are distinctly separate.

A survey of the chemical structures of the various types of lipids indicates the general nature of the lipophilic and hydrophilic groups present. The phospholipids and sphingolipids that appear in many membranes possess two long-chain hydrocarbon portions. We will consider only the more commonly encountered members of these classes.

A structural characteristic of the glycerophospholipids is the presence of fatty acid esters in the 1 and 2 position of glycerol and phosphoric acid esterified in the 3 position. A great diversity of fatty acid chains has been reported in the hydrocarbon portions, but for most of the lipids the fatty acid found in the 1 position is saturated and in the 2 position, unsaturated. The chain lengths vary from 14 to 20 carbon atoms. The smaller chains may possess 1 or 2 unsaturations, but the larger chains may have as many as 5 or 6 double bonds.

Among the phosphocholines, phosphoethanolamines, and phosphoserines, another form of hydrocarbon grouping is encountered and the resulting compounds are called plasmalogens. They possess a fatty acid ester in the 1 position of glycerol but, unlike the diacyl glycerides, the hydrocarbon chain at the 2 position is linked to glycerol by an ether oxygen. In addition, an unsaturation is present. They are therefore vinyl ethers. These plasmalogen forms are most common in the phosphatidly ethanolamine lipids.

In many of the membrane lipids a hydrophilic region is present in a terminal position, and of these the orthophosphoric acid esters are of major importance. In phosphatic acid, two ionizable protons are available, but in other phospholipids, where two ester groups are united to the phosphoric acid, only one ionizable proton remains. Phosphatidyl-L-serine contains a carboxyl group of the serine as well as the phosphoric acid. Basic groups are also represented among the lipids. Lecithins and sphingomyelin contain choline, which is basic by virtue of the quaternary amine. Ethanolamine lipids and serine lipids contain a basic amine group united to the phosphate. A small group of lipids are related to sulfuric acid; of these, sulfatide is of special interest, since it is present in brain tissue.

Nonionic polar groups are also common. The phosphatidyl inositides contain the inositol group esterified to the phosphate. Cerebrosides contain galactose united to the sphingolipid structure. Glucose is found in other lipids.

III. Forces Acting between Lipid Molecules

Lipid molecules in water show a remarkable ability to form micellar aggregates, lamellar structures, or surface films. For many lipids there is no indication that single molecules split away and exist in a simple dissolved state. This characteristic of lipids is the result of forces acting between the lipid molecules themselves and between the lipid molecules and water. Among the forces that could act to produce these effects are:

(1) Coulombic or electrostatic (ionic) forces;
(2) ion–dipole interactions;
(3) London–van der Waals dispersion forces; and
(4) hydrogen bonding.

A. Ionic Interactions

The importance of the ionogenic groups of lipid molecules has not yet been fully explored. It is not difficult, however, to see how the forces and

interactions that could result from these groups influence the characteristics of lipid systems. We need only compare two lipids such as cerebroside
(7) and sulfatide (8) whose structures are closely related. These lipids
have similar apolar regions and the galactose molecule is present in both.
The sole difference is the sulfuric acid portion present in sulfatide. Cerebroside does not contain any ionic group but does contain the polar groups of
galactose. An immediate difference detected between these two molecules
is the ability of sulfatide to form stable dispersions in water with low
turbidity, while cerebroside does not readily form dispersed aqueous
systems.

The forces between ions of charges q and q' separated by a distance D
between their centers vary inversely with D^2 and with ϵ, the effective
dielectric constant of the medium. The effective dielectric constant is
drastically different from the value determined for the same medium in
bulk. When the distance between the ions is large, they are in fact separated by the medium, which then exhibits its normal, macroscopic properties. When the distance between the ions is small, the medium does not
separate the ions and it is influenced and altered by the ions. Pressman
et al. (1) have suggested that for distances less than 10 Å the effective
microscopic dielectric constant of water is given by the equation $\epsilon =
6D_{\text{Å}} - 11$. Thus, at small distances, the forces between ions can be very
significant because of the reduced value for the dielectric constant.

Salem (2) has presented an analysis of the forces acting between lipids
and proteins in water. For the reaction of anionic O^-, radius 1.4 Å, and
a basic ion such as $N^+(CH_3)_3$, radius 3.5 Å, there is a charge separation of 5 Å. Since the macroscopic dielectric constants of lipids and
proteins are roughly 2, the microscopic dielectric constant of water in the
presence of these charged groups at a separation of 5 Å might be 15 or
lower. The electrostatic energy of interaction of two univalent ions under
these conditions would be 4.1 kcal/mole.

In contrast to this large energy resulting from the interaction of two
ionic groups, Salem shows that the energy of the bond resulting from
polarization forces is quite small. Polarization forces would result from
the interaction of a charged ion with a polarizable portion of a neighboring
molecule, and could take place between an acid group of a lipid and the
CH_2 groups of another lipid or protein. Similar but weaker bonds can
arise from the polarization induced by a polar group such as OH. For
the bond formed between an ion with unit charge and a CH_2 group at
a separation of 5 Å, the energy of polarization is calculated as 0.002
kcal/mole.

B. London–van der Waals Dispersion Forces

In considering the structures formed by aggregates of lipid molecules, sufficient attention must be given to the long hydrocarbon chains. In an aqueous medium many lipids form layered structures in which the chains are directed away from the water and are oriented parallel to each other. Although these hydrocarbon groups are neutral, fluctuations in the charge density of one molecule will induce electric moments in a neighboring molecule. For two identical groups in vacuum, this energy is given by

$$W = \frac{3\overline{\Delta E}}{4D^6}\alpha^2$$

where α is the polarizability of the groups and $\overline{\Delta E}$ is usually the ionization potential of the systems. This is taken as the average electronic excitation energy. For two interacting CH_2 groups, the energy is given approximately by $W = -1340/D^6$ kcal/mole; thus, when the separation between the groups is 5 Å, the energy of interaction is about 0.1 kcal/mole. For two linear hydrocarbon chains containing n CH_2 units aligned parallel to each other, the dispersion energy of attraction varies with n and inversely with D^5.

The introduction of a methyl side chain in the hydrocarbon structure decreases the energy from -8.4 kcal/mole in stearic acid to -2.8 kcal/mole for isostearic acid. This large change in bonding energy clearly demonstrates the effects that can result from changes in structure that alter the closeness with which molecules approach each other. For the hydrocarbon chains of lipids, the greatest attractive energies will be encountered with the long chains that are fully saturated. Among the unsaturated hydrocarbons, the cis-unsaturated ones cannot be packed tightly because of the kinks in the chains. This is evidenced by a lowering of the melting point as we go from a saturated to a cis-unsaturated fatty acid of identical chain length. Similarly, the trans-unsaturated fatty acid melts at a higher temperature than the cis isomer (3). The distortion of the chain structure is less in the trans form, where the packing of the molecules is closer.

C. Hydrogen Bonding

The fact that lipids possessing strongly hydrophilic ionic or polar groups are able to form stable micellar aggregates or lamellar structures when mixed with water points to a decreased free energy for these systems.

A detailed, quantitative analysis of the thermodynamics of these systems must await a more complete knowledge of their structural details. These structures are ordered, and form, in spite of unfavorable entropy changes, because of compensating energies associated with the hydrogen bonding of the −OH and other polar groups with water. The hydration energies of ionogenic groups probably are major factors in maintaining the favorable free energy change. In addition, the polar groups of the lipid cause a breakup of the water structure at the interface and this possibly could produce an increased entropy for the entire lipid–water system.

The polar groups of membrane lipids are undoubtedly involved in hydrogen bonding in many ways. Intermolecular bonding between phosphate groups has been pointed out in the phospholipids (4). Hydrogen bonding between ionized and unionized carboxyl groups in neighboring molecules is another instance of such forces between molecules. A number of lipids contain either inositol, galactose, or some other moiety that is rich in −OH groups. The exact manner by which these OH units form hydrogen bonds with water or with other lipids is not known. However, this type of bonding is undoubtedly important in the formation of stable micelles or lamellar structures.

The role of the structure of liquid water in the bonding of the hydrocarbon regions of solute molecules is an area of current theoretical interest, and some of these concepts have been reviewed by Scheraga (5). The presence of hydrocarbons in water produces a greater degree of hydrogen bonding and increased order. However, the formation of a water cluster around the hydrocarbon leads to a lower energy, resulting from the interaction of water and the hydrocarbon. This change in the water structure around the nonpolar hydrocarbon groups becomes a major contribution to the formation of hydrophobic bonds.

In our description of specific lipid characteristics we will encounter other ionic interactions that serve to bind lipid molecules. Thus, the action of divalent cations on anionic lipids is to form bridges between negative charge sites on the lipid surface. The interaction of oppositely charged ionic groups on neighboring molecules is also important.

IV. Organization of Amphipathic Lipids in Water

When a liquid hydrocarbon is agitated with water, a stable dispersion does not form. The van der Waal-type forces between the hydrocarbon chains draw the molecules together in close association, while the hydrogen bonding of the water tends to squeeze out the nonpolar molecule from the aqueous phase. In contrast to this behavior, when hexose is agitated in water, hydrogen bonding of water with the polar −OH groups of the hexose is sufficiently energetic to disrupt both the water bonds and also the weak van der Waal forces of the CH and CH_2 members. Complete solubility on a molecular level takes place for the hexose in water. Intermediate conditions between that of the insoluble hydrocarbon and the soluble hexose arise when the molecule is a hydrocarbon chain at the end of which is a highly polar group, such as OH, or an ionic unit, such as an acidic group or a basic amine. In these cases, stable systems with a free energy minimum can form when the polar groups are directed into the water phase and the nonpolar hydrocarbon chains, associated with each other, are directed away from the aqueous region. The stable arrangements of these amphipathic molecules that are of interest to this discussion are monolayers, micelles, and lamellar aggregates.

V. Lipid Monolayers

The study of monolayers formed by a great number of lipids and other amphipathic compounds has been a valuable tool in determining some of the important characteristics of these compounds. When small amounts of an amphipathic substance are added to a clean water surface, the forces between the water molecules and the polar or ionic groups of the amphipathic molecules cause a spreading of the molecules along the water surface. When the hydrocarbon chains are sufficiently long, their cohesive forces prevent the molecules from dissolving in the water, and a monomolecular film forms. The introduction of the polar groups into the surface of the water lowers the surface tension of water. The difference in surface tension between a clean water surface and the water surface coated with a monomolecular layer of the amphipathic compound is measured as the surface pressure (π).

Experimentally, two devices are favored. The Langmuir–Adams film balance uses a float suspended from a torsion wire to measure the difference in surface tension between the clean water on one side of the float and the monolayer on the other side. Another device, named for Wilhelmy, employs a glass slide or platinum foil dipping part-way into the water. The plate hangs suspended from a torsion balance. When a monolayer is formed on the water, the surface tension is reduced and the force acting on the torsion balance is decreased. When either of these procedures is used, the area of the monolayer is decreased by moving a paraffin-coated glass slide along the surface of the water, and the relation between the surface pressure and the area of the film is measured. The surface pressure is expressed as the force acting per unit length of the float and the area is expressed as the cross-sectional area in $Å^2$ of the molecules making up the monolayers. The latter can be computed from the knowledge of the mass of substance spread on the surface, its molecular weight, and the measured area of the compressed film.

The study of monomolecular films at the air–water interface very often includes measurements of the potential across the surface film. The procedure employed most frequently is to mount a metal probe at a small distance above the liquid surface. Some polonium or radium on the lower surface of the probe emits α particles, which ionize the air space between the probe and the liquid surface. A silver–silver chloride reference electrode dips into the liquid and an electrometer is connected in the circuit to measure the dc potential between the probe and the liquid surface. When a film is formed on the surface the potential between the probe and the liquid differs from that measured for the clean surface. This difference in potential (ΔV) between the potential for the substrate alone and the substrate coated with a film is related to the number of charges per square centimeter of surface (N), the dipole moment (μ), and the angle of inclination of the dipole to the perpendicular by the equation

$$\Delta V = 4\pi N\mu \cos \theta$$

It is assumed that the dielectric constant of the film is unity and that the water structure at the surface of the substrate is the same with and without the overlying film. The restructuring of water along the surface produced by the polar groups of the film quite understandably alters the potential at the interface, and this change influences the surface potential.

VI. Properties of Lipid Monolayers

Monolayers formed by lipids at the air–water interface have been studied extensively. This review is necessarily limited to some of the results that have added significantly to our knowledge of the properties of lipids. An important area of investigation has been concerned with the reactions at such interfaces, and these include the reactions with acid or base or with dissolved ions in the aqueous substrate. The kinetics of such reactions as ester hydrolysis or enzyme activity were followed by measuring the change in surface pressure or potential as the film reacted with a component in the aqueous phase. The reactions of lipid films with a large number of compounds of biological interest have also been investigated. In these studies a lipid film is formed on the aqueous substrate which, in some cases, contains added acid, base, buffer, or other ionic compounds. The substance of biological interest (proteins, anesthetics, or antibiotics) in an appropriate solvent is injected behind the movable float into the aqueous phase. By stirring, it is brought to the region below the film where it may react, and changes in surface pressure or potential measure the extent of the reaction.

A. SIMPLE LIPIDS

We will first review some of the studies made with simple lipids and then some of the more complex. Purified fatty acids when dissolved in petroleum ether can be spread on carefully distilled water or salt solutions. A monolayer forms on evaporation of the solvent. For saturated fatty acid chains containing 16 or more carbon atoms, the pressure–area (π–A) curves show the characteristics of the liquid-condensed state when the neutral molecules are compressed. In a study of monolayers of fatty acids with 14–18 carbon atoms, Spink (6) reported on the effects of pH on the film characteristics. Using 0.01 M NaCl substrates with pH increasing from 2.0 to approximately 8.0, π–A curves show a decrease in the molecular areas of the carboxy acids. The acids with shorter chains become soluble with increasing pH. Myristic acid becomes soluble above pH 5.5 and palmitic acid above pH 8.5. For stearic acid, which could be brought to a high pH, a contraction was followed by an expansion at pH > 10. At this high pH, the carboxylic acid is fully ionized and repulsions act between the charged head groups expanding the layer. At lower pH levels a mixed monolayer of both ionized and unionized species is present. The contraction in the film is attributed to ion–dipole interaction between these two forms as well as increased hydrogen bonding between them.

Measurements of surface potential at different pH levels for these films at a constant molecular area resemble a titration curve. The midpoint corresponds to a substrate with a pH of 8.7. The dissociation constant of the acid in a monolayer (pK_s) is then almost four units greater than in the bulk phase. If the ionization of the carboxyl groups is assumed to have the same characteristics in the surface film as it has in true solution, then the apparent difference in pK values is attributed to a difference in pH of the surface and the solution. This difference results from the negative potential (ψ) of the ionized surface which holds H^+ ions in the region of the surface. The surface pH and bulk pH (pH_s and pH_b) values are related by the equation

$$pH_s = pH_b + \frac{e\psi}{2.3kT}$$

At 20° we can evaluate $2.3kT/e$ as 59 mV; then $pH_s = pH_b + (\psi/59)$. When $\psi = 0$ for an uncharged surface, $pH_s = pH_b$. For anionic surfaces, ψ is negative and $pH_s < pH_b$. Using the equation of Guoy, a value for ψ can be calculated for the incompletely ionized surface:

$$\psi = -\frac{2kT}{e} \sin h^{-1} \frac{(134\alpha)}{(A\sqrt{c})}$$

Assigning a value of 0.5 for the degree of dissociation, an area A of 21 Å², and an electrolyte concentration c of 0.01, the value of ψ is found to be -213 mV, from which $pH_s = 5.1$. Although some questions arise concerning the applicability of Guoy's equation to highly charged surfaces, the value for pH_s is in good agreement with the ionization constant for stearic acid. This supports the view that the dissociation constant of an acid at an interface does not differ greatly from the value in a bulk phase, although the pH of the surface and the bulk solution do vary.

The reactions of monolayers of fatty acids with metallic ions in the aqueous subphase have interested a number of investigators. Goddard and Ackilli (7) measured π–A and ΔV–A changes for stearic acid monolayers at various pH levels. From pH 3 to 8.5, the π–A curves were not influenced by changes in proton activity. In this pH range the curves show a linear section running at an angle from approximately 3 to 26 dynes/cm before the beginning of the steeply rising curve characteristic of the compact film. This linear region is ascribed to "close-packed head groups." At pH levels above 10.6, however, the curve bent and rose sharply at an area of 20 Å (Fig. 1).

Interesting effects were found on adding Ca^{2+} or Mg^{2+} to the sub-solution at pH 10. Here, the divalent metal ions bound together adjacent acid molecules. In 10^{-6} M $CaCl_2$ the films contracted sharply at low surface pressures with the disappearance of the close-packed head region. On adding pyrophosphate or other complexing agents, the return of this part of the curve indicated the absence of free alkaline earth metal in the subsolution. Adding an excess of calcium or magnesium to the solution again caused a sharp contraction of the film.

Experiments in the author's laboratory showed additional changes that can influence this property of the π–A curves (8). When the concentration of Ca²⁺ was increased to 10^{-4} M, the "close-packed head" portion of the

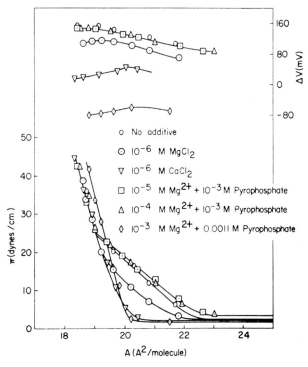

Fig. 1. π–A isotherms of stearic acid monolayers at 25° on solutions at pH 10 containing 0.01 M NaCl and various additives. [Reproduced from E. D. Goddard and J. A. Ackilli, *J. Coll. Sci.* **18,** 588 (1963), by permission of publisher.]

curve was not present, giving way to the compact film at pH levels above 8.0. For lower concentrations of Ca²⁺, higher pH values were needed to produce the compact film. When the subsolution contained 1×10^{-3} M NaHCO₃ as well as 1×10^{-4} M Ca²⁺, this condensed film could be produced at pH levels as low as pH 5.5. This effect of bicarbonate may be explained by the formation of some subsurface "structure" with the Ca²⁺ layered between the carboxylate ions of the film and bicarbonate ions in solution leading to an effective neutralization of charges and reduced free energy.

The binding of carboxylic acid monolayers and metallic ions was studied in a direct manner by analysis of the film. Bagg et al. (9) allowed stearic acid monolayers to react with Ca^{2+} in subsolutions at known pH levels. The film was skimmed and analyzed by infrared spectroscopy for stearic acid and calcium stearate. These could be identified by measuring the intensity of the unionized COOH absorbance and that of the COO^-. Since the absorbance of CH groups should be unchanged by the ionization of the acid group, the intensities of both groups were related to the CH absorbance. The variation of Ca^{2+} content with pH resembled a titration curve.

Similar studies (10) of the reactions of other alkaline earth metals with fatty acid monolayers show a strong reaction of beryllium with the carboxyl group, with considerable covalency. Ba^{2+} forms an ionic bond with the lowest stability constant for the series. Calcium, in an anomalous manner, binds more strongly than magnesium. This result is not in keeping with the decrease in binding that is usually associated with increasing ionic radius of the metal.

In a related area, Sears and Schulman (11) measured π–A and ΔV–A curves for stearic acid films on solutions of alkali metal hydroxides. In the temperature range 15–37°, the molecular area of the metallic soap varied in the order $K > Na > Li$. This sequence is the order of the radii of the alkali cations.

Since many of the lipids of biological interest contain phosphate, the study of related amphiphilic phosphates is of considerable value. In a study of monolayers formed from mono- and dioctadecyl phosphoric acid, Parreira (12) investigated the effect of pH and inorganic ions on these films. In the pH range from 4.0 to 6.0, the molecular area increased with pH. No changes occurred below pH 4.0, while dissolving of the film precluded the use of solutions above pH 7.5. The limiting area for the unionized molecules was 25.5 $Å^2$, ascribed to the area of the phosphate head group. At pH 6.0, this area increased to 27.5 $Å^2$. Films formed from dioctadecyl phosphoric acid showed an expansion from pH 3.0 to 9.0. The molecular area at low pH was 39 $Å^2$, that of two fatty acid chains oriented almost vertically to the surface. Measurement of surface potential at constant area showed a variation with pH that resembled a titration curve. An apparent surface pK was obtained from the inflection point of the curve. The measurements of surface potential permitted calculations of the degree of ionization of the surface molecules, and from this pK_s, the true pK for the surface acid, was found. For mono-octadecyl phosphoric acid, the pK_s for the first ionization is 2.8, and for dioctadecyl phosphoric acid it is 1.7.

In a later study of long-chain alkyl phosphates, Gershfeld and Pak (*13*), in addition to studying π–A and ΔV–A changes in monooctadecyl phosphate films, also measured the surface viscosity. This increased with time and with decreased pH, changes attributed to the formation of surface aggregates through hydrogen bonding of the phosphate groups.

B. Electron Microscopy of Films

Ries and Walker (*14*) have obtained electron micrographs of a variety of compounds spread as monomolecular films. In their method a glass plate is raised vertically through the surface film and a portion of the monolayer is transferred onto a collodion covered screen carried on the plate. The film is then shadow-cast with metal and studied in the electron microscope. π–A measurements can be made at the same time.

In one study they used binary mixtures of compounds—one molecule was oriented horizontally on the surface, the other, vertically. A typical system was an equi-weight mixture of stearic acid and polyvinyl acetate. A film of stearic acid alone has an area of 20 Å^2/molecule with a film thickness of 25 Å. Polyvinyl acetate films indicate a molecular area of 27 Å^2 with a thickness of 5 Å, pointing out the horizontal alignment in this case. For the mixed film the π–A curve shows combined areas that are smaller than calculated for the ideal system and indicates a tighter packing of the two types of molecules. At intermediate pressures a plateau in the π–A curves results from the squeezing out of the polyvinyl acetate from the film leaving stearic acid. The collapse pressure for the mixed film is very much greater than for either component, indicating strong forces between the molecules.

Films containing *n*-hexatriacontanoic acid were also studied. These longer hydrocarbon chains give good results in the micrography of transferred films. At low surface pressures the fatty acid forms islands with empty regions between them. As the pressure increases, the empty regions decrease, until homogeneous, continuous areas are formed at high pressures. At collapse pressures, long platelets of bimolecular layers are observed resting on a continuous monolayer structure.

C. Phospholipids

Regarding the lipids present in biological membranes, relatively few experimental procedures have been available for study of the structural and chemical properties of their aqueous systems. Without question, the study of their monolayers has given us an important understanding of some of these properties. It is unfortunate that much of the early work was done with preparations of doubtful purity. Only in the more recent studies, using either synthetic compounds or products from natural sources purified by modern chromatographic methods, do we have data that can be pieced together to give a more complete and meaningful picture.

The most abundant and readily available compound of this class of lipids is lecithin (phosphatidyl choline) (2). Obtained from natural sources, the fatty acid chains vary in length and degree of unsaturation. Many of the experimental studies have been carried out with egg lecithin, which is a mixture of compounds containing almost 50 mole% saturated C_{16} and C_{18} chains and an almost equal percentage of C_{18} mono- and di-unsaturated chains.

Recent work has preferred the use of synthetic compounds of precise structure. A thorough investigation of the effect of the hydrocarbon chain on the π–A characteristics of monolayers formed from phospholipids was carried out by Van Deenen and co-workers (15). For lecithins containing saturated acyl groups the molecular area of closest packing was approximately 60 Å² for the 1,2-didecanoyl-L-phosphatidyl choline, and this decreased with increasing chain length to 40 Å² in 1,2-ditetracosanoyl-L-phosphatidyl choline. This indicates that the intermolecular van der Waals' forces between CH_2 groups of neighboring molecules are additive. The shorter chains also showed greater compressibility before reaching the closest packed configuration. Similarly, the introduction of one unsaturated chain of identical length results in an expansion of the film. In this case, the expansion is caused by the "kinking" of the carbon chain.

The π–A, ΔV–A, and surface viscosities of 1,2-dipalmitoyl-L-phosphatidyl choline monolayers were recently reinvestigated by Vilallonga (16). The π–A curves at 25, 35, and 45° indicate an increased area with increasing temperature, whereas the compressibility–pressure curves showed sharp discontinuities at 25 and 35°. These are attributed to phase transitions that take place at these temperatures. No such change was observed at 45°. When fine crystalline particles of this lecithin were dropped onto a clean water surface, at temperatures up to 40°, no spreading on the surface occurred and no increase in surface pressure was observed. At 45°, however, spreading took place rapidly until an equilibrium pressure was attained. The relation between these observations and the physical properties of lecithins are discussed on p. 62. From the measurements of surface potentials of the films, calculations were made of the surface dipole moment per molecule: $\mu_1 = (A\Delta V)/12\pi$, where A is the molecular area in $Å^2$. The dipole moments are lower at 35° than at 25°. The general range of these values is consistent with a model in which the two charged groups of the phosphoryl choline are coplanar and parallel to the interface.

The phase transition that takes place in the monolayers of lecithin is further discussed by Phillips and Chapman (*17*). They point to a close similarity between the phases in the monolayer and the states detected in aqueous dispersions of lecithin. The hydrocarbon chains in the gel state of the dispersion are like those in the condensed monolayer, both resembling the crystalline form of the hydrocarbon. In aqueous dispersions of lecithin, the smectic state (consisting of bimolecular lamella structures) and in the liquid-expanded state of films, the hydrocarbon chains are in a molten state with ability to flow along a two-dimensional surface.

It is of interest to compare monolayers of lecithins with some of the other major phospholipids. Studies by Van Deenen and his group (*15*) compared synthetic 1,2-distearoyl-L-phosphatidyl choline with 1,2-distearoyl-L-phosphatidyl ethanolamine. The lecithin gave a molecular area of 40 $Å^2$ on closest packing, while the ethanolamine lipid was smaller with an area of 36 $Å^2$. When one of the hydrocarbon chains was unsaturated (oleoyl replacing stearoyl), both lecithin and phosphatidyl ethanolamine films had expanded areas, with the lecithin again approximately 4 $Å^2$ larger. These differences cannot easily be reconciled. The limiting area of 36 $Å^2$ for the saturated diacyl phosphatidyl ethanolamine agrees with that of two fatty acid chains. The somewhat larger area for the lecithin having the same hydrocarbon structure could be explained by reasoning that the geometry and dimensions of the phosphorylcholine (including the bulky trimethyl group) require larger intermolecular spacing than the phosphoryl ethanolamine. However, at elevated temperatures when thermal effects have increased the average spacing between the carbon chains, the difference in the areas of lecithin and phosphatidyl ethanolamine should not persist.

Phillips and Chapman (*17*) also show that the monomolecular areas at 22° of saturated phosphatidyl ethanolamines are smaller than lecithins with the same carbon chains. At this temperature among the lecithins, the π–A curve of dipalmitoyl lecithin (C_{16}) appears to show a phase transition from the liquid-expanded film to the condensed film, while among the phosphatidyl ethanolamines it takes place at this temperature at roughly the same surface pressure with the dimyristoyl (C_{14}) member.

Although this change of state involves the increased ordering produced by the cohesive forces of the hydrocarbon chains, it appears that two (CH$_2$) units are needed to compensate for the greater bulk or greater hydration or greater disorder associated with the choline rather than the ethanolamine moiety. Layered on a subphase of 0.1 M NaCl these lipids are at or very near the isoelectric point and no charge difference can be the basis for their differences in state. It would then appear that for the lipids with saturated diacyl groups the lecithins would remain in the form of the liquid film or aggregates in the smectic mesophase with somewhat longer carbon chains than the phosphatidyl ethanolamines.

Van Deenen et al. (15) further showed that the π–A curves at pH 7.4 of phosphatidic acid and phosphatidyl ethanolamine, both having distearoyl chains, have equal limiting areas. At this pH level, phosphatidic acid possesses a smaller polar group but does have a large negative charge, with repulsive forces tending to increase intermolecular distances. That both these lipids have equal limiting areas gives weight to the view that this area is determined by the hydrocarbon chain dimensions. Phosphatidyl serine, which is also highly negative at this pH level, has a limiting area similar to that of lecithin, which is neutral. Here the area is greater than that of the hydrocarbon chains and is not attributable to the effect of a net charge. We may find some similarity in either the dimensions, orientation, or hydration of the phosphorylcholine and the phosphorylserine groups to be the determinant of the limiting areas for both of these lipids.

The interaction of monolayers of phospholipids with metallic ions has been studied by several investigators. Some of the more recent results will be described here. The π–A and ΔV–A curves were measured by Shah and Schulman (18) for several phospholipids on solutions containing metallic salts. They found that the π–A curves of several types of lecithin as well as those of dicetyl phosphate are not altered by the presence of univalent metal ions in the aqueous phase. The ΔV–A curves, however, do show changes when layered over solutions of divalent cations. Of the lecithins tried, the fully saturated dipalmitoyl showed the greatest increase in surface potential on solutions of divalent ions, while the highly unsaturated yeast lecithin showed very little change. Shah and Schulman explain the differences observed for the different forms of lecithin in terms of the increase in the intermolecular separation produced by unsaturations of the hydrocarbon chains. In the case of dipalmitoyl lecithin, the molecules are sufficiently close so that a Ca^{2+} ion can bridge the negative phosphates of two molecules. This reduces the contribution of the negative phosphate toward the resultant dipole of the molecule, which then becomes more positive. With unsaturated lecithins the phosphate groups are further apart and univalent ions and water enter the intervening spaces, decreasing the effectiveness of the divalent cation.

The binding of calcium by monolayers of phospholipids was also studied by Hauser and Dawson (19). They formed monolayers of various lipids on an aqueous phase containing a low concentration of $^{45}CaCl_2$. The radioactivity detected directly above the surface was measured before and after the monolayer was spread on the surface. The increased activity was related to the $^{45}Ca^{2+}$ adsorbed by the lipid. Their results indicate that at pH 5.5 the only lipids that reacted with calcium were those that were anionic and had a negative charge. It is understandable that phosphatidyl choline, sphingomyelin, and phosphatidyl ethanolamine (which have zero net charge) would not react, nor would cerebroside, which does not possess any ionic groups. Stearic acid reacted with calcium only in the alkaline pH range. After a lipid monolayer had reacted with calcium, the addition of magnesium, or alkali metal salt, resulted in a displacement of calcium by the added ion. No difference was observed in the effectiveness of potassium and sodium in exchanging for calcium. However, differences did exist among the lipids in the concentration of K^+ or Na^+ required to reduce by one-half the amount of calcium bound. This was greatest for dicetyl phosphate and least for ganglioside.

In a related study, mixtures were prepared of lecithin with varying amounts of one of the anionic lipids. These were dispersed in NaCl and the electrophoretic mobilities of the particles were measured. For each of the acidic lipids used, the mobilities of the particles increased with the mole percent of the acidic lipid in the mixture with lecithin. Furthermore the mobilities of the systems containing lecithin mixed with dicetyl phosphoric acid were the same as those with phosphatidyl inositol, since both of these carry one unit of charge when fully ionized. Phosphatidic acid systems had mobilities twice this value and triphosphoinositide, five times as much. These are also the numbers of unit charges of the ionized molecule.

An apparent disagreement exists in the results of Hauser and Dawson and those of Shah and Schulman. Whereas the former found no binding of Ca^{2+} to lecithin films, the latter explain the changes in the ΔV-A curves of lecithin on solutions containing Ca^{2+} as resulting from the interaction of Ca^{2+} with the phosphate. Possible reasons for the disparity may be that Hauser and Dawson used very low concentrations of $CaCl_2$, while Shah and Schulman used 0.01 M solutions. Further, the former group used egg lecithin, which was shown to have a smaller interaction with Ca^{2+} than a saturated lecithin.

A problem that has drawn the attention of several investigators of lipid chemistry has been the relation of cholesterol to phospholipids in membranes. In a number of biological systems the cholesterol and phospholipids are present in approximately equal molar amounts. This has led many workers to the belief that some form of complex between cholesterol and phospholipid aids in the stabilization of the membrane structure. An early study was made by DeBernard (20), who formed monolayers of lecithin and cholesterol in varying amounts and determined the mean molecular areas at a constant pressure for the two compounds. The mean molecular area is obtained by dividing the area of the film by the total number of lipid molecules deposited on the trough. DeBernard found a negative deviation (a condensation) from that predicted for ideal mixing of the lecithin and cholesterol. This suggested some form of bonding between the two compounds. Several investigators have recently returned to this question.

Demel et al. (21) studied monolayers formed from synthetic lecithins or phosphatidyl ethanolamines mixed with cholesterol. They found condensation between lecithin and cholesterol for some systems but not all. Some of these forms of lecithin contained saturated hydrocarbon chains and some contained unsaturated ones. The highly unsaturated lecithins did not show any condensation with cholesterol. The forms of phosphatidyl ethanolamine with hydrocarbon chains analogous to the lecithins exhibited the same behavior with cholesterol as did the lecithins. No simple explanation can be given for these different effects. A possible explanation may involve a phase transition produced by cholesterol for some of the lipids at the temperature of the experiment. For the highly unsaturated lipids this temperature is above the transition temperature, and cholesterol has no condensing effect, while for the long-chain saturated lipids, despite the presence of cholesterol, the transition temperature is above the experimental temperature.

Shah and Schulman (22) also investigated monolayers formed with cholesterol mixed with dicetyl phosphate or lecithin. They found that the ΔV–A curves were a better indicator of any interaction between cholesterol and the phosphate than the π–A curves. The mean molecular areas for mixtures of dicetyl phosphate and cholesterol showed no deviation from additivity, whereas the surface potentials showed a negative deviation as a result of the ion–dipole interaction of the hydroxy group of cholesterol with the phosphate. This effect was less for films on solutions of $CaCl_2$ than on NaCl because of the neutralization of the phosphate charge by

Ca^{2+}. The π–A curves do not show any contraction with added cholesterol because the dicetyl phosphate forms condensed films with tightly packed hydrocarbon chains. The cholesterol must assume spaces between the dicetyl phosphate molecules, causing a progressive expansion of the film. Phosphatidic acid mixed with cholesterol gave effects similar to those found with dicetyl phosphate—there was a small contraction of the mean area per molecule but a large deviation in the surface potential. This again points to the interaction of the phosphate with the cholesterol hydroxy group.

In mixtures of dipalmitoyl lecithin and cholesterol, a reduction in the mean molecular area was found at low surface pressures but not at maximum pressure. Furthermore, the surface potentials of the films showed a linear variation with the molar ratio of the two constituents, indicating that no interaction took place between the cholesterol and the lecithin phosphate. Shah and Schulman explain their results by stating that at low surface pressures, thermal disturbances create cavities or spaces between the molecules of lecithin into which the cholesterol fits. This leads to an observed decrease in the mean area for the two types of molecules. At high pressures these cavities are not present and the mean molecular area varies linearly with the mole ratio of the two components.

Shah and Schulman also formed mixed films with egg lecithin, which contains unsaturated and saturated acyl chains. Egg lecithin has a larger molecular area than dipalmitoyl lecithin, providing cavities for the cholesterol that persist even when the film is highly compressed. The mean surface areas per molecule for the lecithin–cholesterol systems showed a contraction at all surface pressures. The surface potentials of these systems followed the simple additivity rule and showed no interaction between the molecules. The significance of much of this work is that the nonadditivity of the surface potentials showed an interaction between cholesterol and the phosphate groups of dicetyl phosphate as well as phosphatidic acid. The absence of such an effect in the systems with lecithin indicates that the phosphate group is not free and is presumably neutralized by the trimethyl ammonium ion.

VII. Aqueous Dispersions of Lipids

A. Chemical Characteristics of Lipid Dispersions

Phospholipids can be dispersed in water with varying degrees of ease and stability and lecithin dispersions in particular have been studied by a number of investigators. Electron micrographs of lecithin aggregates in which the aqueous and polar regions are made visible by treatment with heavy metals can be prepared. The hydrocarbon portions of the lipid appears as light bands 40–70 Å thick enclosing regions of water or aqueous solution. The lipids that contain an anionic charged group disperse most readily. With mild agitation in water they form highly hydrated particles, the system having a low turbidity. The lipids with zero or small net charge such as lecithin, sphingomyelin, or phosphatidyl ethanolamine form dispersions less readily. These systems are quite turbid with larger particles that can be sedimented by centrifugation.

An illustration of the effectiveness of an anionic group in forming dispersions can be seen by mixing a relatively small amount of phosphatidic acid or dicetyl phosphate with lecithin in an organic solvent; after evaporating the solvent the mixed lipid is dispersed in water. This dispersion is much less turbid and more stable than lecithin alone, except at low pH levels. In this mixed lipid the repulsions of the charged groups as well as the increased hydration of the polar portion of the molecules are major factors in its ability to form dispersions.

B. TITRATIONS WITH ACID OR BASE

Studies made by the author and others that are now bringing to light some additional information on the chemical characteristics of aqueous lipid systems will be described here. A number of the procedures used and general observations made can be illustrated by the studies on phosphatidic acid.

Phosphatidic acid is a relatively simple lipid. It can be prepared from lecithin by simply splitting off the choline portion. As a diprotic acid it forms salts with metallic ions. Combined with univalent cations, the phosphatidate can be dispersed in water (23), but with the calcium salt, dispersion is more difficult or incomplete. Dispersions of sodium phosphatidate have been titrated with acid and base, and a typical titration curve is shown in Fig. 2. When calcium phosphatidate is partitioned with aqueous HCl, the resulting phosphatidate in water has an initial pH of 3.1–3.2. In this form the lipid contains little metallic ion and is therefore predominantly in the acid form.

Systems that had been partitioned with NaOH or KOH contained more than 1 equivalent of metal cation per mole. The initial pH of these systems were above 7. Titrations were also performed with standard NaOH and HCl in the range of pH 3–10. A blank titration of water under like conditions was made and this was subtracted from the titration curve of the lipid–water system so that the number of equivalents of H$^+$ taken up or released by the phosphatidic acid alone at each pH level was obtained. This corrected titration curve showed regions of increased slope in the neighborhood of pH 5.8 and 9.8. These are equivalence points for the conversion of phosphatidic acid to sodium hydrogen phosphatidate and disodium phosphatidate, respectively.

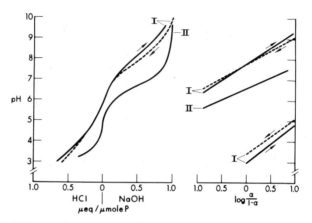

Fig. 2. (I) 18-hr titration of an acid-dialyzed dispersion of phosphatidic acid. The titration curve was continuously monitored during this period. Dispersion contained 13.8 μmoles phosphatidic acid in 5 ml of water and had an initial pH of 3.10. (II) titration of 13.5 μmoles of sodium β-glycerophosphate in 5 ml of H$_2$O. Panel at right shows log ($\alpha/1 - \alpha$) against pH for first and second acid groups of I and second acid group of II. [Reproduced from M. B. Abramson, R. Katzman, C. E. Wilson and H. P. Gregor, *J. Biol. Chem.* **239**, 4068 (1964), by permission of publisher.]

By graphical methods, the changes in pH produced by small volume additions of reagent (ΔpH/ΔV) were plotted against pH, and the pH of maximum slope gave a more accurate evaluation of the equivalence points. In a like manner the region of minimum slope gave the pH of the apparent pK (pK$'$) for the acid group titrated. In water, pK values for the first and second acid ionizations were 3.8 and 8.6, respectively, values that were lowered when titrations were carried out in 0.1 N NaCl or KCl. These results are of interest in that one equivalent of base per mole of phosphatidic acid is required for the titration from the first to second equivalence points. This indicates that although the single lipid molecules are insoluble in water, all of the molecules in the aggregate have their ionizable groups directed toward the aqueous solvent. These molecules are either on the outer surface of the aggregate or have their ionic groups exposed on the surface of the lamellar structures making up the aggregate. The solvent that surrounds the entire particle then must be freely penetrable into the interior portions. These systems then can serve in some studies as models for biological membranes.

Sodium β-glycerophosphate is a water-soluble analog of phosphatidic acid. Its solution was titrated under identical conditions and showed values for the equivalence points and pK that were roughly 1.5 units below those of phosphatidic acid. This difference, which is also observed for soluble and insoluble carboxylic acids, is attributed to the negative surface potential acquired by the insoluble polyelectrolyte as it releases protons. This surface potential impedes the further ionization of the acid groups, effectively decreasing the acid strength of the polyelectrolyte. The presence of salts in the solution decreases the surface potential and thereby lowers the pK of the polyelectrolyte. This was observed in the titrations of other acidic lipids studied.

When phosphatidic acid and lecithin were dissolved in organic solvents and the solvent was evaporated, the solid that formed contained the two lipids intermixed. Moreover, when this solid was dispersed the particles contained both compounds. The titrations of these mixtures showed the reaction with base of all the phosphatidic acid present, but the lecithin did not react. Because of its tightly bound acid and base groups, lecithin shows very little reaction with acid or base in the pH range 3.5–10. The phosphatidic acid molecules, which presumably become oriented within the mixed lamellar surface, are situated so that the acid groups are available for reaction with dissolved ions.

Titration studies of other lipids have been carried out in an analogous manner. Phosphatidylserine possesses a monoprotic phosphate and a carboxylic acid group as well as an amine. The titration of this lipid in water shows an equivalence point in the region of pH 7.5 and a pK' of 3.8 (23). The equivalence point is that of the formation of the mono-sodium salt, while the acid group titrated below this pH is the carboxyl. The phosphate is bound in some manner to the amine. Titrations to pH 10 involve the amine group. This frees the phosphate so that at this pH level the disodium salt is formed with a metallic ion on each acid group. In dilute salt solutions the acid strength of the ionic group is effectively increased and the pK' for the amine was found to be 9.7.

Phosphatidylinositol contains a phosphate esterified to glycerol and to inositol and thus possesses one ionizable hydrogen. As it contains both the highly polar inositol and also an anionic group, this compound readily forms aqueous dispersions. The titrations of this lipid show that the pK' for the acid in water is 3.1 (24). This is lowered in the presence of salts so that in 0.08 M NaCl it is 2.5.

Titration curves for phosphatidylserine and triphosphoinositide in water and in salts of Ca, Mg, and Ni were carried out by Hendrickson and Fullington (25) and an evaluation of the stability constants of the metal complexes was made. It was found that these lipids formed more stable metal complexes than related soluble compounds with similar polar groups. Thus, phosphatidylserine formed more stable metal complexes than O-phosphoserine.

C. REACTIONS WITH CATIONS

Another area of considerable interest in the study of membrane lipids is their reaction with metallic ions and some organic cations. Such reactions alter the surface characteristics of lipid particles and may point to similar changes in biological membranes. There is little doubt that in biological systems, since the membrane lipids are exposed to aqueous media at or near neutral pH, the acidic groups are ionized and are able to react with metallic ions in the surrounding fluid. Studies conducted in the author's laboratory on phosphatidic acid will serve to illustrate concepts applicable to other lipid systems.

Titration studies showed (26) that at pH 7 phosphatidic acid (H_2PA) exists as the acid anion (HPA^-) and the basic ion (PA^{2-}). At this pH level a small concentration of cation is present, due to the addition of base required to bring the acidic form of the lipid to the neutral range. We can visualize the negatively charged lipid particle surrounded by the double layer in the aqueous phase with hydrogen ions, and other cations held in the double layer by the negative surface potential. When small amounts of salt solutions (also at pH 7) are added to such a dispersed system a sharp drop in pH is observed. With further additions of salt the change is less pronounced and then becomes minimal. Three effects are occurring: (1) The addition of the salts causes an increase in the concentration of cations, thus permitting some of these cations to enter the double layer and exchange for H^+; (2) the increased ionic strength of the medium leads to a greater ionization of the lipid; and (3) any specific binding of the lipid anion to a metallic ion decreases the concentration of anions and the surface charge, permitting further ionization of the acid. In order to measure specific cation reactions with the lipid it is desirable to minimize the first two effects. The addition of quaternary ammonium chlorides to the lipid in water releases fewer hydrogen ions than addition of alkali metal chlorides, indicating that little or no specific binding takes place between the lipid and the large organic cation.

Systems were then prepared in 0.10 M tetramethylammonium chloride (TMACl) and the hydrogen ion released by the addition of metallic chlorides was assumed to be the result of a specific binding of metal and lipid. Experimentally, these cation–proton exchange experiments were satisfactorily conducted at a constant pH level. The lipid, with added TMACl, was brought to the desired pH by the addition of 0.10 N tetramethylammonium hydroxide (TMAOH) and the amount of added base was noted. Small serial additions of NaCl, KCl, or other salts, all adjusted to the same pH, were then made. Following each addition, with ensuing release of hydrogen ions, the pH of the system was restored to the initial level by addition of measured volumes of TMAOH. In this way, it was possible to construct curves relating the concentration of added metallic salt and the number of microequivalents H^+ released per micromole of lipid. This permits a comparison of the relative effectiveness of the different metals in their reactions with the lipid. For univalent salts the concentrations studied ranged as high as 0.8 M. In such instances the total ionic strength varied considerably. For divalent metals, concentrations did not exceed 0.003 M and, therefore, the ionic strength did not change significantly from the value of the 0.1 M TMACl. Using such procedures, the apparent stability constants (since activity coefficients are neglected) for the metallic complex with phosphatidic acid were calculated for the reaction $Me^+ + PA^{2-} \rightarrow MePA^-$. The reader is referred to the original papers for the method of calculation of K' from the experimental data (26a).

In a modification of this method the phosphatidic acid in a solution of TMACl was brought to pH 7 by addition of TMAOH. Additions of the appropriate salt were then made until the pH had dropped to either 6.5 or 6.0. From the number of microequivalents of H^+ ion exchanged and data obtained from the titration curve of phosphatidic acid, the stability constants shown in Table 1 were calculated. The first column gives values calculated from the experiments at constant pH 7, while the second and third column give values obtained from pH changes from 7.0 to 6.5 and 7.0 to 6.0 respectively.

These constants point out the much stronger binding of divalent cations than the univalent ones. This is in part the effect of the greater charge but it is also associated with the ability of the divalent ion to bridge two negative charge sites, either on the same or neighboring molecules.

Among the univalent metal ions the binding to phosphatidic acid decreases with increasing ionic radius of the element.

An entirely different procedure was found especially effective for determining the stability constants of calcium complexes with phospholipids (26b). The lipid was dispersed in water and Tris buffer was added to maintain a pH of 7.2 and to serve as a supporting electrolyte. Small additions of $CaCl_2$ were made, producing a sharply rising increase in the turbidity of the systems as the calcium reacted with the lipid anions. The turbidity was measured by a light-scattering photometer by comparing

TABLE 1
STABILITY CONSTANTS

	I	II	III
Ca PA	1.60×10^4	1.17×10^4	1.95×10^4
Mg PA	0.97×10^4	1.49×10^4	1.10×10^4
Li PA$^-$	17.3	20.5	15.0
Na PA$^-$	15.8	12.2	11.0
K PA$^-$	8.9	9.9	6.42

the ratio of the intensity of light scattered at 90° with that of the transmitted beam.

In order to determine the stability constants, the procedure used two equal aliquots of known concentration of phosphatidic acid in Tris buffer. Sodium citrate was added to one system and sodium chloride was added to the other to give equal sodium concentrations. Serial additions of $CaCl_2$ were made to both systems, with measurement of the turbidities after each addition. As shown in Fig. 3, in the presence of citrate ion the concentration of calcium required to produce an increase in turbidity was greater than in its absence. Using these curves, the difference in the concentrations of calcium chloride added gives the concentration of Ca complexed to citrate ion in the formation of CaCit$^-$.

The value for the stability constant of CaCit$^-$ at roughly equal ionic strength is known to be 1.62×10^3. This permits calculation of the concentration of unbound calcium ions in equilibrium simultaneously with citrate and phosphatidate ions. From this concentration the stability constant for calcium phosphatidate was found to be 1.4×10^4 at pH 7.23—in reasonable agreement with the value obtained by proton exchange measurements. Further details of the computation are given in the original paper (26b).

Similar studies were made of the reactions of phosphatidylinositol (PI) with metallic ions. Titrations of this compound showed that at pH levels above 6 it is fully ionized; therefore, to obtain cation–proton exchange reactions, the pH of the system was brought to levels at which some of the phosphate groups are not ionized. The apparent stability constant found this way for the calcium complex $CaPI^+$ was 1×10^3 and for

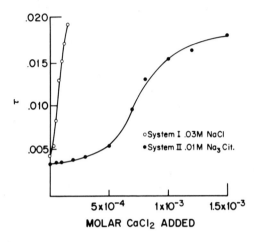

Fig. 3. Turbidities of micellar dispersions of phosphatidic acid with increasing concentrations of $CaCl_2$. Equal aliquots buffered at pH 7.23 in 0.05 M Tris. System I containing 0.03 M NaCl, System II 0.01 M Na_3 citrate. For equal turbidities the difference in the concentrations of $CaCl_2$ gives the concentration of $CaCit^-$ in System II. This permits calculation of the apparent formation constant of CaPA. [Reproduced from M. B. Abramson, R. Katzman and R. Curci, *J. Coll. Sci.* **20**, 777 (1965), by permission of publisher.]

NaPI, 6.9. Using the method based upon turbidimetric measurements, a value of 1.5×10^3 was obtained for the constant for $CaPI^+$.

Flocculation is another useful procedure for determining the relative binding of different cations to a lipid. A mixture of two or more salts in a desired concentration ratio was added to a lipid dispersion and the system brought to a suitable pH level and at a designated temperature. The concentrations of salts were selected so as to cause flocculation of the lipids. After centrifugation, an aliquot of the supernatant solution was analyzed for the concentrations of cations and the solid was also analyzed for cations. A correction was made for solution cations unavoidably entrapped by the solid. In this way the number of microequivalents of each cation per micromole of lipid was found, along with the concentration of cation in the solution at equilibrium with the solid. For two cations (A and B) having the same valence, a separation factor α_A^B of used to describe the relative degree of binding of B compared with is $\alpha_A^B = (\bar{x}_B)(x_A{}^+)/(\bar{x}_A)(x_B{}^+)$. Here, $\bar{x}$ is the number of microequivalents A. metal per milligram of coagulated lipid and x is the cation concentration of that metal in the solution. For phosphatidylinositol it has been found that $\alpha_{Na}^K = 1.4$ and $\alpha_{Mg}^{Ca} = 2.4$. These separation factors are in reasonable agreement with the relative effectiveness of the ions in proton–cation exchange and in producing increased turbidities.

Another lipid studied for its reactions with cations was sulfatide. The sulfate group in this compound is strongly acidic and is fully ionized at pH 4. Cation–proton exchange reactions are not feasible; therefore, turbidimetric titrations and coagulation studies were made. Sulfatide showed a selectivity of K/Na = 2. This greater binding of potassium is the reverse of the order found for phospholipids where sodium is more strongly bound than potassium. Turbidimetric titrations showed the effectiveness in increasing turbidities is K > Na > Li. This order of selectivity follows the order of binding exhibited by sulfonic acid ion-exchange resins (27). Phosphonic and also carboxyl acid resins show the reverse order of selectivity.

These parallel properties show the similarity of the two types of poly-anions present in lipids and ion-exchange resins. To account for these orders of ion behavior, we note that the order of the polarizabilities of anions and water is $PO_4^{3-} > COO^- > H_2O > SO_4^{2-}$. Thus a small ion, Li^+, which produces a large polarization of H_2O, will produce even greater polarization of phosphate or carboxylate ions. It then binds to these anions in preference to water. A large ion does not polarize water or anions greatly and its larger size reduces the ionic attraction for the anion. With sulfate as the polyanion the small metallic ions remain highly hydrated, so that K^+ approaches more closely to the sulfate than Na^+ or Li^+. There is consequently a greater force between sulfate and K^+ than between sulfate and Li^+.

D. X-Ray Studies

Much of the understanding of lipid–water structures has come from x-ray diffraction studies and electron microscopy. Phase diagrams have been constructed for soap–water systems and x-ray diffraction diagrams of the different phases have been made (28). At lower temperatures and especially at high soap concentrations, a gel phase exists. At higher temperatures, liquid crystals are present with a short spacing of 4.5 Å, which remains constant with temperature. In the concentration range of 3–40% soap, the liquid crystalline form is called "middle soap." It is composed of cylindrical micelles in which the hydrocarbon chains are directed inwardly. These cylinders become arranged in a hexagonal pattern with water in the spaces between the cylinders. At higher soap concentrations, several intermediate phases exist until, at a concentration of roughly 60% soap, a transition to the smectic phase or "neat" soap takes place. This liquid–crystalline phase is characterized by a layered structure in which each layer is of bimolecular thickness with the ionic groups situated on the two parallel surfaces with water between the layers. Luzzati and Husson (28) point out that in the liquid–crystalline form the ionic group at the end of the hydrocarbon chain is held at the aqueous interface, while the acyl chains between the surfaces are in a disordered structure similar to a liquid hydrocarbon. In this liquid–crystalline state, mixtures of lipids act as a single component, with the hydrocarbon chains intermixed.

From studies of the phase diagrams of a number of lipids whose structures contain anionic, cationic, or nonionic polar groups, some generalizations can be reached concerning the relation between the chemical structure of a lipid and its phase diagram. A lipid with large hydrophilic groups forms a middle phase that ranges over a wide concentration variation. Lipids that possess bulky hydrocarbon chains form the neat phase over an extended range of concentrations. An example is a compound with two hydrocarbon chains joined to a single ionic group. This compound may form only the neat phase as its liquid–crystalline form.

Similar x-ray studies were made of mixed phospholipids obtained from brain tissue with varying amounts of water added. At high concentrations of lipid the liquid–crystalline structure observed is a hexagonal array of long cylinders in which the water is at the center of the cylinder. This is an understandable result as the area of the two hydrocarbon chains is larger than that of the hydrophilic groups of the phospholipids. At lower concentrations of lipid a lamellar structure appears. This resembles the neat phase of soap–water systems. A difference between the soap–water and phospholipid–water systems is that in the former the middle phase, composed of cylinders in a hexagonal order, occurs at a lower lipid concentration than the neat phase with lamellar structure. In the phospholipid–water system the lamellar structure forms at a lower lipid concentration than the hexagonal structure. It must, however, be recognized that the hexagonal structure in the phospholipid–water system has water at the center of cylinders, while in the soap–water system, water is the continuous phase.

In a study of lecithin–cholesterol–water systems (29), x-ray diffraction patterns gave information on the effect of cholesterol on the structure of phospholipids in water. With increasing mole percent cholesterol in the lipid mixture there is an increase in the dimensions of the long spacing of the lamellar structure—from 64 Å for the pure phospholipid to a maximum of 81 Å at 7.5 mole% cholesterol. At concentrations higher than 50 mole% cholesterol, the cholesterol precipitates and gives its diffraction pattern with a long spacing of 34 Å. Ladbrooke and co-authors describe the structure of 1,2-dipalmitoyl-L-lecithin alone in water as arranged in layers with the hydrocarbon chains of the lecithin molecules aligned at an angle of 58° to the lipid–water interface. With the addition of small concentrations of cholesterol the hydrocarbon chains become oriented perpendicular to the interface, with the cholesterol molecules accommodated amongst the lecithin molecules. At higher concentration of cholesterol the hydrocarbon chains of the lecithin become fluid, indicated by a lowering of the transition temperature and heat of transition of the lipid and an increase in the amount of bound water.

E. Electron Microscopy of Lipid–Water Systems

Numerous electron-microscopic studies have been made of the structures formed by lipid–water systems using lipids of biological origin as well as pure, synthetic compounds. The results of such studies have been described by several workers. Bangham (30) describes the formation of the lamellar structures in the smectic phase of the lipid–water system. A solution of potassium phosphotungstate added to the system deposits metal in the aqueous but not in the hydrocarbon regions. In this way the polar portion of the lipid aggregates appear dark and are distinguishable from the hydrophobic portions. This procedure of "negative staining" is believed to produce little alteration in the structural arrangements of the aggregates.

Lecithin–water systems are characterized by the presence of spherules. These vary in size, but the outer regions of the spherules consist of alternate layers of lipid and water. The lipid layer is approximately 30 Å wide and the water layer, about 20 Å. The central region enclosed by these lamellae is aqueous but contains smaller aggregates of lipid.

Various other lipids can be mixed in moderate amounts with lecithin and the electron micrographs show the same typical spherules containing lamellar outer layers. Amphipathic ionic compounds such as dicetyl phosphoric acid and neutral molecules such as cholesterol can be incorporated into these lecithin structures; in the author's laboratory, mixtures

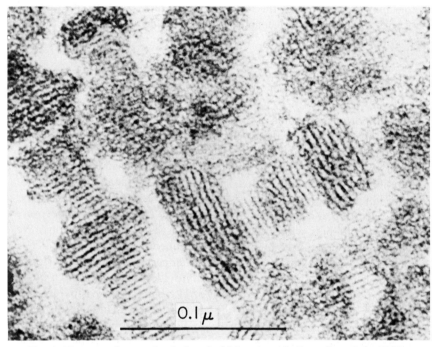

Fig. 4. Electron micrograph of phosphatidyl serine. A dispersion of the lipid was flocculated by the addition of calcium chloride. Particles were stained with osmium oxide, embedded, and sectioned. The marker represents 0.1 μ.

of lecithin with lipids containing anionic groups such as phosphatic acid, phosphatidylinositol, and sulfatide showed the characteristic structures of lecithin. These highly charged lipids alone did not show the remarkable ability of lecithin for forming layered structures and it appears that lecithin may serve as the basis for the formation of a layered structure into which other lipids may be incorporated. However, layered structures have also been formed of phosphatidyl serine and phosphatidic acid after reaction with metal salts. Figure 4 shows an electron micrograph of a dispersion of phosphatidyl serine flocculated by the addition of calcium chloride.

The structures formed by lecithin mixed with other lipids are interpreted differently by Lucy and Glauert (*31*). They suggest that such
structures may be composed of small globular micelles. The micelles
aggregate and are held together by hydrogen bonds as well as electrostatic
attractions. In lecithin–cholesterol systems the globular micelles are
molecular associations of the two lipids with their polar groups directed
outward radially. Arrangement of the micelles into layers of tubular
structures is also postulated.

F. OTHER PHYSICAL CHEMICAL STUDIES OF LIPID–WATER SYSTEMS

The importance of lipids in the structures of biological membranes has
spurred investigations of the properties of lipid–water systems using new
instruments and procedures. One such study (*32*) investigated the phase
transitions with temperature of phosphatidylcholine and phosphatidylethanolamine in the anhydrous form as well as in aqueous systems.
Using differential thermal analysis and differential scanning calorimetry,
the temperature and heats of transition for the mesomorphic states were
measured for lipids with different chain lengths and unsaturation. We
will only refer here to the findings for the aqueous systems.

For the saturated lipids the transition temperature and heats of transition increase with chain length. Dipalmitoyl lecithin undergoes a phase
change at 41° with a heat of transition of 11.8 cal/g. Dimyristoyl lecithin
changes at 23° and distearoyl lecithin changes at 60°. These results show
how temperatures, either above or below mammalian body temperature,
can be major factors in determining the behavior of lipids in biological
systems. For systems containing 0–20% water the heat absorbed in melting
ice at 0° was zero but increased from 20 to 100% water. This indicated
that the systems contained bound water in a weight ratio of 1:4, water to
lipid. Lecithin is known to form a monohydrate, but the nature of this
much larger amount of water (roughly 10 molecules) that appears to be
closely associated with each lipid molecule is not well defined. It could be
anchored to the surface of the lipid by bonds to the ionic and polar groups.

A related study (*29*) of lecithin–cholesterol mixtures containing 50%
water by weight showed that increasing the percentage of cholesterol in
these systems had several related effects. The temperature for the transition from the gel to the liquid crystalline phase decreased and the heat
absorbed in this transition decreased. Furthermore, the heat absorbed in
the ice transition at 0° was a minimum at 50 mole% cholesterol. This

would indicate that the bound water is a maximum at this composition of cholesterol and lecithin. In a number of biological systems the molar quantities of cholesterol and phospholipid are approximately equal and the amount of bound water will probably be large.

In another study of the physical properties of phospholipids, the electrical conductivity of some of these compounds was measured (*33*). This study was devoted chiefly to anhydrous samples of phospholipids and showed that at the temperature at which a mesomorphic transition takes place, a sharp change in conductivity occurs with a characteristic activational energy for each of the mesomorphic states. The mechanism for conduction is considered to be either electronic or protonic involving the polar end groups of the phospholipid. The authors point to some observations made with hydrated samples in which the resistance is much lower than for the anhydrous compounds. Some of the water probably becomes associated by hydrogen bonding with the polar groups of the lipid and provides a pathway for proton or electron conduction. Additional investigations of this property of aqueous systems of phospholipids should be very valuable because of its possible bearing on membrane processes.

A number of workers, in describing the chemical characteristics of membrane compounds, discuss the possibility of ionic interactions between the membrane lipids and neighboring protein molecules and also between different molecules of lipids in the membrane structure. Studies of the titration characteristics of some mixed lipid systems showed that such lipid–lipid interaction could take place (*34*). In these experiments, sulfatide was mixed with either lecithin or sphingomyelin. These latter lipids are alike in containing the phosphorylcholine group in which there is the zwitterion structure composed of an ionic phosphate and the basic quaternary amine of the choline. The titration of sodium sulfatide in water systems shows that sulfatide is a strong acid that does not bind protons except at low pH levels. Sphingomyelin and lecithin also possess a strong acid group. The presence of the basic choline increases the acid strength of the phosphate in these molecules.

As seen in Fig. 5, the titrations of these lipids alone showed very little reaction with protons in the acid range down to pH 3.5. However, when sphingomyelin and sulfatide or lecithin and sulfatide were dissolved together in an appropriate organic solvent and the solvent was then evaporated, a solid mixture of the two lipids remained. This mixture was dispersed in water and titrated. The titration curve showed the presence of a weaker acid group than was found in either sulfatide or the

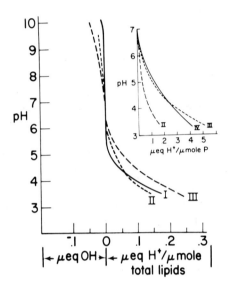

Fig. 5. Increased acid capacity of dispersed lipids, showing the interaction of sulfatide with sphingomyelin. Each system containing the lipid dispersed in 5 ml water was titrated with 0.100 N HCl or NaOH. Curve I, 7.63 μmoles of sodium sulfatide; curve II, 10.0 μmoles of sphingomyelin; curve III, 7.9 μmoles of sphingomyelin + 7.9 μmoles of sodium sulfatide, mixed and dispersed. The inset compares III, a 1:1 complex of sphingomyelin–sulfatide with IV, phosphatidylinositol, and II, sphingomyelin alone. The similarity of III and IV indicates that the ion titrated in the complex is the phosphate of sphingomyelin, made available by the interaction of the sulfate with the positive choline. [Copyright 1968 by the American Association for the Advancement of Science. Reproduced from M. B. Abramson and R. Katzman, *Science*, **161**, 576 (1968), by permission of publisher.]

phosphorylcholine lipid alone. In fact, the titration closely resembled that of a phospholipid containing a monoprotic phosphate group such as phosphatidyl inositol. The explanation given for these results is that the strong acid group of the sulfatide is able to form an ionic bond with the positively charged trimethyl nitrogen of a neighboring molecule of sphingomyelin or lecithin. As a result, the phosphate of the phosphorylcholine is released and reacts as a weaker acid, uniting with protons to a greater extent than when in the zwitterion structure. To explore this reaction further, a series of mixtures was prepared with sphingomyelin and sulfatide in varying mole ratios. When these were titrated, quantitative agreement was obtained between the acid capacity and the number of moles of a complex formed by the two lipids combining in a 1:1 mole ratio.

It was possible to use this interaction between sulfatide and a lipid containing phosphorylcholine to gain information concerning the role of cholesterol in lipid mixtures. Cholesterol was added to the 1:1 molar mixture of sulfatide and sphingomyelin and this mixture of three lipids was titrated in an aqueous system. The reaction of the phosphate group was again evident as in the mixture of sulfatide and sphingomyelin alone, except that not all the phosphate groups appeared to be available for titration. On increasing the mole percent of cholesterol in the mixture, the fraction of the total titratable phosphate decreased. It may be reasoned that the cholesterol occupies spaces between the other two lipids, and with increasing content of cholesterol the average distance between the other two lipids increases. The ionic interaction of the sulfatide with the quaternary amine is decreased and the availability of the phosphate groups for titration is lessened.

REFERENCES

1. Pressman, O., Grossbery, A. L., Pence, L. H. and Pauling, L., *J. Amer. Chem. Soc.* **68**, 250 (1946).

2. Salem, L., *Canad. J. Biochem. Physiol.* **40**, 1287 (1962).

3. Deuel, H., *The Lipids*, Vol. 1, Wiley-Interscience, New York, 1951, pp. 52–54.

4. Abramson, M. B., Norton, W. T. and Katzman, R., *J. Biol. Chem.* **240**, 2399 (1965).

5. Scheraga, H. A., *Ann. N.Y. Acad. Sci.* **125**, 253 (1965).

6. Spink, J. A., *J. Colloid Sci.* **18**, 512 (1963).

7. Goddard, E. D. and Ackilli, J. A., *J. Colloid Sci.* **18**, 585 (1963).

8. Abramson, M. B., unpublished results.

9. Bagg, J., Abramson, M. B., Fichman, M., Haber, M. D. and Gregor, H. P., *J. Amer. Chem. Soc.* **86**, 2759 (1964).

10. Deamer, D. W., Meek, D. W. and Cornwell, D. G., *J. Lipid Res.* **8**, 255 (1967).

11. Sears, D. F. and Schulman, J. H., *J. Phys. Chem.* **68**, 3529 (1964).

12. Parreira, H. C., *J. Colloid Sci.* **20**, 742 (1965).

13. Gershfeld, N. L. and Pak, C. Y. C., *J. Colloid Interface Sci.* **23**, 215 (1967).

14. Ries, H. E. Jr., and Walker, D. C., *J. Colloid Sci.* **161**, 361 (1961).

15. Van Deenen, L. L. M., Houtsmuller, U. M. T., de Haas, G. H. and Mulder, E., *J. Pharm. Pharmacol.* **14**, 429 (1962).

16. Vilallonga, F., *Biochim. Biophys. Acta* **163**, 290 (1968).

17. Phillips, M. C. and Chapman, D., *Biochim. Biophys. Acta* **163**, 301 (1968).

18. Shah, D. O. and Schulman, J. H., *J. Lipid Res.* **6**, 341 (1965).

19. Hauser, H. and Dawson, R. M. C., *Europ. J. Biochem.* **1**, 61 (1967).

20. DeBernard, L., *Bull. Soc. Chim. Biol.* **40**, 164 (1958).

21. Demel, R. A., Van Deenen, L. L. M. and Pethica, B. A., *Biochim. Biophys. Acta* **135**, 11 (1967).

22. Shah, D. O. and Schulman, J. H., *J. Lipid Res.* **8**, 215 (1967).

23. Abramson, M. B., Katzman, R. and Gregor, H. P., *J. Biol. Chem.* **239**, 70 (1964).

24. Abramson, M. B., Colacicco, G., Curci, R. and Rapport, M. M., *Biochemistry* **7**, 1692 (1968).

25. Hendrickson, H. S. and Fullington, J. G., *Biochemistry* **4**, 1599 (1965).
26. Abramson, M. B., Katzman, R., Wilson, C. E. and Gregor, H. P., *J. Biol. Chem.* **239**, 4066 (1964).
26a. Abramson, M. B., Katzman, R., Gregor, H. and Curci, R., *Biochemistry* **5**, 2207 (1966).
26b. Abramson, M. B., Katzman, R. and Curci, R., *J. Colloid Sci.* **20**, 777 (1965).
27. Bregman, J. I., *Ann. N.Y. Acad. Sci.* **57**, 125 (1953).
28. Luzzati, V. and Husson, F., *J. Cell Biol.* **12**, 207 (1962).
29. Ladbrooke, B. D., Williams, R. M. and Chapman, D., *Biochim. Biophys. Acta* **150**, 333 (1968).
30. Bangham, A. D. in *Advances in Lipid Research* (R. Paoletti and D. Kritchevsky, eds.), Vol. 1, Academic, New York, 1963.
31. Lucy, J. A. and Glauert, A. M., *J. Mol. Biol.* **8**, 727 (1964).
32. Chapman, D., Williams, R. M. and Ladbrooke, B. D., *Chem. Phys. Lipids* **1**, 445 (1967).
33. Leslie, R. B., Chapman, D. and Hart, C. J., *Biochim. Biophys. Acta* **135**, 797 (1967).
34. Abramson, M. B. and Katzman, R., *Science* **161**, 576 (1968).

3

Adsorption of Biological Analog Molecules on Nonbiological Surfaces. Polymers

B. J. FONTANA

Chevron Research Company
Richmond, California

I. Introduction

A considerable amount of information on the adsorption of macromolecules from solution onto solid surfaces has accumulated since systematic study first began about 20 years ago. The major part of this information, however, concerns polymers that are of interest because of their practical applications. Common examples are the polymers of butadiene, isobutylene, styrene, vinyl esters, and of acrylic acid and its derivatives. In general, these polymers consist of several hundred to a few thousand segments, which may or may not be all identical, arranged in an inherently linear and flexible array. For the purpose of this chapter, this latter description will be the primary claim to analogy with polymers of biological origin.

Biological polymers in general, because of the bulky nature of the segments, may be more inflexible than most of the nonbiological types for which adsorption studies have been made. In addition, biopolymer inflexibility will be extreme in those cases where highly specific intra- or intermolecular interactions, such as hydrogen bonding, occur. The helical secondary or tertiary structures of the biopolymers that result are in fact rigid-rodlike. However, such structures apparently often unfold upon adsorption at an interface (1). It will be pertinent to note where possible the effects on the adsorption behavior caused by macromolecular rigidity or chain extension, albeit for reasons other than helix formation.

In vivo biological phenomena occur primarily in aqueous media. In view of the motivations of polymer applications, the vast majority of data on adsorption of nonbiological polymers, however, were obtained from nonaqueous solutions. This is apparent in the two major reviews on polymer adsorption available to date by Patat et al. (2) and Kipling (3). Some features of the adsorption behavior of polymers in nonaqueous solvents should be directly translatable to the behavior in aqueous solutions. To assist in extrapolating adsorption phenomena from nonaqueous to aqueous systems, the role of the solvent must be emphasized. A brief description of the fundamental properties of polymer solutions is therefore given in the following section. In addition, virtually all of the relatively meager information on the adsorption of nonbiological polymers from aqueous media will be discussed.

The formation of hydrogen bonds is a common occurrence and has important consequences in biological systems. Aside from being the energizing factor for helix formation, hydrogen bonding must also be of primary importance for interfacial interactions at biosurfaces. Thus polymers and surfaces capable of hydrogen bond formation are of interest here. Examples of this kind are available in both aqueous and nonaqueous solutions and will be described. This is, however, just part of the more general problem of the effect of the magnitude of the polymer/surface interaction energy on the adsorption characteristics.

Studies at the gas–liquid interface, especially the air–water interface, have been much used to yield very useful structural information about insoluble monolayers of both biological and nonbiological polymers (4, 5). However, as will be discussed later, behavior at gas–liquid interfaces is not typical of adsorption on a solid surface. The discussion here will be restricted to adsorption at a solid interface from the liquid phase. Most biological processes that involve *adsorption* must occur at solid–liquid interfaces.

Some of the unique consequences of the way polymers behave at the liquid–solid interface will be considered in conclusion. It is hoped that such examples will assist in making the extrapolation from nonbiological to biological problems.

Thus it is expected that the basic concepts arrived at in the study of the adsorption of nonbiological macromolecules will be useful in the interpretation of the behavior of some biological systems. It is outside the scope of this chapter, and the ken of the author, to suggest where this application may be possible. Also, it does not appear pertinent to present an exhaustive coverage of the literature but only such references (preferably recent) as serve to illustrate the various facets and concepts of polymer adsorption behavior.

II. Polymer Solutions

Since we are concerned here with the equilibrium process

Polymer in solution ⇌ Polymer adsorbed at the surface

it is pertinent to consider first the nature of the polymer solution. The physical state of the dissolved macromolecule is strongly dependent upon the relative properties of polymer and solvent and it is reasonable to expect that the state of the polymer in solution might also be reflected in some way in the adsorbed state. In addition, the nature (or behavior) of

the polymer solution can yield information on inherent structural characteristics of the macromolecule, particularly its flexibility.

The unique physical properties of polymer molecules in solution (or
adsorbed at a solid surface) are the direct result of their flexibility. This
flexibility arises from the freedom of rotation of one chemical bond
about another at a constant angle. Thus a long, chain-like molecule
containing many hundreds of bonds in the backbone can assume an
enormous number of different configurations. The resulting large configurational entropy *per mole* then favors the formation of a random,
tangled coil rather than an extended state. This condition is a dynamic
one that may undergo drastic conformational alteration under the influence
of external forces. Such alteration could be the result, for example, of a
change in the solvent medium or of adsorption at a surface.

A convenient and much used method of determining the physical state
of a dissolved macromolecule is the study of the hydrodynamic properties,
particularly the viscosity, of dilute polymer solutions. Viscosity behavior
has the virtue of a physical representation that is relatively easily visualized
intuitively. If η is the viscosity of a dilute suspension of spherical incompressible particles in a solvent of viscosity η_0 and ϕ is the volume
fraction of the particles, the relative increase of the viscosity is given by
the Einstein equation

$$\frac{\eta - \eta_0}{\eta_0} = 2.5\phi \tag{1}$$

The solvent enmeshed in a macromolecular coil is considered to be
virtually all trapped and carried along during rotational or translational
motion of the polymer molecule (6). Thus the hydrodynamic size of a
polymer molecule can be approximated as an "equivalent" incompressible
sphere of volume V_e and the molal hydrodynamic volume $\bar{V}_e = NV_e$,
where $N = $ Avogadro's number. If the concentration C of the polymer
is expressed in g/ml, then

$$\phi = C\bar{V}_e/M \tag{2}$$

where M is the molecular weight of the polymer. Substituting Eq. (2)
in Eq. (1) and rearranging gives the expression for the so-called intrinsic
viscosity $[\eta]$ in the limit at infinite dilution—i.e., as $C \to 0$

$$[\eta] = \left(\frac{\eta - \eta_0}{\eta_0 C}\right)_{C \to 0} = 2.5\frac{\bar{V}_e}{M} \tag{3}$$

$\bar{V}_e/M$ is the apparent density of the polymer coil. For a suspension of
solid particles of density $= 1$, $[\eta] = 2.5$. Thus the ratio $[\eta]/2.5$ for a

dissolved polymer can be considered to be an approximate gauge of the voluminosity or degree of swelling of the macromolecule in the particular solvent medium. In a poor solvent for the polymer, solvent–polymer interactions decrease, and hence the macromolecular coil will collapse. This results in a decrease in the value of $\bar{V}_e$ and hence of the intrinsic viscosity. Under these latter conditions the limiting size of the collapsed coil is largely determined by the chain flexibility—that is, by the nature of the hindrance to internal rotation around single bonds, the so-called short-range interactions. For most polymer systems the ratio of $[\eta]$ in a good and poor solvent is from 2 to 10. A less flexible macromolecule will remain more extended in the "collapsed" state. This should result in a smaller decrease in $\bar{V}_e$ and hence $[\eta]$.

Comparisons of polymer coil dimensions, however, should not be made in a "poor" solvent but in a so-called theta solvent. When the temperature T is equal to the Flory theta temperature θ (6), the excess free energy of mixing ΔF_m of polymer and solvent is zero, i.e.,

$$\Delta F_m \propto \left(\frac{\theta}{T}\right) - 1 \qquad (4)$$

θ is the theoretical critical temperature of miscibility for a polymer of infinite length. In a θ-solvent the entropy contributions (due to long-range interactions between widely separated segments on the polymer chain) and the enthalpy contributions (due to solvent–polymer segment interactions) to the free energy of mixing just cancel each other. Under these conditions, in the effective absence of long-range interactions, the polymer coil is considered to be "unperturbed."

The end-to-end distance r of the linear chain that constitutes the polymer coil can be used to express the coil dimensions. In particular, the value r_0 for unperturbed coils gives a relative measure of the chain flexibility. The value of r_0 can be determined from the viscosity–molecular weight relationship in a θ-solvent, among other methods (7). In addition, theoretical values r_{0_f}, can be calculated for a freely rotating unperturbed macromolecular chain of given structure. The ratio r_0/r_{0_f} is a measure of the effect of steric hindrance.

Examples of values of the intrinsic viscosity and unperturbed dimensions of some polymers in organic and aqueous solvents are given in Table 1. The polymers are arranged in the apparent order of increasing rigidity and/or steric hindrance. Some examples of the effect on $[\eta]$ of θ and "good" solvents are also shown. In order to make the comparison in Table 1 somewhat more meaningful, the calculations of $[\eta]$ and r_0 were

TABLE 1
INTRINSIC VISCOSITY AND UNPERTURBED DIMENSIONS OF POLYMERS[a]

Polymer and solvent	Temp., °C	Mol. wt., $M \times 10^{-5}$	$[\eta]^{b}$ cm³/g	r_0/r_{0_f}	$r_0,^{b}$ Å
Poly(ethylene oxide)					
0.45 M K₂SO₄, aq. (θ)	35	1.1	43	1.46	260
Poly(propylene)					
Isoamyl acetate (θ)	34	1.6	67	1.76	329
Poly(hexamethylene adipamide)					
90% Formic acid, aq., 2.3 M KCl (θ)	25	1.2	88	1.85	349
Poly(methyl methacrylate)					
Butanone/isopropanol 1:1 vol. (θ)	25	3.7	36	2.08	389
Chloroform (good)	25	3.7	137		
Poly(styrene)					
Butanone/methanol 89:11 vol. (θ)	25	3.9	45	2.22	416
Toluene(good)	20	3.9	105		
Poly(vinyl pyrrolidone)					
Water/acetone 33:67 vol. (θ)	25	4.1	47	2.22	417
Water(good)	25	4.1	242		
Poly(vinyl sulfonic acid)					
0.65 M KCl, aq. (θ)	26	4.0	50	2.31	433
0.35 M NaBr, aq. (good)	30	4.0	586		
Poly(n-dodecyl methacrylate)					
Pentanol (θ)	30	9.4	34	2.59	485
Cellulose tricaproate					
Dimethyl formamide (θ)	41	4.4	162	2.65	646

[a] Calculated with viscosity constants K and a, and ratio r_0/r_{0_f} from the compilation by M. Kurata, M. Iwama, and K. Kamada in Ref. (8).

[b] Calculated for the indicated molecular weights, corresponding in every case to a polymer backbone length of 7400 atoms.

made for polymers that all had about the same linear chain length. The intrinsic viscosities are calculated from the Mark–Houwink–Sakurada equation

$$[\eta] = KM^{a} \tag{5}$$

where K and a are viscosity constants. It may be noted, in conclusion, that in θ-solvents K generally increases with polymer backbone rigidity. The value of a reflects the effect of solvent power, $a = 0.50$ in θ-solvents, and $a \geqslant 0.8$ in good solvents.

III. Polymer Adsorption Characteristics

A. PRIMARY TRAITS

The adsorption of a flexible macromolecule can be expected to differ from that of ordinary low-molecular-weight compounds because of two unique factors. First, the inherent polyfunctionality of the polymer suggests that if one segment of the molecule adsorbs onto a solid surface, the probability of the adsorption of neighboring segments will be greatly increased. Second, the unusual configurational behavior of a macromolecule, as evinced by the properties of polymer solutes discussed in the previous section, suggests that the intramolecular configuration (as well as the intermolecular arrangement) of the adsorbed molecules will be an important aspect of the adsorption process. Thus it is likely that simple monolayer, or ordered multilayer, adsorption may not occur.

The first of the above factors is made manifest in the effect of the polymer concentration in solution on the amount adsorbed. Typically, the equilibrium amount adsorbed per unit surface area increases extremely rapidly with polymer concentration. Indeed, the initial steep rise in adsorption often occurs at concentrations too small to be readily measured at all. A plateau value (with respect to concentration) is thus rapidly reached and this does not change appreciably with further increase in the equilibrium concentration of the polymer solute. Thus the plateau often extends virtually over the entire experimentally measurable range of concentration. These effects are illustrated in Fig. 1 for the adsorption of polymethyl methacrylate onto Pyrex glass and iron powders from benzene (9).

The adsorption behavior described above must be the result of the cooperative effect of an adsorbed polymer segment upon a neighboring but as yet unadsorbed segment. This effect has been demonstrated unequivocally in a study of the adsorption of simple polyfunctional esters (10). The adsorption on silica from n-dodecane of the mono-, di-, and tridodecyl esters derived from acetic, succinic, and tricarballylic acid, respectively, is illustrated in Fig. 2. These three compounds are the first members of poly(dodecyl-carboxy-methylene):

$$\left[\begin{array}{c} C_{12}H_{25}-O-C-C- \\ \quad\quad\quad\quad \| \\ \quad\quad\quad\quad O \end{array} \right]_\nu$$

where $v = 1$, 2, and 3, respectively. It is apparent that marked increase in adsorption occurs immediately with the bifunctional member of this series. Thus the very rapid approach to complete coverage, expected of a macromolecular polyester, occurs with these simple polyfunctional compounds. The tremendous enhancement of the adsorption of a macromolecule thus is not surprising. It is perhaps possible then that

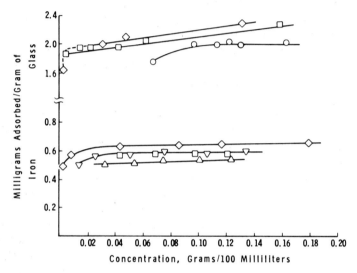

Fig. 1. Typical polymer adsorption isotherms. The adsorption of poly(methyl methacrylate) by glass and iron from benzene at 30°C. Molecular weight of the polymer $\times$ 10^{-5}: $\bigcirc$, 0.23; $\triangle$, 1.18; $\square$, 5.32; $\triangledown$, 10.80; $\diamondsuit$, 22.70. [Redrawn from Ref. (9) by courtesy of John Wiley and Sons, Inc., N.Y.]

the sensitivity of some biological processes to extremely low solute concentrations may be ascribed to a similar factor if the adsorption of biopolymer at a surface is involved.

The earliest studies of polymer adsorption suggested that the configuration of the adsorbed polymer was indeed not analogous to the monolayers typical of simple surfactants. Amborski and Goldfinger (11) estimated that the layer of GR-S rubber (butadiene–styrene copolymer) adsorbed on carbon black from xylene solutions was 150–200 Å thick. These values were obtained by determining the effective volume of the particles from viscosity measurements on the carbon suspensions and application of Eq. (1) (modified with an additional ϕ^2 term to allow application to higher concentrations).

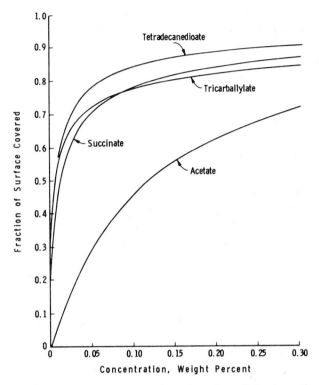

Fig. 2. Effect of polyfunctionality on adsorption. Smoothed adsorption isotherms for *n*-dodecyl esters on silica from *n*-dodecane at 25°C. [Reproduced from Ref. (*10*) by courtesy of the American Chemical Society.]

Jenckel and Rumbach (*12*) noted that the plateau values of polymer adsorbed per unit surface area were of such a magnitude that it did not appear that all of the polymer segments could be accommodated in the surface layer. The spatial requirements of the polymers were assumed to be as estimated from the pure solid polymer densities. The results for poly(methyl methacrylate), polystyrene, and poly(vinyl chloride) adsorbed onto aluminum, quartz, or glass suggested to these authors that the polymers were attached to the surface by relatively few segments in long loops extending out into the solvent phase. More reasonable estimates of the surface area requirements for adsorbed poly(vinyl acetate) based on actual measurements at an air–liquid interface with the Langmuir film balance, or on adsorption studies with monomeric ethyl acetate, led to a similar conclusion (*13*). It may be noted here that the character of the adsorption isotherms (*14*) (e.g., Fig. 1) indicates that it is unlikely that the thick polymer layers are the result of multilayer adsorption (i.e., one polymer molecule adsorbing on top of another).

Thus it is apparent that adsorption data alone serve to yield a picture, albeit indirect and crude, of the unusual nature of the adsorbed macromolecule. More definitive studies of the conformation of the adsorbed polymer molecule will be discussed in Sect. V.

B. Effect of Molecular Weight

The study of the effect of molecular weight on the adsorption of synthetic polymers is complicated by the typically broad molecular weight distribution. The problem of dealing with molecular weight distribution is only heightened by the fact that even the average of the distribution may be expressed in a number of different ways (*7*) and depends upon whether the method of measurement yields a number average M_n (e.g., from osmotic pressure), weight average M_w (e.g., from light scattering), or other more complex average (e.g., from viscosity M_v). Thus fractionation of polymers is highly desirable but has rarely been done in adsorption studies. Coupled with the wide variety of polymer compositions and solvent, adsorbent, or rate effects (to be discussed in Sect. C, E, and F), the net result is that the effect of molecular weight is perhaps the least well-defined characteristic of polymer adsorption behavior. It appears unjustified at present to do more than very broadly generalize on the available data.

Polymer/solvent/adsorbent systems are found where the saturation adsorption is virtually independent of molecular weight over a wide molecular weight range or, in some cases, only above some high molecular weight. More usually, however, the adsorption increases with increasing molecular weight but with a wide range of dependence. It appears quite definite that a decrease in adsorption with increasing molecular weight never occurs on a nonporous adsorbent. All of the discussion in this section will pertain, unless otherwise noted, to nonporous adsorbents only. The special effects caused by porous adsorbents will be discussed in Sect. III.F on the effect of the nature of the adsorbent surface.

The saturation value A_s of polymer adsorbed per unit surface area is often found to fit the relation

$$A_s = K'M^n \qquad\qquad (6)$$

where K' and n are constants and M is molecular weight. This equation was first suggested by Koral et al. (13) from considerations of the possible relation of the space occupied by an adsorbed macromolecule to the coil size in solution. The parallelism of Eq. (6) and Eq. (5)—the molecular weight dependence of intrinsic viscosity—is obvious.

The data collected in Table 2 illustrate the range of molecular weight dependence mentioned above. The saturation adsorption of poly(isobutylene) from benzene onto carbon black (18) as a function of molecular weight is shown in Fig. 3. Here, the molecular weight dependence, initially high, becomes virtually negligible above a molecular weight of about 10^6. In addition, preferential adsorption of the high-molecular-weight species is observed. Thus at equilibrium between the solution and solid phases the average molecular weight of polymer remaining in solution is smaller than that in the initial solution. The carbon adsorbent used in this latter study (Vulcan 3) has an apparent-to-true particle diameter ratio of 1.55, calculated from N_2 adsorption and electron microscopic examination, respectively (19), and is probably porous. This would be expected to cause the opposite effect (see Sect. III.F). The preferential adsorption of high molecular weight polymers is also evident when chromatographic methods are used to fractionate polymers. A carefully studied case is that of polystyrene (20).

The more tenacious adsorption of the higher molecular weight polymers is to be expected because of the increased polyfunctionality discussed above. However, the simple increase in saturation value A_s with molecular weight need not be due to or necessarily caused by preferential adsorption. Geometrical considerations alone can lead to a variety of values for the molecular weight dependence n of Eq. (6) as discussed by Perkel and Ullman (17). If the polymer segments all lie in the plane of the surface (i.e., as a "monolayer"), then $n = 0$. The value of n would also be zero if the polymer adsorbed as a layer of contiguous, rigid (noninterpenetrating) spheres. This latter state is that of a random macromolecular coil in solution in a poor solvent. However, if the adsorbed macromolecules interpenetrate freely, and if the segment density as a function of distance from the surface is the same as that from the center of mass of a free random macromolecular coil in solution, then $n = 0.5$. In other words, the molecular weight dependence will be determined by the equilibrium polymer configuration at the surface. This configuration must depend primarily on the balance between the two opposing effects of the polymer-surface interaction energy and the entropy effects due to the compression and/or interpenetration of the macromolecular coils.

TABLE 2

MOLECULAR WEIGHT DEPENDENCE OF POLYMER ADSORPTION ON NONPOROUS SOLIDS

Polymer	Adsorbent	Solvent	Molecular weight range, $\times 10^{-5}$	n^a	Ref.
Poly(vinyl acetate)	Fe	Carbon tetrachloride	$4.7-2.8(M_w)$	0.43	(13)
	Fe	Carbon tetrachloride	$9.1-1.4$	0.39	
	Fe	Benzene	$9.1-2.5$	0.02	
	Fe	1,2-Dichloroethane	$9.1-2.5$	0.10	
Poly(methyl methacrylate)	Fe	Benzene	$22.7-2.3(M_w)$	0.04	(9)
	Glass	Benzene	$22.7-2.3$	0.00	
Poly(ethylene oxide)	Silica	Benzene	$2.6-0.3(M_n, M_v)$	$0.11-0.07^b$	(15)
Poly(styrene)	Al	Cyclohexane	$3.7-0.7(M_w)$	$0.13-0.16^c$	(16)
Poly(dimethyl siloxane)	Fe	Benzene	$5.4-0.3(M_n)$	0.43	(17)
	Glass	Benzene	$5.4-0.3$	0.40	
	Fe	n-Heptane	$5.4-0.3$	0.23	
	Glass	n-Heptane	$5.4-0.3$	0.35	

[a] See Eq. (6).
[b] For two different grades of aerosil silica adsorbent.
[c] At 34.7 and 50°, respectively.

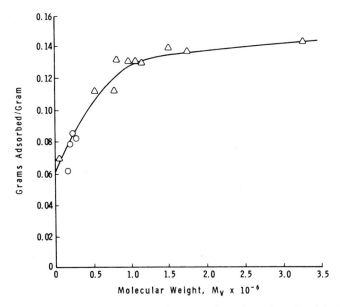

Fig. 3. Effect of molecular weight on the saturation adsorption of polyisobutylene △ and polyisobutylene-isoprene copolymer ○, by carbon black from benzene at 25°C. [Redrawn from Ref. (*18*) by courtesy of John Wiley and Sons, Inc., N.Y.]

C. SOLVENT EFFECTS IN NONAQUEOUS MEDIA

A discussion of the general effects of solvents on polymer adsorption is mostly limited to organic solvents, not only because of the paucity of data in water but because water-soluble polymers are rarely soluble in other common solvents for comparison. The nature of the solvent, as might be expected, has a very marked effect on the adsorption of polymers. These effects can usually be readily interpreted qualitatively in terms of two factors: solvent–polymer interactions and solvent–adsorbent interactions. Either of these effects may predominate, or they may also function together to either intensify or inhibit the adsorption of polymer.

In the absence of appreciable solvent–adsorbent interactions, polymers are usually adsorbed to a greater degree from poor solvents than from good solvents. Solvent power is assessed here, of course, specifically with respect to the solubility of the particular polymer species under consideration. Convenient and often used measures of the relative solvent power are the intrinsic viscosity or the theta temperature. An example of such a

TABLE 3
EFFECT OF SOLVENT ON THE ADSORPTION OF POLY(VINYL ACETATE) BY IRON[a]

Molecular weight, $M_w \times 10^{-5}$	Solvent	Intrinsic viscosity $[\eta]$, cm^3/g	Saturation adsorption A_s, mg/g
2.5	Chloroform	138	0.35
2.5	1,2-Dichloroethane	110	0.54
2.5	Benzene	94	0.68
2.5	Carbon tetrachloride	33	1.54
9.1	1,2-Dichloroethane	368	0.61
9.1	Benzene	314	0.70
9.1	Carbon tetrachloride	111	2.74

[a] Data taken from Ref. (13).

comparison is afforded by the data in Table 3 for the adsorption of poly(vinyl acetate) on iron (13). An expedient method of studying such effects is that of using solvent mixtures. Table 4 shows data for the adsorption of polystyrene from methyl ethyl ketone–methyl alcohol mixtures (21). In both of these cases adsorption increases as the solvent power for the polymer decreases. By way of illustrating the specificity of the solvent–polymer interaction, note that the change in polarity of the solvent in going from good to poor solvent for poly(vinyl acetate) is reversed in the poly(styrene) case.

TABLE 4
ADSORPTION OF POLYSTYRENE ON CHARCOAL FROM SOLVENT MIXTURES[a]

Vol. % methyl alcohol in methyl ethyl ketone	Saturation adsorption, A_s
0[b]	34
0.5	35
1.0	48
1.5	46
2.5	81
5	144
10	144
12.3	(Polymer precipitates)

[a] Data taken from Ref. (21).
[b] Intrinsic viscosity, $[\eta] = 136\ cm^3/g$.

The effect of polymer solubility illustrated above may be in part due to the close parallelism between the solubility of a solute and its absolute activity. This relationship was noted by Hansen et al. (*22, 23*) in a study of the adsorption of low-molecular-weight, monofunctional solutes. In a homologous series, where the adsorbent acts on the same functional groups, congruency of the adsorption isotherms was observed if the amount adsorbed per unit surface area was plotted against the ratio of the equilibrium concentration of the solute to its solubility, C/C_0. The analogous use in gas adsorption of the reduced pressure, P/P_0 (where P_0 is the saturation vapor pressure), was also pointed out.

However, in the case of polymer adsorption it is probable that the effect of solubility primarily reflects the effect of the solvent on the macromolecular coil dimensions (see Sect. II). In the absence of other strong interactions such as between polymer and surface it is reasonable to expect that the polymer–solvent interactions will be partially responsible for the conformation of the adsorbed polymer, i.e., the more compact polymer coil in a poor solvent can form a more densely packed adsorbed layer than from a good solvent and vice versa. More detail on such behavior will be given in Sect. V. In passing, it may be noted that the ability of part of a macromolecule to be essentially *in solution* although some segments are attached to a surface is in some respects analogous to the behavior of block and graft copolymers in selective solvents (*24*). Such polymers have sequences of different homopolymers in the same molecule and display colloidal properties in solvents in which one of the homopolymers is soluble and the other is not.

When a wide variety of solvents are studied it is apparent that the simple relationship between solubility and adsorption does not hold. In the absence of strong polymer–solvent interactions the solubility of a polymer should be highest in a solvent whose solubility parameter is closest to that of the polymer (*25*). The adsorption of poly(methyl methacrylate) on iron, however, from solutions in benzene, toluene, and ten halogenated hydrocarbons (*9*) shows a poor correlation with solubility parameter, or with "polarity" as assessed by the dielectric constant, dipole moment, or polarizability. It is apparent that the lowest adsorption in this case, observed in chloroform, is the result of strong specific interaction (hydrogen bonding) to the polymer and/or the surface.

It is particularly apparent that when strong solvent adsorption can occur, any effects of polymer solubility can become negligible. For example, in the study of poly(vinylacetate) adsorption on iron described above (Table 3), when acetonitrile was used as solvent, where $[\eta] = 81$, no polymer adsorption whatsoever occurred $(A_s = 0)$. The adsorption onto glass of poly(dimethyl siloxane) from a 10% acetonitrile–90% benzene (v/v) solvent mixture was found (*17*) to be about half that from the better solvent benzene. The mixed solvent also completely prevented adsorption of the poly(dimethyl siloxane) onto iron.

A good survey of solvent interactions is given by the critical studies made by Howard and McConnell of the adsorption of poly(ethylene oxide) onto silica (*15*), carbon (*26*), and nylon (*27*) from benzene, dioxane, chloroform, methanol, dimethylformamide, and water. A complex ranking of solvent dependence was found, which was different for each of the three adsorbents. For example, no polyether adsorption was observed from dioxane and dimethylformamide on silica; adsorption from all six solvents occurred on carbon; and only from benzene on

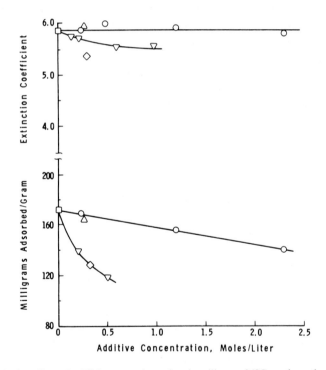

Fig. 4. The effect of additives on adsorption by silica at 24°C, and on the carbonyl infrared extinction coefficient of polyvinylacetate in trichloroethylene. Additives: □, none; ○, acetonitrile; ▽, cyclohexanol; △, dioxane; ◇, methanol. [Redrawn from Ref. (*28*) by courtesy of the American Chemical Society.]

nylon. Solvent adsorption was assessed in some cases by vapor sorption measurement. No explanation could be advanced for the failure to adsorb on silica from dioxane.

In addition to the possibilities of the solvent competing with the polymer for surface sites or affecting the polymer coil dimensions, it is apparent that the solvent may also compete *with the surface* for the functional groups of the polymer that are directly involved in the polymer attachment to the surface. Thies (*28*) has studied the effect of low concentrations ($\leqslant 10\%$ weight/v) of acetonitrile, methanol, cyclohexanol, and dioxane in trichloroethylene on the adsorption of polyvinyl acetate ($M_n =$ 465,000) by silica. All of the added solutes were found to adsorb on silica in separate adsorption experiments, and all caused a reduction in the adsorption of polymer (Fig. 4). However, the greatest reduction occurred for those solutes that interacted most strongly (via hydrogen bonding) with the polymer ester groups. The relative extent of the latter interactions was assessed from the change in the extinction coefficient of the ester carbonyl infrared absorbance (Fig. 4).

D. Effect of Temperature

The effect of temperature on polymer adsorption is not simple and temperature coefficients for polymer adsorption that range from large negative values to small positive values have been observed. Examples of systems displaying all of these degrees of behavior are collected in Table 5.

The interaction of the functional group of a solute with a surface site will normally be exothermic. It follows then that the effect of increasing temperature would normally be to decrease the extent of adsorption, as is usually the case for simple, low-molecular-weight solutes. Thus the negative temperature coefficients for polymer adsorption are to be expected. In addition, since the solubility of a polymer increases with temperature, this also will have the effect of decreasing adsorption with increasing temperature (Sect. III.C).

However, the magnitude and sign of the temperature coefficient of adsorption can be highly solvent dependent. Such a case has been described by Davidson and Kipling (*29*) for the adsorption of polyisobutene on carbon black. Some of their results are described in Table 6. These authors point out that the variation in the *negative* coefficient can also be explained in terms of polymer solubility when account is taken of the theta temperature. It will be recalled from Sect. II that the excess free energy of

TABLE 5

Effect of Temperature on Polymer Adsorption

Polymer	Molecular weight, $\times 10^{-5}$	Adsorbent	Solvent	Temp., °C	Saturation adsorption A_s, mg/g	Ref.
Poly(vinyl acetate)	2.5 (M_w)	Iron	Carbon tetrachloride	30.4	1.54	(13)
				69.5	1.66	
	9.1	Tin	Benzene	30.4	0.221	
				51.5	0.228	
Poly(methyl methacrylate)	12.0 (M_v)	Aluminum	Toluene	25	0.426	(12)
				50	0.477	
	12.0	Silica	Toluene	25	0.154	
				50	0.164	
Polystyrene	5.1	Glass	Toluene	25	0.400	
				50	0.400	
Poly(vinyl chloride)	0.37	Glass	Toluene	25	0.373	
				50	0.264	
Polyisobutylene	2.9 (M_v)	Carbon black[a]	Benzene	25	0.080	(18)
				75	0.027	
	11.6	Carbon black[a]	Benzene	25	0.125	
				75	0.036	
Poly(dimethyl siloxane)	8.0 (M_w)	Glass	Heptane	30	3.13	(17)
				80	2.04	
				103	0.51	

[a] Vulcan 3, a porous carbon black; see Ref. (19).

mixing of polymer and solvent ΔF_m is a function of the temperature T and theta temperature θ as given by Eq. (4)

$$\Delta F_m \propto (\theta/T) - 1$$

Thus the polymer solubility, and hence its adsorption, can be expected to change markedly when the working temperature T is near the theta temperature and less so when T is much higher than θ. This can explain

TABLE 6
EFFECT OF SOLVENT ON THE TEMPERATURE COEFFICIENT OF ADSORPTION[a]
POLYISOBUTENE ON CARBON BLACK (GRAPHON)

Solvent	Theta temperature,[b] °K	Molecular weight, $M_v \times 10^{-5}$	Saturation adsorption[c] A_s, mg/g at	
			293°K	313°K
Benzene	297	3.2	120	85
		46.0	95	75
Cyclohexane	126	3.2	57	57
Diisobutene	84	46.0	78	78
n-Hexane	0	3.2	66	77
		46.0	61	70

[a] Compiled from data in Ref. (29).
[b] For polyisobutene in the corresponding solvent.
[c] Approximate values taken from figures in Ref. (29).

the behavior in benzene, cyclohexane, and diisobutene solvents (Table 6) but not the positive temperature coefficient observed in n-hexane. Davidson and Kipling suggest that in n-hexane the increased flexibility of the polymer chain with increasing temperature now outweighs the effect on solubility because of the extreme gap between T and θ. Thus the polymer can coil more tightly with increase in temperature. The temperature coefficients of the viscosity of polyisobutene in benzene, cyclohexane, diisobutene, and n-heptane apparently are in accord with this explanation of the adsorption behavior.

The more general observation of zero or small positive temperature coefficients of adsorption (Table 5) appears to require a different explanation than that suggested above. The desorption of a large number of solvent molecules required in order to accommodate a single polymer molecule at the surface must result in an increase in entropy. This increase has been suggested to explain the very strong adsorption of polymers in spite of the "heat of adsorption" of virtually zero (*13*). Such values of the heat of adsorption ΔH are ordinarily obtained by application of the Clausius–Clapeyron equation

$$\Delta H/R = d \ln C/d(1/T) \tag{7}$$

to the values of the equilibrium solute concentrations C (at different temperatures T), which correspond to equal amounts of adsorbed polymer. This can only yield an *apparent* heat of adsorption, however.

A relevant characteristic, which has yet to be measured, is the temperature dependence of the number of polymer segments attached to the surface. Also, there appears to be no reason to expect the heat of adsorption of a functional group (normally negative, i.e., exothermic) to approach zero as the number of such groups becomes very large as in a macromolecule (*10*). It seems more reasonable that polymer adsorption results in a net decrease rather than increase in entropy. The decrease in configurational entropy of a macromolecule that occurs upon adsorption must be somewhat analogous to that observed during the freezing of a polymer melt. Large entropy decreases *per segment* occur upon freezing, which Kirshenbaum (*30*) has shown to be largely due to the rotational contributions to the macromolecular conformations.

Thus small temperature effects are probably observed for polymer adsorption because the equilibrium

Polymer solute ⇌ Adsorbed polymer

is always shifted far to the right as indicated by the characteristic adsorption isotherms (Sect. III.A). The observed temperature effects may be due only to changes in the number of attached segments or in the surface area required per attached segment. Silberberg's theoretical analysis (*31*) predicts that the number of attached segments will decrease with increasing temperature, i.e., the number of segments in a free loop will increase. The data of Rowland and Eirich (*32*), which will be discussed in Sect. V.A, also indicate that loop size increases with increasing temperature. If the area per attached segment does not increase correspondingly, this would have the effect of increasing the amount of adsorbed polymer and a positive coefficient of adsorption would result. A somewhat analogous effect is the increased polymer adsorption observed when the number of adsorbing segments is decreased by the introduction of a few more polar segments into a copolymer molecule (*13, 33*). These effects will be further discussed in Sect. V.C.

E. RATE OF ADSORPTION AND DESORPTION

In most studies of polymer adsorption the rate of adsorption has usually been determined at relatively high surface coverage. This rate can be very slow, requiring many hours or even days in some cases. Slow attainment of equilibrium is probably always the result of adsorption

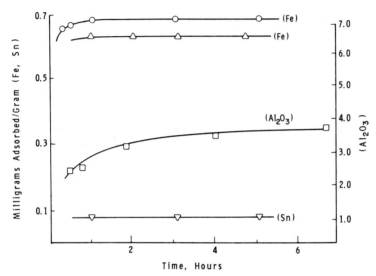

Fig. 5. The effect of adsorbent porosity on the rate of adsorption of polyvinylacetate. Molecular weight × 10^{-5} (solvent, temperature): ○, 2.50 (benzene, 30.4°C); △ 9.05 (1,2-dichloroethane, 69.5°C); □, 1.40 (carbon tetrachloride, 30.4°C); ▽, 1.40 (1,2-dichloroethane, 30.4°C). [Reproduced from Ref. (13) by courtesy of the American Chemical Society.]

on a porous surface. (For a discussion on the effect of the nature of the surface see Sect. III.F). For example, the adsorption of polyvinyl acetate on nonporous iron or tin appears complete in less than one hour, while the amount adsorbed is still slowly changing on porous aluminum oxide after seven hours (13), Fig. 5.

The finite rate of equilibration even on a nonporous surface at relatively high surface coverage is apparently due to conformational changes of the adsorbed macromolecule. However, the fundamental rate at which a polymer molecule in solution becomes attached to a solid surface appears to be rapid and not appreciably different from that observed for low-molecular-weight, monofunctional solutes. The adsorption of poly(vinyl acetate) from 0.02 wt. % solution in benzene onto chrome ferrotype plate is about 70% complete in 20 sec, as measured using radioactively labeled polymer (*34*). Similar studies have been made of the rate of adsorption of polystyrene from very dilute solutions in cyclohexane onto chrome ferrotype plate (*35*). The rate of adsorption is very dependent on the polymer concentration, as depicted in Fig. 6. In addition, the plateau

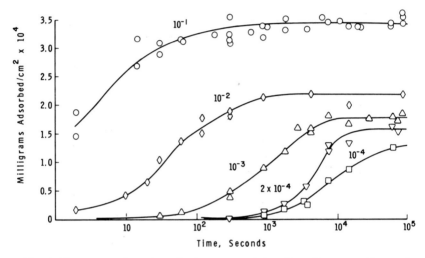

Fig. 6. The rate of adsorption of polystyrene and the effect of concentration on the conformation on chrome ferrotype plate from cyclohexane at 30°C. Numbers on the graph are the concentrations in mg/ml. [Reproduced from Ref. (*35*) by courtesy of the authors and the National Bureau of Standards.]

value or apparent saturation adsorption A_s is seen to be also highly concentration dependent in this region of very dilute solute concentration. Surface coverage is considered to be complete at each of the plateaus, and the differences in A_s then are due to changes in the conformation of the adsorbed macromolecule. Thus the adsorbed polymer is extended on the surface in a relatively flat conformation from the most dilute solutions. With increasing concentration, the polymer attaches to the surface with a decreasing number of segments and hence extends farther out into the solution phase. These conclusions were borne out by ellipsometric estimation of the thickness of the adsorbed polystyrene films (*36*).

Baret (*37*) has pointed out that apart from the energy barrier near the interface that must be surmounted by a solute molecule in order to adsorb at the solid surface, the delay in the establishment of equilibrium is the result of two opposing factors: the rate of diffusion to the interface and the rate of desorption. The diffusion coefficient of a macromolecule [for example, poly(methyl methacrylate) of molecular weight of several hundred thousand to a million in acetone (38)] is of the order of 1/20 to 1/50 of that of a low molecular weight solute (such as glucose in water). The slow rate of diffusion of polymer to the interface must be largely offset, however, by the extremely low rate of desorption.

In general, the desorption of polymers from systems lacking strong solvent–surface interactions is extremely slow; indeed, the adsorption in many such cases is often deemed "irreversible." The latter behavior has been cited as evidence that in such cases the adsorption is not an equilibrium process. The polymer can usually be desorbed, however, by changing to an appropriate solvent. Silberberg (31) showed that the difficulty is only one of degree and that the adsorption process can always be treated by equilibrium thermodynamics.

As examples of such desorption behavior, poly(vinyl acetate) is very slowly and incompletely desorbed from iron by carbon tetrachloride (13); and polystyrene, not at all from carbon black by methyl ethyl ketone (20) or toluene (39). The reason for such behavior lies in the fact that in order for a macromolecule to desorb, *all* of the large number of adsorbed segments must detach from the surface and remain detached long enough for the molecule to diffuse away from the surface. The dependence of the desorption rate on the number of segment attachments was shown in the kinetic study previously cited (35), Fig. 6. Here the rate of desorption into pure solvent (cyclohexane) was found to decrease markedly to virtually nil for adsorbed films prepared from successively lower concentrations of polymer (corresponding to the plateau values of Fig. 6). The latter films, it will be recalled, are apparently attached to the surface in increasingly flat conformations with correspondingly greater numbers of segments attached to the surface. The "irreversibly" adsorbed polymers can invariably be desorbed with solvents that interact strongly with the surface. Better solvency for the polymer may also be a contributing factor but of less importance. In the two previously cited cases, the poly(vinyl acetate) could be completely desorbed with acetonitrile (13) and the polystyrene, about 77% desorbed with tetralin (20).

A study of the adsorption of polymer mixtures by Thies (40) illustrates both the reversibility of the adsorption process and the role of preferential adsorption in the desorption process. It was found that polystyrene was not adsorbed from trichloroethylene solutions onto silica previously saturated with poly(methyl methacrylate). Consistent with this effect, the addition of poly(methyl methacrylate) to silica previously saturated with polystyrene resulted in the complete desorption of the polystyrene.

That the difficulty of desorbing a macromolecule is a matter of degree is also shown by the observation that simple di- or triesters are very rapidly and reversibly adsorbed onto or desorbed from silica in the non-polar solvent n-dodecane (*10*).

F. Effect of the Nature of the Surface

The complex nature of solid surfaces (*41*), perhaps more than any other single factor, contributes to the poor reproducibility of adsorption studies, whether of low- or high-molecular-weight adsorbates. The difficulty lies primarily in the inability to accurately characterize two pertinent properties—the nature of the active adsorption sites and the porosity of the solid medium.

The effect of porosity on polymer adsorption manifests itself in two ways. First, the rate of attainment of equilibrium may decrease markedly

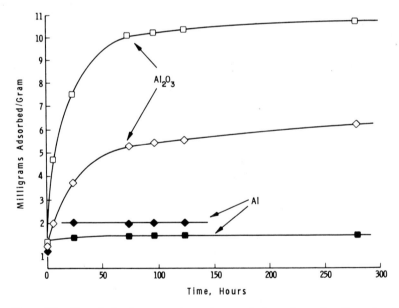

Fig. 7. The effect of adsorbent porosity on the rate of adsorption and on the molecular weight dependence. Polystyrene from cyclohexane at 50°C. Molecular weight $\times$ 10^{-5}: $\Diamond$, $\blacklozenge$, 3.70; $\Box$, $\blacksquare$, 0.67. [Redrawn from Ref. (*16*) by courtesy of the American Chemical Society.]

over that observed on nonporous solids. Figure 5 demonstrated this effect for poly(vinyl acetate) on porous aluminum oxide as compared with nonporous metal powders. A similar effect was observed for the adsorption of polystyrene fractions from cyclohexane by aluminum oxide as compared with aluminum metal (*16*) and is depicted in Fig. 7. These latter data also illustrate the second effect of porosity: the *apparent*

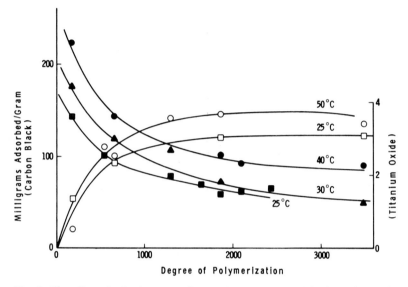

Fig. 8. The effect of adsorbent porosity on the molecular weight dependence of the adsorption of polyvinyl alcohol from water by nonporous titanium dioxide, ○, □; and by porous carbon black, ●, ■, ▲. [Reproduced from Ref. (*42*) by courtesy of the author and the Association for Science Documents Information, Tokyo.]

reversal of the usual effect of polymer molecular weight upon adsorption by the porous oxide. In another example (Fig. 8) adsorption of poly(vinyl alcohol) from water by a nonporous titanium oxide increases with molecular weight, but this dependence is reversed on a porous carbon black (*42*).

Adsorption is usually independent of, or increases with, the molecular weight. Any reversal of this must be due to the ability of the lower-molecular-weight species to penetrate into porous areas of the solid adsorbent not accessible to larger molecules. Thus the reversal is "apparent" in the sense that it is the result of effects other than the actual process of attachment of the macromolecule to the solid surface. The selective retention of low- versus high-molecular-weight macromolecules by a porous matrix is the basis of the technique of gel permeation chromatography for the characterization of the polydispersity of polymers. The permeation elution volume has been shown (43) to be inversely proportional to the logarithm of the molal hydrodynamic volume $\bar{V}_e$, or to log $[\eta]M$ as in Eq. (3). This relation appears to be independent of the shape of the macromolecule, whether a flexible coil or rigid and rodlike.

The very slow rate of attainment of equilibrium on porous solids appears difficult to explain on the basis of rate of diffusion alone. Conformational changes of polymer adsorbed within the pores may be partly responsible. It is also possible that the effect is largely due to the slow displacement of gas trapped in the pores, as has been suggested to explain slow attainment of equilibrium in some gel permeation studies (44).

The chemical nature of the adsorption sites will obviously influence the specificity of interactions with the polymer and/or the solvent. Some instances of such specificity have already been described in the discussions of solvent effects (Sect. III.C) and desorption (Sect. III.E). Particularly striking in the former discussion was the different solvent ranking for adsorption of polyethylene oxide on silica, carbon, or nylon adsorbents (15, 26, 27).

The relative adsorption of aliphatic and aromatic polymers or solvents illustrates the complex effects of the nature of the surface sites. Thus polyisobutylene adsorbs strongly from benzene onto graphitized carbon blacks but not at all on more polar, oxygenated carbon blacks or on silica (18, 29). Polyisobutylene is also not adsorbed from benzene by aluminum, but polysytrene is adsorbed from cyclohexane by aluminum (16). It should be noted that both of the latter polymer–solvent systems were theta mixtures; hence, the difference in behavior must be due to the extent of interaction with the adsorbent. In addition, the polystyrene cannot be readily desorbed from aluminum with cyclohexane but is desorbed efficiently by benzene.

Equal amounts of polyvinyl acetate are adsorbed from carbon tetrachloride per unit area of surface of cellophane, cellulose triacetate, or glass (45). Pretreatment of the cellophane with water, however, increases the adsorption fortyfold. A similar effect of the water treatment is observed on the adsorption from benzene. This effect was shown to be due not to an increase in porosity but apparently to the great increase in hydrogen bonding capacity of the surface.

The present review is restricted to adsorption onto solid surfaces; however, brief note should be taken of the relation to polymer adsorption at a liquid interface (i.e., gas–liquid or liquid–liquid) in view of the large number of studies that have been made, particularly at the water–air interface. The results have been invaluable for the estimation of molecular size and shape in oriented monolayers. Such studies on polymers are in fact best limited to cases where it is quite certain that the polymer mono-layer is completely spread, i.e., with all segments at the interface (46). It is difficult to interpret the behavior of polymer films at the liquid–gas interface in which some of the segments are in the bulk phase. It is evident then that most of the studies do not relate readily to the behavior of random three-dimensional systems typical of polymers adsorbed at the liquid–solid interface. In addition, there may be a fundamental difference in the adsorption of a polymer onto a liquid adsorbent as compared to a solid surface. Stromberg and co-workers have found the conformation of polystyrene adsorbed on mercury (from cyclohexane near the theta temperature) to be highly extended and independent of time or molecular weight (47), quite unlike the behavior at solid metal surfaces (48).

IV. Adsorption from Aqueous Solutions

A. The Role of Water in Adsorption

Biological processes occur primarily in aqueous media, but the dis-cussion up to this point has of necessity been largely concerned with nonaqueous solutions. It is important to note any differences or similarities between aqueous and nonaqueous systems in order to increase the likeli-hood that the information in the preceding sections will be of use to the biologist. Water is a liquid having many unique physical properties, but of primary importance here are the different types of solute–water inter-actions. These interactions are of three fundamental types (49), resulting from ionization of electrolytes, hydrogen bonding, and hydrophobic hydration. Although the interactions of ionic species in water can be of considerable consequence, they are not necessarily overriding. For example, micelle formation of ionic surfactants in aqueous solutions is believed to be due primarily to the negative excess entropy arising from the water structure promoting effect (hydrophobic hydration) of alkyl groups. As another example, it has also been suggested that the classifi-cation of surface active agents with respect to their biological effects is in better accord with their hydrogen bonding rather than their ionic prop-erties (50).

It is to be expected, particularly when hydrogen bonding can occur, that water will compete strongly with the polymer solute at the solid surface. However, because of the amphoteric character of water (and its bifunctionality as a hydrogen donor), the adsorption of water at the surface may either inhibit or enhance subsequent polymer adsorption. This dual possibility is shown in studies of the effect of water on polymer adsorption from organic solvents. For example, small amounts of water very effectively inhibit adsorption of poly(dimethyl siloxane) on iron or glass (*17*) from benzene. The effect is more pronounced on iron than glass and is less so in less polar solvents than benzene. The adsorption of nitrocellulose from acetone by starch is greatly decreased by the addition of a few percent of water (*51*). The effect is about 180 times greater than that of methyl alcohol. The adsorption of poly(vinyl acetate) by cellophane from benzene or carbon tetrachloride, however, is increased about fortyfold after treatment of the cellophane with water (*45*). In another case the adsorption of polystyrene on active carbon from methyl ethyl ketone was found to be decreased about twofold after drying the adsorbent over phosphorus pentoxide (*21*).

It is probable that these inverse effects of adsorbed water are in part the result of the orientation dependence of hydrogen bonding, unlike ionic or hydrophobic hydration interactions. Thus the solid surface, after adsorbing a water layer, may present either a better or poorer geometrical arrangement for subsequent polymer adsorption via hydrogen bonding to the water layer. In this connection, note that correlations have been found between the chemical or biological reactivity of polyfunctional organic compounds and the match of the spacing between electron donors in the molecule to the next nearest neighbor spacing in ice (*49*).

B. Nonionic Polymers

It should be apparent from the discourse on polymer adsorption characteristics given in Sect. III that a valuable tool throughout has been the study of the effect on any given variable of a change in the solvent medium. Polymers that are soluble in both water and a variety of common organic solvents are obviously rare. This perhaps explains in part the paucity of fundamental studies of adsorption of, particularly, nonionic polymers from solutions in water. However, there does not appear to be any reason to expect any unusual characteristics of nonionic polymer adsorption from water. All of the variations that depend on the relative

effects of polymer–solvent, polymer–surface, and solvent–surface interactions can be expected.

Before proceeding to some examples, it is of interest to note that the
very rapid approach by simple, low-molecular-weight, polyfunctional
compounds to the strong adsorption characteristic of macromolecules
(illustrated in Fig. 2) has been observed to occur from water solutions also.

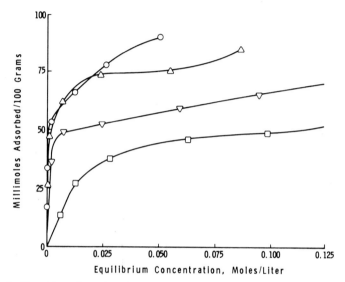

Fig. 9. The effect of polyfunctionality on adsorption by hydrogen montmorillonite
from water at 20°C. Adsorption of glycine, □; glycylglycine, ▽; diglycylglycine,
△; and triglycylglycine, ○. [Reproduced from Ref. (52) by courtesy of the authors
and the Faraday Society.]

Greenland et al. (52) have found the analogous behavior for the adsorption of glycine and its di-, tri-, and tetrapeptides from water by montmorillonite clay (Fig. 9). Glycine and its peptides are of course electrolytes.
However, note that they are all only dipolar electrolytes and that the
polyfunctionality arises via the nonionic amide linkages. Thus even such
simple derivatives presage the possible overshadowing of ion effects by
hydrogen bonding in the adsorption process.

Poly(ethylene oxide) polymers are conveniently soluble in water and a
wide variety of organic solvents. Various aspects of the behavior of
poly(ethylene oxide) have been described in Sect. III.B, C, F. On nonporous silica the adsorption from water is independent of molecular
weights below $M_n = 18,000$, indicating an extended, flat conformation

at the surface (*15*). The adsorption characteristics from organic solvents indicated that moderate looping occurred. It is pertinent to note, however, that Howard and McConnell (*15*) report that adsorption results could not be obtained for high-molecular-weight ($M_v = 190,000$) poly-(ethylene oxide) from water because the silica suspensions became too stable to be sedimented by centrifugation. As will be discussed Sect. VII.A, the mechanism of colloid stabilization by nonionic macromolecules involves adsorption on the particle with long loops of polymer chain extending into the solvent medium. This "steric protection" of aqueous gold sols by poly(ethylene oxide) was shown by Heller and Pugh (*53*) to increase with molecular weight.

Although water is strongly adsorbed on silica, such a surface is still obviously capable of adsorbing the polyether. In fact, it is found that the adsorption from benzene is greater on undried than dried silica. No adsorption of poly(ethylene oxide) occurs on nylon from water (*27*) but the polyether is adsorbed by nylon from benzene. Copolymers (either block or random) of ethylene oxide and propylene oxide will adsorb on nylon from water, the adsorption becoming greater as the proportion of the water-insoluble propylene oxide segments increases. As might be expected, the reverse effect occurs from benzene. Adsorption of poly-(ethylene oxide) from water also occurs on activated or graphitized carbon (*26*) and is greater than from a number of organic solvents. The latter apparently interact strongly with carbon to inhibit the polyether adsorption.

The adsorption of poly(vinyl alcohol) from water by porous carbon and nonporous titanium oxide (*42*) has been discussed in Sect. III.F. The molecular weight dependence of the adsorption on the nonporous oxide (Fig. 8) is in accord with the view that the macromolecules adsorb with long free loops of polymer chain protruding from the surface. On the porous carbon a reversed molecular weight dependence resulted because of the inability of the larger polymer coils to adsorb within the pores. Adsorption of poly(vinyl alcohol) on sodium montmorillonite (hydrous aluminum silicate), however, occurs readily at both the external and lamellar-internal surfaces (*54*). The interlamellar spacing of 20.5 Å, determined by x-ray diffraction, requires that the adsorbed polymer be highly extended on the surface, with perhaps half the segments in contact. Thus it appears if the polymer–surface interactions are sufficiently great that the adsorption of a macromolecule can occur in a highly restricted space. The adsorption isotherm for the poly(vinyl alcohol) on the sodium montmorillonite (Fig. 10) shows the extremely rapid rise to the plateau value at very low concentrations typical of macromolecular adsorption (cf. Fig. 1). It is interesting to note that the poly(vinyl alcohol) did not desorb from the surfaces discussed above.

The examples given above suffice to show that the basic principles for adsorption of nonionic polymers from water are the same as for organic solvents involving interplay of the polymer–solvent–surface triad. In conclusion it may be noted that the moieties that cause a nonionic polymer (or an un-ionized ionic polymer) to be water-soluble will also very likely be responsible for very strong interactions with an appropriate surface.

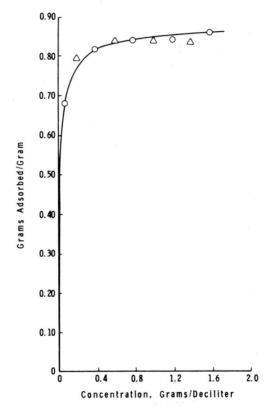

Fig. 10. Adsorption of polyvinyl alcohol ($M_v = 100,000$) by dilute suspensions of sodium montmorillonite from water at 23°C. Weight percent of clay: ○, 0.25; △, 0.50. [Redrawn from Ref. (54) by courtesy of Academic Press, Inc., N.Y.]

C. POLYELECTROLYTES

Singular behavior of polymer adsorption from aqueous solutions occurs when the macromolecular solute contains segments capable of dissociating into ions. Although there is the obvious relevance of polyelectrolytes to many biopolymers and practical interest as flocculating agents, the fundamental nature of their adsorption at solid surfaces has been the least studied facet of polymer adsorption. In addition, as might be expected, the solution behavior of a polyelectrolyte solute is complex and has some unique characteristics (*55*). In the present discussion the effect of the degree of ionization on the polymer solute coil dimensions is of primary interest.

Because of the long range of electrostatic forces, the repulsion of like charges on the polyion results in considerable extension of the macromolecular coil. For example, polymethacrylic acid ($M_v = 84,000$) expands with increasing ionization (*56*), Fig. 11. The maximum size at

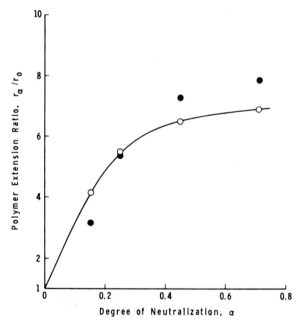

Fig. 11. Ratio of the polymer chain end-to-end distance r of polymethacrylic acid in water as a function of the degree of neutralization, α. From measurements of light scattering, $\bigcirc$; and viscosity, $\bullet$. [Redrawn from Ref. (*56*) by courtesy of the American Chemical Society.]

about 70% neutralization corresponds to a sevenfold increase in the polymer chain end-to-end distance r. The polymer molecule, however, is not completely extended at this maximum. The degree to which the ionized molecule is extended is reduced in the presence of added electrolyte (simple salts) because of the shielding of the electrostatic repulsions between the charged polymer segments. Under these conditions the polyelectrolyte solution characteristics approach those of nonionic polymers and in many cases theta solvent behavior can be attained. The requisite choice of added electrolyte, especially the counterion, to achieve the theta condition is critical, however. It should also be noted that by their very nature polyelectrolytes must contain functional groups that are highly polar even in an un-ionized state.

The earliest studies of polyelectrolyte (nonbiological) adsorption were prompted by their function as efficient flocculating agents. Of particular interest was the observation that flocculation (and the inferred adsorption) could occur with charged polyions and surfaces with the same sign. Michaels (57) observed the flocculation of kaolinite (a clay with negatively charged surface) with partially hydrolyzed polyacrylamide to be most effective when about one-third of the segments were hydrolyzed. Adsorption studies (58) on kaolinite of a 21% hydrolyzed polyacrylamide and of sodium polyacrylate showed that adsorption increased with decreasing pH, i.e., with decreasing degree of ionization. Neither polymer was adsorbed at sufficiently high pH of 7–8. Thus it is apparent that adsorption, via hydrogen bonding of carboxyl (un-ionized) and/or amide segments to the surface, can occur up to a certain limit, in spite of the repulsion of the anionic carboxylate segments from the surface. The anionic segments also have the effect, as discussed above, of causing increased extension of the polymer coil away from the surface in this case. This latter effect favors the bridging of adsorbent particles with polymer, which is the flocculation mechanism. Thus a critical ratio of ionized to un-ionized segments is necessary to achieve the proper balance of adsorption and particle-bridging for the maximum flocculating effect. Partially hydrolyzed polyacrylamide has been shown (59) to adsorb on the negatively charged surface of silica (quartz) particles also. However, here the adsorption was so weak as to be apparently shear dependent.

A case of adsorption of a polyelectrolyte by a surface, with respective charges of opposite sign, is that of polymethacrylic acid on titanium dioxide (anatase) in the pH range 1–3.8 (60). The adsorption capacity rises to a maximum by about a factor of five upon increasing the pH from 1 to 2. This result is presumably due to the effect of increasing the number of very strong ionic bonds to the surface. It is interesting to note that the number of charges per macromolecule at pH 2 was estimated to be only five and that the molecular weight of the polymethacrylic acid was 190,000 (M_w), from which the total number of segments per molecule must be about 2200. The fraction of segments attached to the surface is not known, although an estimate of at least 30% may be ventured (see Sect. V.B). In any case it is apparent that the effect on the adsorption of a very small fraction of very strongly held segments is marked. Qualitatively similar effects occur in completely nonionic polymers in nonpolar solvents (9, 13).

Upon further increasing the pH from 2 to 3.8, however, the poly-methacrylic acid adsorption is about halved. It does not appear possible to decide with any certainty among the several possible causes of this reduction. Expansion of the polymer to a more voluminous coil by the intermolecular replusion of ionic segments may be the primary cause. Also, the amphoteric anatase surface must become less positively charged and eventually negative at sufficiently high pH. Indeed, at pH 12 the polymethacrylic acid is completely desorbed. A third possibility is that as the number of very strongly held segments increases, at some point the polymer may start to be adsorbed in a flatter conformation at the surface.

The effect of charge repulsion was also observed with a series of vinyl acetate–crotonic acid copolymers (61). In the pH range 7.6–5 the adsorption on anatase increased markedly with decreasing pH or with decreasing crotonic acid content (42% to 13%). These data are shown on Fig. 12, where it is apparent that a constant shift in pH units for the individual copolymer curves will cause all of the points to fall on a single curve. Analysis of this result led to the conclusion that the primary variable determining the maximum amounts adsorbed (and the solubility limit in the solvent as well) was the ratio of repelling carboxylate ions to adsorbing vinyl acetate groups. It appeared that the adsorptive capacity was uninfluenced by the number of un-ionized carboxylic acid groups. This seems difficult to reconcile with the adsorption behavior of poly-methacrylic acid described above (60). Again, with regard to the electro-static interactions responsible for the reduction in adsorption, the relative contributions of the surface–segment and segment–segment repulsions are not apparent.

A more definitive assessment of the relative contributions of the re-pulsive interactions is that of Peyser and Ullman (62) for the adsorption of poly-4-vinylpyridine onto a silica glass. The charge on the silica is always negative, yet the polymer adsorption always decreased with increasing amount of cationic segments (produced by neutralization with hydrochloric acid). In addition, the adsorption increased with increasing concentration of added electrolyte. No difference in the latter effect was observed whether $MgCl_2$ or $NaCl$ was used as the added electrolyte, although the silica carries a higher negative charge in $NaCl$ than in $MgCl_2$ (as determined from ζ-potential measurements). All of these effects are depicted in Fig. 13 for a polymer of 360,000 molecular weight (M_v) in 50% methanol–water solvent. The degree of ionization produced by varying degrees of neutralization by HCl was determined by ultraviolet spectroscopy. The ionic strength was controlled by adding inorganic electrolyte. These results are in accord with the view that the effect of

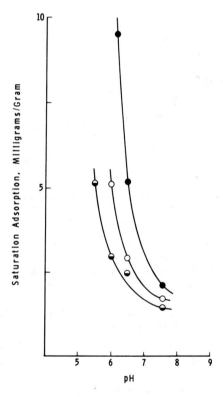

Fig. 12. Effect of pH on the adsorption of vinyl acetate–crotonic acid copolymers by titanium dioxide from water at 30°C. Percent crotonic acid in the copolymer (molecular weight, $M_w \times 10^{-5}$): ●, 12.9 (0.61); ○, 26.1 (0.31); ◑, 42.4 (0.18). [Reproduced from Ref. (61) by courtesy of the American Chemical Society.]

degree of ionization and of added electrolyte on the polymer coil dimensions is the primary factor controlling the extent of adsorption, i.e., the size of the polymer solute is reflected in the spatial requirements (area per molecule) at the surface.

Virtually all of the available information on nonbiological polyelectrolyte adsorption has been described above. The information at present

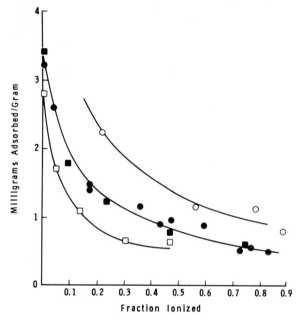

Fig. 13. Effect of the degree of ionization and of added electrolyte on the adsorption of poly-4-vinylpyridine by silica from 50% methanol–water at 30°C. Total Cl⁻ normality (nature of electrolyte): ○, 0.2 (NaCl); ●, 0.04 (NaCl); ■, 0.04 (MgCl$_2$); □, 0.007 (NaCl). [Reproduced from Ref. (62) by courtesy of John Wiley and Sons, Inc., N.Y.]

suggests that charge interactions at the liquid–solid interface play a minor role in polyelectrolyte adsorption. Only at extremely small fractions of ionized segments and over a very narrow range of composition does electrostatic attraction between polymer and surface contribute uniquely to polymer adsorption.

It is pertinent at this point to include a brief reference to a polymer of biological origin in order to reinforce the above views. Gelatin, a polypeptide, is a much studied polymer whose adsorption characteristics (63) are in accord with those of nonbiological polymers. In particular, the adsorption is found to be greatest in the region of the isoelectric point of the gelatin, for example, on silica (63) and on silver bromide (64).

A final case of polyelectrolyte adsorption having a unique consequence is that of the adsorption of diallyldimethylammonium chloride–SO_2 copolymer on sodium bentonite (65). The latter is a negatively charged clay capable of ion exchange. The polymer adsorption and its effect on

TABLE 7

ADSORPTION OF A POLYELECTROLYTE ON BENTONITE
EFFECT ON THE ION-EXCHANGE CAPACITY[a]

Total adsorbed polymer, A	Cation- exchange capacity, B	Anion- exchange capacity, C	Bound polymer, $D = A - C$	Total cation- exchange sites, $B + D$
0	62.4	0	0	62.4
8.6	49.5	0	8.6	58.1
18.5	44.6	0	18.5	63.1
33.6	30.7	1.0	32.6	63.3
40.7	24.5	3.9	36.8	61.3
43.5	23.9	5.3	38.2	62.1
47.9	21.4	6.7	41.2	62.6
46.6	23.7	6.4	40.2	63.9
57.7	15.8	13.2	44.5	60.3
58.5	14.4	15.1	43.4	57.8
63.0	12.8	15.5	47.5	60.3
64.3	11.1	18.2	46.1	57.2

[a] All amounts in milliequivalents per 100 g bentonite. Polyelectrolyte: diallyldimethylammonium chloride–SO_2 copolymer of molecular weight 167,000. Data taken from Ref. (65).

the ion-exchange capacity are shown in Table 7. Adsorption of the polycation decreased the cation-exchange capacity of the clay (B). Furthermore, although the original bentonite had no anion-exchange capacity, at high polymer adsorption appreciable anion-exchange capacity appeared (C). The latter must be due to the unattached polymer segments in the free loops of the adsorbed polymer. Thus it was possible to calculate the amount of bound segments (D) and then the total number of cation-exchange sites ($B + D$) on the original bentonite. The constancy of this latter value is evidence that the polymer adsorption is via ion exchange, $Na^+ \rightleftarrows$ polycation. It should be noted that the ratio D/A is the apparent fraction of attached polymer segments. It is thus evident, in accordance with previously discussed behavior (p. 104), that the polymer is completely extended and attached at all segments at low surface coverage. When the surface is saturated with polymer, some looping occurs, freeing about one-fourth of the segments.

It is evident that the study of polyelectrolyte adsorption is usually more difficult than that of nonionic polymers in general because of the greater difficulty of characterizing the solute and the surface. The chemical nature of both the solute and the surface, as well as the physical nature of the solute, is often seen to vary drastically with changes in the aqueous environment (pH, added electrolyte).

V. Conformation of the Adsorbed Polymer

A. THICKNESS OF THE ADSORBED LAYER

It should be apparent at this point that an appreciable amount of qualitative information regarding the probable conformation of adsorbed macromolecules can be and has been deduced from adsorption data alone. An estimate of the film thickness from adsorption data must be based on assumptions regarding the surface area required per attached segment and on the state of the free loops of unattached segments. Obviously only direct measurement of the adsorbed layer thickness can be entirely satisfactory. Several methods have been devised for this purpose, but each method has its own limitations and idiosyncrasies.

The first direct measurements were made by Öhrn (66), who examined the anomalous values of the apparent reduced specific viscosity of very dilute polymer solutions obtained from measurements with capillary viscometers. The anomaly is evidently due to the reduction in effective diameter of the capillary by the adsorbed macromolecular film. The maximum apparent thickness of adsorbed polystyrene ($M \approx 500,000$) in toluene was thus found to be about 1300 Å—a result that we will see is too high.

Since the work of Öhrn, other such estimates of adsorbed polymer thickness have been made, usually as an adjunct to viscometry studies. The method does not appear well suited to the purpose and unrealistically high values often result. Discussions of the sources of error in capillary viscometry are given by Tuijnman and Hermans (67) and by Öhrn (68). It is interesting to note that one complication is the change in concentration caused by the strong polymer adsorption in very dilute solutions (see Fig. 1).

The difficulties in utilizing this reduction in effective cross section of flow for systematic study of polymer adsorption have been circumvented by Rowland and Eirich (69) by the use of porous discs rather than a single capillary. A systematic study of poly(vinyl acetate), poly(methyl methacrylate), and polystyrene on Pyrex glass from several solvents was made by this method (32). The thickness of the adsorbed layers was found to

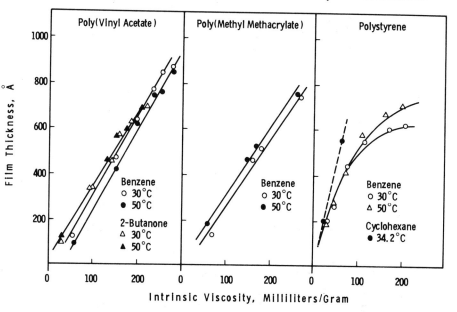

Fig. 14. Polymer film thickness at high surface coverage on Pyrex glass as a function of intrinsic viscosity. From measurements of flow rates through porous glass disks. [Reproduced from Ref. (32) by courtesy of John Wiley and Sons, Inc., N.Y.]

vary in a way closely related to the polymer coil dimensions in solution. This is illustrated in Fig. 14, where the thickness is shown as a function of intrinsic viscosity. However, deformation of the coils must occur upon adsorption. A measure of this, called the reduced film thickness, was taken to be the ratio of the film thickness to twice the radius of gyration R_G of the free coils in solution. (The equivalent hydrodynamic volume V_e discussed in Sect. II, is proportional to R_G^3.) The ratio should be about 0.9 for a coil adsorbed without distortion. On this basis Fig. 15 shows that weakly adsorbing polystyrene adsorbs almost undeformed at low molecular weights. At high molecular weights the deformation approaches that of the strongly adsorbing poly(methyl methacrylate).

A related viscometric method of film thickness determination is to measure the effective increase in volume of dispersed particles when coated with a layer of adsorbed polymer. This was first done by Amborski and Goldfinger (*11*) and involves use of the Einstein viscosity equation [Eq. (1)] modified with an additional ϕ^2 term for application at higher concentrations. Rothstein (*70*) applied this method to alkyd resins (fatty

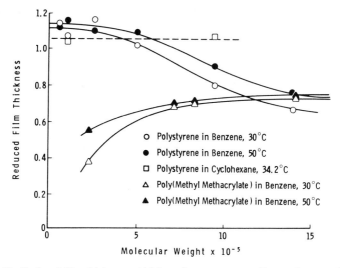

Fig. 15. Reduced film thickness at high surface coverage on Pyrex glass as a function of molecular weight. Molecular weights are M_v for polystyrene and M_w for poly(methyl methacrylate). [Reproduced from Ref. (*32*) by courtesy of John Wiley and Sons, Inc., N.Y.]

acid-modified glycerol phthalates, $M_v \approx 4000$ to 14,000) adsorbed on titanium dioxide particles from methyl ethyl ketone. The apparent thickness of the adsorbed film increased markedly with increasing molecular weight and was a linear function of the intrinsic viscosity, analogous with Fig. 14. The measured thicknesses ranged from 75 to 280 Å. It has been suggested, however, since a small degree of flocculation greatly increases the dispersion viscosity, that erroneously high values can result (*71*). It is further suggested that this effect can be eliminated by appropriate extrapolation of viscosities to infinite shear rate.

Another method allied to those discussed above is that of measuring the influence of the adsorbed polymer upon the sedimentation rate of a particle of known size (*72*). Particles of a narrow size distribution must be used in order to obtain sharply defined sedimentation boundaries. Rather limiting is the necessity of using a solvent of about the same density as the polymer so that the polymer film will contribute nothing to the effective mass of the sedimenting particle. A thickness of 25 Å for poly(lauryl methacrylate) on carbon black in fluorotoluene was obtained. This is in accord with other evidence to be discussed in Sect. V.B, which indicates that this polymer is highly extended on polar surfaces.

It must be noted that all of the methods discussed thus far yield an effective hydrodynamic thickness. Although the relationship between these values and the true molecular configuration at the surface is not known, the values certainly are of considerable practical consequence.

A method of measurement suggested by Stromberg et al. (36) which has the advantage of not perturbing the adsorbed layer is that of ellipsometry. In this method, changes in the state of polarization of light after reflection from the surface (while immersed in the equilibrium solution) are measured. Unfortunately the precision is lowered when the refractive index of the adsorbed film is close to that of the solution, as may be expected if the polymer film is swollen with solvent. To date, only polystyrene has been studied by this technique (36, 48). The thickness at saturation on chrome ferrotype from cyclohexane (at the θ-temperature, 34°C) was found to be a linear function of the square root of the molecular weight, at least up to 19×10^5 (M_n or M_w) as shown in Fig. 16. This relationship suggests that the polymer is adsorbed as a random coil under these conditions. It is pertinent to recall from Eq. (5) that at the θ-condition the intrinsic viscosity of a random coil is proportional to the square root of the molecular weight. The polystyrene extension was found to be approximately independent of the type of metal surface, even for gold, as compared with the oxidized chromium ferrotype surface.

The only results suitable for comparison amongst determinations of polymer film thickness are those for polystyrene by porous disk viscometry (69) and by ellipsometry (48). These two studies agree very well in spite of the different surfaces, solvents, and temperatures used. The two methods might also be expected to yield basically different effective thicknesses.

An additional single determination for polystyrene is available by a novel method used by Peyser and Stromberg (73). The method, employing attenuated total reflection (ATR or internal reflection spectroscopy) in the ultraviolet, is closely related in many respects to the ellipsometric technique. Polymer adsorption takes place on the ATR prism, which must therefore be transparent to ultraviolet light. For the polystyrene of $M_v = 76,000$ a film thickness of 240 Å on quartz from cyclohexane at 34°C was obtained. The same polystyrene fraction had previously given a value of 225 Å on chrome ferrotype by ellipsometry (48), and a polystyrene fraction of $M_v = 110,000$ measured 200 Å by porous disk viscometry (69) on Pyrex glass, all from cyclohexane at 34°C. This is good agreement albeit for three different surfaces.

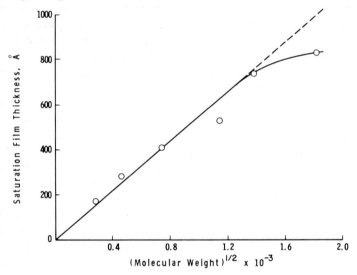

Fig. 16. Effect of molecular weight on the film thickness of polystyrene adsorbed on chrome ferrotype plate from cyclohexane at 34°C. From ellipsometric measurements. [Reproduced from Ref. (48) by courtesy of the American Chemical Society.]

B. Number of Attachments to the Surface

A complete description of the conformation of an adsorbed macromolecule must also include the number of attachments to the surface. Again, qualitative estimates can be made from the adsorption and particularly the desorption behavior (Sect. III.E). The only direct method of counting the number of chain segments attached to surface sites is the spectroscopic method of Fontana and Thomas (72). When the functional groups responsible for adsorption interact with surface sites, a frequency shift in the characteristic infrared absorption of the functional group will occur. The spectrum of unattached groups on the adsorbed macromolecule is unchanged from that of the free molecule in solution. The fraction of attached segments can then be determined from their infrared absorbance by comparison with that of adsorbed monomeric analogues (72) or from the relative absorbances for bound and unbound groups, as done by Thies et al. (74). The method has been applied thus far only to adsorption on silica of very small particle size (0.02 μ) for experimental convenience. However, recently devised techniques (75, 76) for the spectroscopic study of adsorption at liquid–solid interfaces would appear to allow a variety of solid phases to be utilized.

In the study of the adsorption of poly(lauryl methacrylate) on silica (71) the effect of the adsorption on the infrared spectrum also confirmed that the mechanism of adsorption was via hydrogen bonding of ester carbonyl to surface hydroxyl. The fraction of attached segments was found to be about 0.36 at high surface coverage, only slightly dependent on surface coverage, and independent of a change in molecular weight (M_v) from 330,000 to 1,190,000 and a 1.3-fold change in polymer solute voluminosity. These results are in accord with the 25 Å film thickness determined from sedimentation rate (as discussed in the previous section) and with expectations for strong polymer–surface interactions.

The results of similar studies of the adsorption of poly(methyl methacrylate), poly(4-vinyl pyridine), and polystyrene on silica (74) are shown in Fig. 17. In agreement with the results described above, the fraction of poly(alkyl methacrylate) segments attached to the surface—about 0.30 in this case—is not appreciably affected by the extent of surface coverage. On the other hand, the poly(4-vinyl pyridine), although probably more strongly bound to the silica surface than are the poly(alkyl methacrylates), undergoes an appreciable decrease in the fraction of bound segments from 0.45 to 0.20 with increasing adsorption. The more weakly interacting polystyrene also exhibits decreasing fractions of attachment with increasing surface coverage to a minimum of 0.02 to 0.05. The latter value was observed to increase to 0.10 to 0.13 for longer periods of contact, however, without any change in the amount of polymer adsorbed.

As with most other features of polymer adsorption it is not surprising that the fraction of attachments to the surface must be determined by the complex relations between a number of factors such as polymer–solvent–surface interactions and polymer flexibility. Considerable evidence has been presented (Sect. V.A) to suggest that the thickness of the adsorbed layer is related to the polymer solute coil dimensions. Appreciable deformation of the coil must ordinarily occur, however, to account for the number of segments attached to the surface [in accord with the "reduced film thickness" concept of Rowland and Eirich (69), Fig. 15].

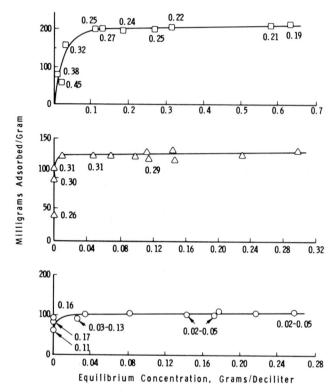

Fig. 17. Adsorption isotherms on silica and the fraction of bound segments (numbers on the graph) by infrared spectroscopy. $\bigcirc$, polystyrene ($M_v = 92,000$), from trichloroethylene at 25°C; $\triangle$, poly(methyl methacrylate) ($M_v = 230,000$), from chloroform at 25°C; $\square$, poly(4-vinyl pyridine) ($M_v = 1,690,000$), from chloroform at 30°C. [Redrawn from Ref. (74) by courtesy of Gordon and Breach Science Publishers, Inc., N.Y.]

Thies et al. (74) note that two possible configurations of an adsorbed polymer with the same fraction of surface-bound segments can result in widely different effective thicknesses or loop lengths, depending upon whether the attached segments are relatively uniformly distributed along the chain or whether they occur in long consecutive sequences.

In concluding this section, the special case of the adsorption of a polyelectrolyte on a surface capable of ion exchange (65) should be recalled (Sect. IV.C, Table 7). Here, the measurement of the effect of polymer adsorption on the ion exchange capability resulted in a count of the number of attached segments. Virtually all the segments were bound at low surface coverage and about three-fourths at saturation of the surface.

This latter effect then appears to be quite general, i.e., upon saturating a surface with polymer, even when segment–surface interaction is strong, the crowding always results in a very appreciable number of unattached segments. It is particularly interesting to note that these "free" loops of polymer segments, except for molecular mobility, behave in many respects as though a solute. The evidence for this lies in the unchanged infrared absorption frequencies (72) and in the segment ion exchange capability (65).

C. Adsorption of Copolymers

The majority of polymer adsorption studies have been made with homopolymers. Of particular interest here to the consideration of conformation is the adsorption of copolymers where one kind of relatively infrequently occurring segment is preferentially adsorbed rather than the other more abundant segments. Under these circumstances the formation of free loops that determine the film thickness is exaggerated. This effect has important practical consequences, especially for colloid stabilization, and these will be discussed in Sect. VII.A.

Such behavior was first noted by Koral et al. (13) for the adsorption of partially hydrolyzed polyvinyl acetate (Table 8). The hydroxyl groups produced by hydrolysis are evidently preferentially adsorbed rather than the ester segments. The longer free loops that result are reflected in the large increase in adsorption. It is pertinent to note that the most pronounced effect occurs for the first few hydroxyl groups introduced into the molecule. In fact it would be expected that the adsorption would start to decrease beyond some optimum fraction of strongly adsorbing segments. Note also that the effects shown in Table 8 are not caused by changes in polymer solubility (as denoted by the intrinsic viscosity). An exactly analogous and even more marked effect was observed for the adsorption of partially hydrolyzed poly(methyl methacrylate)—to give more strongly adsorbing carboxyl groups in this case—on iron (9), as shown in Table 8.

Incorporation of about 17 mole% N-vinyl-2-pyrrolidone into poly-lauryl methacrylate) as a copolymer increased the adsorbed film thickness on carbon black, as determined from sedimentation measurements (72), to about 210 Å as against about 25 Å for the homopolymer. The infrared

TABLE 8

ADSORPTION OF PARTIALLY HYDROLYZED POLYMERS ON IRON

Polymer	Molecular weight, $M_W \times 10^{-5}$	Intrinsic viscosity,[a] ml/g	% Hydrolyzed	Saturation adsorption, mg/g	Reference
Poly(vinyl acetate)[a]	1.40	73	0	0.35	(13)
		75	13	0.96	
		53	26	1.07	
Poly(methyl methacrylate)[b]	5.32	—	0	0.77	(9)
		—	0.9	2.09	
		—	1.5	2.17	

[a] In 1,2-dichloroethane at 30.4°C.
[b] In benzene at 30°C.

absorption frequency shift for the adsorbed species on silica from *n*-dodecane confirmed that the vinyl pyrrolidone segments interact with the surface more strongly than do the ester segments. In a spectroscopic study (*33*) made with a 20:1 mole ratio alkyl methacrylate–polyglycol methacrylate copolymer adsorbed on silica, it was possible to show that the fraction of attached ester segments was reduced to as low as 0.03 at high surface coverage, against 0.40 for the ester homopolymer.

Similar studies (*28*) with an 88:12 mole ratio poly(ethylene–vinyl acetate) copolymer showed a very high fraction (0.94) of the *vinyl acetate segments* attached to the silica surface from cyclohexane at low adsorbance, decreasing to 0.50–0.60 (i.e., to 0.06–0.07 of the total number of segments) at high adsorbance. This was in contrast to the poly(vinyl acetate) homopolymer, where the fraction of attached segments changed very little (at about 0.41) with surface coverage. Thus the copolymer can evidently more readily undergo configurational changes to satisfy the changing spatial requirements at the surface. Virtually equal saturation adsorbance values were observed, although the copolymer molecular weight was about one-tenth that of the homopolymer. The copolymer chain looping thus compensates for the disparity in the molecular weights.

An interesting consequence of this kind of adsorption behavior of copolymers was revealed by Steinberg (*77*) in studies of the co-adsorption of a homopolymer—poly(lauryl methacrylate)—with a copolymer—poly(alkyl methacrylate-vinyl pyridine)—by iron from toluene or decane. It was found that adsorption of the homopolymer did not occur on a surface already covered with the copolymer. The copolymer, however, not only could adsorb over the previously adsorbed homopolymer, but under these conditions an optimum adsorbance could be attained where the total amount of the two polymers adsorbed was about twice that possible with either polymer alone.

VI. Polymer Adsorption Theory

A. HOMOPOLYMERS

Theoretical analysis of the adsorption of macromolecules has required rather formidable statistical–mechanical treatment. The numerous available treatments are consequently diverse in their methods and conclusions. The problem is in some ways akin to that of accounting for the excluded volume of macromolecules for the determination of the thermodynamic properties of polymer solutions (7), which has yet to be completely solved. Rowland and Eirich (32) note that comparison of existing theories of polymer adsorption with the experimental results is difficult because of the limitations of both theory and experiment. Most of the theoretical treatments have of necessity dealt with the "isolated" macromolecule, i.e., with very dilute solutions and low degrees of surface coverage. This is a region especially difficult for experimentation. In addition, many of the parameters used in the theoretical treatments are not capable of rigorous definition or evaluation.

Stromberg (78) gives a good discussion of the major theoretical treatments from the original attempt by Simha et al. (79) to early 1965, including the first two papers of Silberberg (80, 81), and summarizes the conflicting conclusions. Recently, Silberberg (82, 83) has greatly expanded his earlier study, taking into consideration subsequent criticisms and extensions of the theory made by others. In particular, the treatment now takes into greater account the role of solvent–polymer and solvent–surface interactions as well as that of polymer–surface, and extends the conditions to finite solute concentrations. The model that arises is that of a macromolecule attached to the surface by contiguous sequences of segments alternating with free loops extending away from the surface. It is concluded that macromolecules are never adsorbed in the random coil form in which they exist as solutes in solution. The concentration of segments in physical contact with the surface (i.e., the fraction of surface sites occupied by polymer segments) is primarily determined by the polymer-surface interaction parameter. However, the concentration of free segments in the attached polymer phase and the associated free loop size are found to be primarily a function of the polymer solute concentration. In view of the model, it is surprising that the relation of adsorbed layer thickness (proportional to the loop size) to molecular weight in a θ-solvent (with appropriate choice of a polymer flexibility parameter) is almost exactly that found by Stromberg et al. (48), Fig. 16. Thus it appears that the experimentally observed relation of film thickness to the square root of the molecular weight, or to the intrinsic viscosity (Fig. 14), need not be taken literally as evidence for adsorption as random coils. Such relationships apparently may only indicate a functional dependence of the film thickness on parameters related to properties of the polymer as a solute.

Silberberg's treatment (83) (again, with proper choice of a flexibility parameter) also gives values for the fraction of attached segments p that fall in the experimentally observed range (Sect. V.B). Theoretically

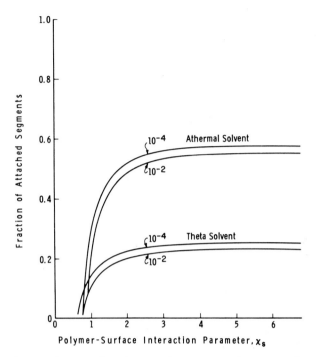

Fig. 18. Theoretical values of the fraction of the number of segments of an adsorbed macromolecule in contact with the surface as a function of the polymer–surface interaction parameter (in units of kT). Total number of segments $= 10^6$. Numbers on the graph are the equilibrium concentrations (volume fraction) of the polymer solute. [Redrawn from Ref. (*83*) by courtesy of the author and the American Institute of Physics.]

calculated values of p for a polymer with 10^6 segments versus the polymer–surface interaction parameter x_s are shown in Fig. 18 for the cases of a good (athermal) and a poor (θ) solvent. The parameter x_s is the negative of the energy change for a polymer segment displacing a solvent molecule on a surface site; adsorption increases with increasing x_s.

B. Polyelectrolytes and Copolymers

There have as yet been no attempts to treat theoretically the adsorption of polyelectrolytes. It will be recalled (Sect. IV.C) that in solutions of high ionic strength, where long-range electrostatic interactions are effectively reduced, the polyelectrolytes behave in a manner analogous to nonionic polymers. Increase in ionic strength decreases the quality of solvent for the polyelectrolyte and increases the amount of polymer adsorbed. Silberberg (*83*) notes that under conditions of high ionic strength his theoretical treatment should fit all these observations.

Detailed theoretical treatment analogous to that discussed above for homopolymers has not been applied to the case of adsorption of copolymers having segments of different adsorbability. The first paper by Silberberg (*80*) considers the case of a polymer where only certain segments are adsorbable. It was concluded that for such a copolymer the free loop length would be considerably increased over that for a homopolymer, in accord with experimental observation (Sect. V.C).

An interesting computer simulation of the adsorption process for such copolymers has been carried out by Clayfield and Lumb (*84*). Because of limitations of computer time only relatively short chain lengths and short runs of segment movements could be considered. Thus the results were expected to provide only broad indications of the configurational behavior. The results suggest that a critical balance of the adsorption energy and the frequency of occurrence of the adsorbing segments in the

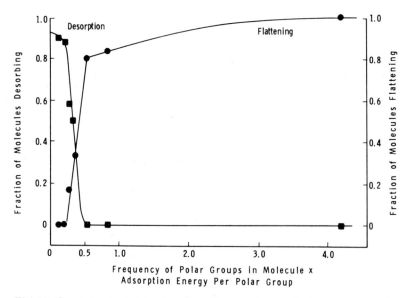

Fig. 19. Computer-simulated adsorption of a copolymer of 128 segments at low surface coverage. Influence of the strength of adsorption on the configurational behavior. Fraction of the surface covered = 0.14. [Reproduced from Ref. (*84*) by courtesy of the American Chemical Society.]

chain is necessary in order to avoid either flattening of the chain on the surface or desorption. This result is shown in Fig. 19 (which must not be taken too literally), where the product of the adsorption energy and segment frequency is used as a numerical rating of the strength of adsorption of the macromolecule as a whole.

C. ADSORPTION ISOTHERM EQUATIONS

Discussion concerning equations for polymer adsorption is largely academic at present. There appears to be no available equation that can be readily applied to experimental data to obtain meaningful values of adsorption parameters. The isotherm equations that arise from statistical–mechanical theory are intractible for this purpose. The experimental isotherm (Fig. 1) lends itself poorly to mathematical expression because the experimentally accessible region ordinarily consists of the slowly changing plateau region. The large changes in adsorption occur at concentrations ordinarily too low to be readily measured. The fit of polymer adsorption data to the Langmuir equation

$$\theta/(1 - \theta) = kC \tag{8}$$

where θ is the fraction of surface covered, C is the concentration of solute, and k is a constant) has been often shown. However, this appears to be the fortuitous result of the insensitivity of the equation to the limited breadth of the measured absorbance. This is borne out by the difficulty as shown by Frisch et al. (39) in differentiating from the experimental data between the Langmuir equation and the Simha–Frisch–Eirich equation (85, 86)

$$\theta/\nu(1 - \theta)^\nu = kC \tag{9}$$

where ν is the number of attached segments per macromolecule Equation (9) was obtained by simplification of a more complex equation by ignoring terms relating to lateral interactions between polymer segments. It is interesting to note that Silberberg (81) states that his isotherm agrees in form with Eq. (9) but with the difference that ν should represent the total number of segments per molecule.

A simple kinetic derivation of Eq. (9) has been given (39) using arguments analogous to those commonly applied in kinetic derivations of the Langmuir equation, Eq. (8). However, no physical behavior particularly unique to macromolecules is invoked, which suggests that the equation must be too oversimplified to apply to macromolecular adsorption.

Interestingly, though, the kinetic derivation and hence Eq. (9) should be valid for simple polyfunctional solutes (e.g., dimer and trimer). This has been shown to be the case (*10*) for the derivatives illustrated in Fig. 2. In this study there was no ambiguity regarding the meaning of *v*, since it was found for such simple compounds that all of the segments of adsorbed molecules were always attached to the surface.

VII. Effects of Adsorbed Polymers

To conclude this chapter, some of the practical consequences of adsorbed polymers will be discussed. The unique aspects of such effects are the result of the polyfunctionality of the macromolecule and its tendency to adsorb with free loops of segments extending out into the solution phase. Some of these effects have been touched upon in Sect. IV.B.C and V.A.C but will be reemphasized and amplified briefly here. It is hoped that this discussion may encourage and assist the application of the knowledge on the adsorption of nonbiological polymers to problems concerning biological systems.

A. Colloid Stabilization

The classical mechanism for retarding the aggregation of dispersions of colloidal particles in aqueous solutions is that of electrostatic-charge repulsion. This is usually accomplished by the adsorption onto the particles of ionic surfactants. In nonpolar media such ionic effects are difficult to bring about; van der Waarden (*87*) showed, however, that the stability of carbon black dispersions in hydrocarbon solvents could be increased by the adsorption of *n*-paraffin chain-substituted aromatics. Theoretical analysis has shown (*88, 89*) that the repulsive forces operating in such a case result from the decrease in configurational entropy that occurs when the paraffin chains adsorbed on approaching particles intermingle. The effectiveness of the stabilization increases with increasing thickness of the adsorbed oleophilic film; this can also be shown from theoretical considerations alone of the van der Waals attractive interactions of the particles, adsorbed film, and solvent (*90*). Thus suitable adsorbed macromolecules would be expected to function very well in this respect, and such is found to be the case. In polymer adsorption studies, in fact, it is occasionally reported that difficulty was experienced in sedimenting the solid adsorbent phase away from the solution phase for analysis due to the formation of stable dispersions. This is most likely to happen with high-molecular-weight fractions, as would be expected. An example of this occurred in the study of the adsorption of polyethylene oxide polymers onto silica in aqueous solutions (15). It should be noted at this point that the steric repulsion mechanism functions equally well in aqueous media, provided only that the adsorbed polymer film be hydrophilic in this case.

The above-mentioned theoretical treatment of Mackor and van der Waals (88, 89) was suited only to short, rigid molecular chains. Recent attempts by Meier (91) and Napper (92) to evaluate the mutual repulsion of adsorbed polymer chains assume that the polymer films obey the configurational statistics characteristic of polymer solutes (6). This implies that the polymer film is swollen with solvent. Their results indicate that at high surface coverage in good solvents for the polymer two factors are responsible for the repulsion of polymer-coated particles. First, and apparently most important, is again the reduction in configurational entropy due to the volume constraints of the interpenetrating polymer chains. In addition, the change in free energy of mixing of polymer and solvent as the density of segments increases operates to inhibit particle aggregation.

The free energy of mixing, however, becomes an attractive force in poor (θ) solvents for the polymer. Under these latter conditions polymer segment–segment interactions become energetically more favorable than segment–solvent interactions. Napper has shown experimentally that this is the case for suitable polymer-stabilized dispersions both in nonpolar organic (92) and aqueous (93) solutions.

The model used in the above theoretical treatments was that of a polymer adsorbed at one end. The block and graft polymer dispersants used by Napper (92, 93) fit such a model. Meier (91) comments that in principle the theoretical treatment could be applied to the more common case where random segments are adsorbed, leaving loops of segments projecting into solution, but that the additional complexities would be formidable. The computer simulation of dispersant action attempted by Clayfield and Lumb (94, 84) is designed to give only broad indications of purely entropic repulsive behavior.

As discussed in Sect. V.C, the preferential adsorption of less populous, more polar segments in a random copolymer can result in a configuration more extended away from the surface. Thus properly designed copolymers can be expected to be very efficient dispersants (33). Such is found to be the case in actual practice (95, 96).

B. Colloid Aggregation

The very strong adsorption of macromolecules at very low concentrations (and hence very low surface coverage) makes possible an effect that is the reverse of the stabilization discussed in the previous section. It has been found that at very low concentrations polymers can be extremely efficient coagulating or flocculating agents. Colloidal dispersions that are stabilized by charge repulsion (in polar media) can be coagulated by increasing the ionic salt concentration, thereby electrostatically screening the ionic interactions. Specific ion adsorption can also cause reduction in the net charge on the particle. Homopolyelectrolytes of charge opposite to that on the dispersed particles were found by Pugh and Heller (97) to be very effective coagulants at concentrations far smaller than required for tri- or quadrivalent inorganic coagulants. The systems studied were gold and iron sols coagulated by polyvinylbenzyltrimethylammonium chloride (molecular weight about 100,000) and polystyrene sulfonate (molecular weight about 111,000), respectively. In both systems, as the polymer concentration was increased beyond that required for coagulation, the sign of the charge on the particles was observed to reverse and the dispersion became stable once again. In this second polymer-stabilized stage the dispersions were far more stable toward coagulation by added salt. Thus it appears that steric as well as charge repulsion is functioning at this point. In these systems utilizing homopolymers none of the described effects occurred when the sol was treated with the polyelectrolyte of the same sign.

It is not clear in the above studies with homopolymers whether the aggregation effects are caused only by the reduction in net charge. A novel mechanism of the flocculation of dispersions became apparent in studies (57, 58) utilizing copolymers consisting of both ionic and non-ionic segments. As discussed in Sect. IV.C, such copolymers can adsorb onto charged particles having the same sign as the ionic segments via the nonionic segments. The charged segments serve only to increase the polymer extension away from the surface and thus enhance the bridging of adsorbed polymer from one particle to another (the flocculation mechanism). Again, when the concentration of polymer exceeds that required for optimum flocculation, the flocs redisperse. The rate of floc formation has been shown (98) to reside in the factor $\theta(1 - \theta)$, where θ is the fraction of the solid surface covered by adsorbed polymer, and hence $(1 - \theta)$ is the fraction of uncovered surface. The factor $\theta(1 - \theta)$ reflects the bridging mechanism, whence optimum flocculation occurs at

$\theta = 0.5$. The flocculated state is metastable, however, with respect to the final state of the system where $\theta = 1$, and is shear dependent (99). This appears to reflect the fact that the usual equilibrium conformation of adsorbed polymer at low surface coverage is relatively flattened on the surface, but a finite time is required to arrive at this state after the initial attachment to the surface is made (Sect. III.E).

Since the thickness of the adsorbed polymer film increases with molecular weight, it is expected and observed that the flocculant effectiveness also increases with molecular weight. Walles (100) has shown that this can be accounted for by the increase in effective particle collision frequency. An effective collision requires only attachment of an adsorbed polymer segment to the surface of another particle.

As expected, nonionic homopolymers are usually not very effective flocculants. The effectiveness of certain naturally occurring macro-molecules, such as starch, must be due to the relatively high rigidity of the polymer backbone and hence to the inherently extended configuration. The latter must also be true for the polyelectrolytes, discussed above (97), but the extended state in this case is due to the intramolecular ionic segment repulsion.

C. RESTRICTED CAPILLARY FLOW

The determination of adsorbed polymer film thickness by measurement of the restriction of flow in capillaries due to the effective decrease in radius has been discussed in Sect. V.A. It remains only to give some idea of the magnitude of the effect that can be expected. The basic equation describing the effect is obtained from Poiseuille's law (66, 69) and may be written

$$f/f_0 = (\eta_0/\eta)(1 - \Delta r/r)^4 \tag{10}$$

where f and f_0 are the flow rates, η and η_0 the true viscosities of the solution and the pure solvent, respectively, r is the capillary radius, and Δr is the effective thickness of the adsorbed polymer film. For a 0.1 wt. % solution of polystyrene ($M_v = 154,000$) in benzene flowing through capillaries of about 0.3 μ average radius, a 27% reduction in flow rate occurs beyond that due to the increase in viscosity ($\eta/\eta_0 = 1.092$) as shown in Fig. 20 (69). This corresponds to a calculated effective thickness of 0.031 μ for the polystyrene film.

At the low polymer concentrations usually employed in these visco-metric studies it is not likely that intermolecular polymer entanglement effects between adsorbed and solute macromolecules are contributing to the effective hydrodynamic thickness of the adsorbed film. However, because of the voluminosity of macromolecular solutes, especially in good solvents, the critical concentration at which the onset of overlapping of spheres of hydrodynamic action occurs can be at quite reasonable levels. For example, for polystyrene of 600,000 molecular weight (M_v) in methyl ethyl ketone ($[\eta] = 102$ cm³/g) the critical concentration is

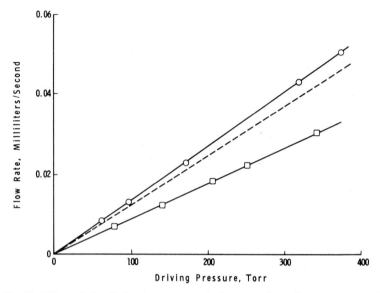

Fig. 20. Effect of adsorbed polystyrene on the flow rate through a porous glass disk. ○, pure solvent, benzene; □, 0.1 wt.% polystyrene in benzene; (– – –), calculated solution flow rate from the relative viscosity. [Redrawn from Ref. (69) by courtesy of John Wiley and Sons, Inc., N.Y.]

calculated to be 1.3 wt.% (101). The linear relation of the reduced specific viscosity versus concentration is observed to start a marked upward curvature near this concentration. Thus it might be expected that the capillary flow restriction caused by adsorbed polymers would show related increased effects for polymer solutions above the critical concentration. Study of this possibility has apparently not been made and may be difficult because of the high viscosities involved.

It should be noted that there is some evidence that multilayer polymer adsorption can occur at high solute concentrations (102). These results may possibly be more a manifestation of entanglement effects above the critical concentration than of true adsorption because of the unique static method of measurement used. The apparent adsorption was determined from the effect on the bouyancy of the adsorbing surface (for example, a metal foil).

D. ADHESION

Adhesion between surfaces by polymeric agents ordinarily involves polymer melts and is thus outside the realm of the subject of this chapter. However, it is fitting to conclude this discussion on effects of polymer adsorption by merely noting that the adhesion caused by polymers involves an extension or magnification of previously discussed adsorption behavior. The efficacy of macromolecular compounds as adhesive agents must be in large part the result of the two most common and outstanding facets of their adsorption behavior. First is the very strong adsorption resulting from the polyfunctional character of the macromolecule. Second is the tendency to adsorb with long, free loops of segments extended away from the surface. This latter effect then can involve the primary adsorbed layer with the intramolecular entanglement that is characteristic of polymer melts or concentrated solutions (103). The details of the mechanism of adhesion, however, are much more complex than suggested by the above two factors alone (104).

E. CONCLUSION

In spite of the generic gap between nonbiological and biological polymers, this chapter has been written with the premise that the adsorption behavior of one is pertinent to that of the other. In this connection, it is interesting to note recent studies that have shown synthetic nonbiological polymers to have biological activity. Antibody-inducing properties have been demonstrated for vinyl polymers (105), and lymphocyte output has been observed to be increased by polymethyacrylic acid (106). Whether the mechanism of these effects involves interfacial activity or not, such effects encourage the comparison of properties of biological and nonbiological macromolecules.

With regard to the application of the behavior of polymers in nonaqueous media it may be noted that evidence exists for hydrophobic fluid regions in some biological membranes, which can behave like a hydrocarbon solvent (107). It may appear that such regions are not large enough to accommodate a macromolecular solute; however, in

Sect. IV.B we have seen how polyvinyl alcohol can penetrate and adsorb in an extended configuration within the 20 Å lamellar spacing of montmorillonite (54). Such considerations cannot be pursued here and are left to the biologist.

REFERENCES

1. Malcolm, B. R., in *Surface Activity and the Microbial Cell*, S.C.I. Monograph No. 19, Gordon and Breach, New York, 1965.
2. Patat, F., Killmann, E. and Schliebener, C., *Fortschr. Hochpolymer. Forsch.* **3**, 332 (1964).
3. Kipling, J. J., *Adsorption from Solutions of Non-electrolytes*, Ch. 8, Academic, New York–London, 1965.
4. Gaines, G. L., Jr., *Insoluble Monolayers at Liquid–Gas Interfaces*, Wiley-Interscience, New York, 1966.
5. Loeb, G. I., "Surface Chemistry of Proteins and Polypeptides," *U.S. Naval Res. Lab. Report 6318*, Nov. 30, 1965.
6. Flory, P. J., *Principles of Polymer Chemistry*, Cornell, Univ. Press, Ithaca, N.Y., 1953.
7. Tanford, C., *Physical Chemistry of Macromolecules*, Wiley, New York–London, 1961.
8. Brandrup, J., and Immergut, E. H., *Polymer Handbook*, Wiley-Interscience, New York, 1966.
9. Ellerstein, S., and Ullman, R., *J. Polym. Sci.* **55**, 123 (1961).
10. Fontana, B. J., *J. Phys. Chem.* **70**, 1801 (1966).
11. Amborski, L. E., and Goldfinger, G., *Rubber Chem. Technol.* **23**, 803 (1950).
12. Jenckel, E., and Rumbach, B., *Z. Elektrochem.* **55**, 612 (1951).
13. Koral, J., Ullman, R., and Eirich, F. R., *J. Phys. Chem.* **62**, 541 (1958).
14. Giles, C. H., and MacEwan, T. H., *Proceedings of the 2nd International Congress on Surface Activity*, Vol. 3, Butterworths, London, 1957, p. 457.
15. Howard, G. J., and McConnell, P. M., *J. Phys. Chem.* **71**, 2974 (1967).
16. Burns, H., Jr., and Carpenter, D. K., *Macromol.* **1**, 384 (1968).
17. Perkel, R., and Ullman, R., *J. Polym. Sci.* **54**, 127 (1961).
18. Gilliland, E. R. and Gutoff, E. B., *J. Appl. Polym. Sci.* **3**, 26 (1960).
19. Cabot Corporation, *Carbon Black Pigments*, Technical Brochure.
20. Yeh, S. J., and Frisch, H. L., *J. Polym. Sci.* **27**, 149 (1958).
21. Hobden, J. F., and Jellinek, H. H. G., *J. Polym. Sci.* **11**, 365 (1953).
22. Hansen, R. S., Fu, Y., and Bartell, F. E., *J. Phys. Chem.* **53**, 769 (1949).
23. Hansen, R. S., and Craig, R. P., *J. Phys. Chem.* **58**, 211 (1954).
24. Molau, G. E., and Wittbrodt, W. M., *Macromol.* **1**, 260 (1968).
25. Small, P. A., *J. Appl. Chem.* **3**, 71 (1953).
26. Howard, G. J., and McConnell, P., *J. Phys. Chem.* **71**, 2981 (1967).
27. Howard, G. J., and McConnell, P., *J. Phys. Chem.* **71**, 2991 (1967).
28. Thies, C., *Macromol.* **1**, 335 (1968).
29. Davidson, E., and Kipling, J. J., Int. Symp. *Macromol. Chem., Prague* (*1965*), Preprint P 193.
30. Kirshenbaum, I., *J. Polym. Sci.* **A3**, 1869 (1965).
31. Silberberg, A., *J. Phys. Chem.* **66**, 1884 (1962).

32. Rowland, F. W., and Eirich, F. R., *J. Polym. Sci.* A-1, **4**, 2401 (1966).

33. Fontana, B. J., *J. Phys. Chem.* **67**, 2360 (1963).

34. Peterson, C., and Kwei, T. K., *J. Phys. Chem.* **65**, 1330 (1961).

35. Stromberg, R. R., Grant, W. H., and Passaglia, E., *J. Res. Nat. Bur. Std.* (*U.S.*) **68A**, 391 (1964).

36. Stromberg, R. R., Passaglia, E., and Tutas, D. J., *J. Res. Nat. Bur. Std.* (*U.S.*) **67A**, 431 (1963).

37. Baret, J. F., *J. Phys. Chem.* **72**, 2755 (1968).

38. Meyerhoff, G., and Schulz, G. V., *Makromol. Chem.* **7**, 294 (1952).

39. Frisch, H. L., Hellman, M. Y., and Lundberg, J. L., *J. Polym. Sci.* **38**, 441 (1959).

40. Thies, C., *J. Phys. Chem.* **70**, 3783 (1966).

41. Brunauer, S., *Pure Appl. Chem.* **10**, 293 (1965).

42. Nakanishi, K., *Rep. Progr. Polym. Phys.* (Japan), **5**, 52 (1962).

43. Grubisic, Z., Rempp, P., and Benoit, H., *J. Polym. Sci.* **B5**, 753 (1967).

44. Yau, W. W., in discussion at Pacific Conf. on Chemistry and Spectroscopy, Symp. on Gel Permeation, San Francisco, Nov. 7, 1968.

45. Weatherwax, R. C., and Tarkow, H., *J. Polym. Sci.* **A2**, 4697 (1964).

46. Gaines, G. L., *Insoluble Monolayers at Liquid–Gas Interfaces*, Wiley-Interscience, New York, 1966.

47. Stromberg, R. R., and Smith, L. E., *J. Phys. Chem.* **71**, 2470 (1967).

48. Stromberg, R. R., Tutas, D. J., and Passaglia, E., *J. Phys. Chem.* **69**, 3955 (1965).

49. Franks, F., *Chem. Ind.* 560 (1968).

50. Nash, T., *Surface Activity and the Microbial Cell*, S.C.I. Monograph No. 19, Gordon and Breach, New York, 1965.

51. Brooks, M. C., and Badger, R. M., *J. Amer. Chem. Soc.* **72**, 4384 (1950).

52. Greenland, D. J., Laby, R. H., and Quirk, J. P., *Trans. Faraday Soc.* **58**, 829 (1962).

53. Heller, W., and Pugh, T. L., *J. Polym. Sci.* **47**, 203 (1960).

54. Greenland, D. J., *J. Colloid Sci.* **18**, 647 (1963).

55. Strauss, U. P., *Characterization of Macromolecular Structure*, Publication 1573, Nat. Acad. Sci., Washington, D.C., 1968, p. 148.

56. Oth, A., and Doty, P., *J. Phys. Chem.* **56**, 43 (1952).

57. Michaels, A. S., *Ind. Eng. Chem.* **46**, 1485 (1954).

58. Michaels, A. S., and Morelos, O., *Ind. Eng. Chem.* **47**, 1801 (1955).

59. Healy, T. W., *J. Colloid Sci.* **16**, 609 (1961).

60. Lopatin, G. and Eirich, F. R., Proc. 3rd Int. Congr. Surface Activity **2**, 97 (1960).

61. Schmidt, W., and Eirich, F. R., *J. Phys. Chem.* **66**, 1907 (1962).

62. Peyser, P., and Ullman, R., *J. Polym. Sci.* **A3**, 3165 (1965).

63. Kragh, A. M., *J. Photographic Sci.* **12**, 191 (1964).

64. Curme, H. G., and Natale, C. C., *J. Phys. Chem.* **68**, 3009 (1964).

65. Ueda, T. and Harada, S., *J. Appl. Polym. Sci.* **12**, 2395 (1968).

66. Öhrn, O. E., *J. Polym. Sci.* **17**, 137 (1955).

67. Tuijnman, C. A. F., and Hermans, J. J., *J. Polym. Sci.* **25**, 385 (1957).

68. Öhrn, O. E., *Ark. Kemi.* **12**, 397 (1958).

69. Rowland, F. W., and Eirich, F. R., *J. Polym. Sci.* A-1, **4**, 2033 (1966).

70. Rothstein, E. C., *J. Paint Technol., Official Digest* **36**, 1448 (1964).

71. Doroszkowksi, A., and Lambourne, R., *J. Colloid Interface Sci.* **26**, 214 (1968).

72. Fontana, B. J., and Thomas, J. R., *J. Phys. Chem.* **65**, 480 (1961).

73. Peyser, P., and Stromberg, R. R., *J. Phys. Chem.* **71**, 2066 (1967).

74. Thies, C., Peyser, P. and Ullman, R., *Proceedings of the 4th International Congress on Surface Activity Substances, Brussels, 1964*, Vol. 2, Gordon and Breach, New York, 1967, p. 1041.
75. Low, M. J. D., and Hasegawa, M., *J. Colloid Interface Sci.* **26**, 95 (1968).
76. Hasegawa, M., and Low, M. J. D., *J. Colloid Interface Sci.* **29**, 593 (1969).
77. Steinberg, G., *J. Phys. Chem.* **71**, 292 (1967).
78. Stromberg, R. R., in *Treatise on Adhesion and Adhesives* (R. L. Patrick, ed.), Dekker, New York, 1967.
79. Simha, R., Frisch, H. L. and Eirich, F. R., *J. Phys. Chem.* **57**, 584 (1953).
80. Silberberg, A., *J. Phys. Chem.* **66**, 1872 (1962).
81. Silberberg, A., *J. Phys. Chem.* **66**, 1884 (1962).
82. Silberberg, A., *J. Chem. Phys.* **46**, 1105 (1967).
83. Silberberg, A., *J. Chem. Phys.* **48**, 2835 (1968).
84. Clayfield, E. J., and Lumb, E. C., *Macromol.* **1**, 133 (1968).
85. Frisch, H. L., *J. Phys. Chem.* **59**, 633 (1955).
86. Frisch, H. L., and Simha, R., *J. Chem. Phys.* **17**, 702 (1957).
87. van der Waarden, M., *J. Colloid Sci.* **5**, 317 (1950); **6**, 443 (1951).
88. Mackor, E. L., *J. Colloid Sci.* **6**, 492 (1951).
89. Mackor, E. L., and van der Waals, J. H., *J. Colloid Sci.* **7**, 535 (1952).
90. Vold, M. J., *J. Colloid Sci.*, **16**, 1 (1961).
91. Meier, D. J., *J. Phys. Chem.* **71**, 1861 (1967).
92. Napper, D. H., *Trans. Faraday Soc.* **64**, 1701 (1968).
93. Napper, D. H. *J. Colloid Interface Sci.* **29**, 168 (1969).
94. Clayfield, E. J., and Lumb, E. C., *J. Colloid Interface Sci.* **22**, 269, 285 (1966).
95. Lyman, A. L., and Kavanagh, F. W., *Proc. Amer. Petrol. Inst. Sect.* III, **39**, 296 (1959).
96. Fowkes, F. M., Schick, M. J. and Bondi, A., *J. Colloid Sci.* **15**, 531 (1960).
97. Pugh, T. L., and Heller, W., *J. Polym. Sci.* **47**, 219 (1960).
98. Smellie, R. H., Jr., and LaMer, V. K., *J. Colloid Sci.* **13**, 589 (1958).
99. Healy, T. W., and LaMer, V. K., *J. Colloid Sci.* **19**, 323 (1964).
100. Walles, W. E., *J. Colloid Interface Sci.* **27**, 797 (1968).
101. Weissberg, S. G., Simha, R. and Rothman, S., *J. Res. Nat. Bur. Stand. (U.S.)* **47**, 298 (1951).
102. Patat, S., Killmann, E. and Schliebener, C., *Makromol. Chem.* **49**, 200 (1961).
103. Porter, R. S., and Johnson, J. F., *Chem. Rev.* **66**, 1 (1966).
104. Huntsberger, J. R., in *Treatise on Adhesion and Adhesives* (R. L. Partick, ed.), Dekker, New York, 1967.
105. Gill, T., and Kunz, H., *Proc. Nat. Acad. Sci. U.S.* **61**, 490 (1968).
106. Ormai, S., and DeClercq, E., *Science* **163**, 471 (1969).
107. Hubbell, W. L., and McConnell, H. M., *Proc. Nat. Acad. Sci. U.S.* **61**, 12 (1968).

4

The Adsorption of Surfactants at Solid–Water Interfaces*

D. W. FUERSTENAU

Department of Materials Science and Engineering
University of California, Berkeley, California

I. Introduction

The biologist is encountering an increasing number of phenomena in which surface-active molecules and ions adsorb at interfaces. Physiologists

* Support of the Miller Institute for Research in Basic Science, University of California, Berkeley, is gratefully acknowledged.

who have considered in detail adsorption at the walls of the biological cell, on lung tissue, and on bone and teeth are now interested in adsorption phenomena on metallic and plastic implant materials. The biochemist is concerned with interaction of surface-active materials in their standard extractive and preparative techniques of chromatography. All these groups of workers are now involved with the mechanism of surface reactions in rejection phenomena, associated with implants.

The aim of this chapter is to delineate the basic principles of adsorption of organic surface-active agents at solid–aqueous solution interfaces. Most of the basic surface chemistry has been carried out with non-biological surfaces and simple organic surface-active agents (surfactants). The latter serve as analogs of the many complex biosurfactants, and research on such model systems provides the basis for understanding the mechanism of adsorption in the wide variety of complex biological systems. Although the concept of the hydrophobic bond was invoked to explain biological structures, there were many prior examples of this type of bonding in nonbiological systems. In the latter the conditions and energetics of the bonding can be more readily defined and studied. This analysis of surfactant adsorption on nonbiological surfaces will be concerned with the role of the solid surface, specifically its charge and solvation, and the role of the polar group and the hydrocarbon chain of the surfactant.

The adsorption of surfactants is intimately associated with their amphipathic character, i.e., the balance between the polar characteristics of the head group and the nonpolar character of the hydrocarbon chain. The present discussion is concerned mostly with simple anionic and cationic surfactants, such as alkyl sulfonates, sulfates, amine salts, soaps, and fatty acids. Specific physicochemical properties of these surfactants can be found in standard monographs such as that of Shinoda (28).

In assessing the nature of the adsorption of these surfactants at the solid–aqueous solution interface, it is first necessary to consider the detailed nature of the solid surface. While many surfaces such as those of oxides and the salt-type minerals are essentially hydrophilic and others such as graphite and polyethylene are hydrophobic, a number of materials will be shown to exhibit both hydrophobic and hydrophilic characteristics. Adsorption on hydrophilic materials is generally controlled by coulombic interactions between the ionic head group of the surfactant and the electrostatic charge of the solid surface, or in some instances by the formation of covalent chemical bonds between the surface and the surfactant. On the other hand, adsorption of surfactants on hydrophobic solids occurs principally because of hydrophobic bonding between the hydrocarbon chain of the surfactant and the nonpolar surface. It will be shown that the adsorption of surfactants on many solid materials is controlled by a combination of hydrophobic and hydrophilic phenomena. This chapter will develop the basic principles involved in the adsorption of surfactants at solid–aqueous solution interfaces and will attempt to catalog adsorption in a number of systems in terms of the types of inter-actions that can occur between the adsorbent, the adsorbate, and the solvent, water. As a result, this should provide a basis for interpreting adsorption phenomena in many of the more complex biological systems.

II. The Electrical Double Layer at Solid–Water Interfaces

Since adsorption phenomena at solid–water interfaces are controlled in most cases by the electrical double layer, we must be concerned with factors responsible for the charge on the solid surface and with the behavior of ions that adsorb as counter ions to maintain electroneutrality. Figure 1 presents a schematic representation of the electrical double layer at a solid–water interface (1), showing the charge on the solid surface and the diffuse layer of counter ions extending out into the liquid phase. This figure also shows the drop in potential across the double layer, neglecting the potential due to dipole effects. The closest distance of approach of counter ions to the surface (δ) is called the Stern plane. Depending on whether ions remain hydrated or are dehydrated upon adsorption, there can exist an inner and an outer Stern plane, but for our purposes we shall not differentiate between an inner and an outer Stern plane. The surface potential is ψ_0 and that at the Stern plane is ψ_δ; from the Stern plane on out into the bulk of the solution the potential drops to zero.

In the case of the ionic solids such as $BaSO_4$, CaF_2, AgI, and Ag_2S, the surface charge arises from a preference for one of the lattice ions to reside at the solid surface as compared with the aqueous phase. Equilibrium is attained when the electrochemical potential of these ions is constant throughout the system. Those particular ions that are free to pass between both phases and therefore establish the electrical double layer are called *potential-determining ions*. In the case of AgI the potential-determining ions are Ag^+ and I^-. For a solid such as calcite, $CaCO_3$, the potential-determining ions are Ca^{2+} and CO_3^{2-}, and also H^+, OH^-, and HCO_3^- because of the equilibria between these latter ions and CO_3^{2-}.

Similarly for hydroxyapatite, the potential determining ions are Ca^{2+}, PO_4^{3-}, and OH^-, with the other hydrolysis products also functioning in this role because of the complex equilibria involved in this system.

For oxides, hydrogen and hydroxyl ions have long been considered to be potential-determining (2, 3) although there still remains a difference of opinion as to how pH controls the surface charge on oxides. Since oxide

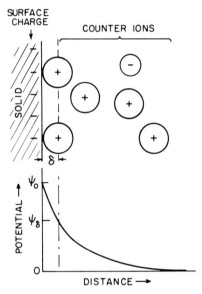

Fig. 1. Schematic representation of the electrical double layer and the potential drop in the double layer at a solid–water interface.

minerals form hydroxlated surfaces when in contact with water vapor (4), a hydroxylated surface should be expected when the solid is equilibrated with an aqueous solution. Adsorption-dissociation of H^+ from the surface hydroxyls can account for the surface charge on the oxide by the following mechanism (5, 6):

$$MOH_{(surf)} \rightleftharpoons MO^-_{(surf)} + H^+_{(aq)}$$

$$MOH_{(surf)} + H^+_{(aq)} \rightleftharpoons MOH^+_{2(surf)}$$

Parks and deBruyn (7, 8) have postulated a different mechanism for the charging of oxide surfaces, involving partial dissolution of the oxide and formation of hydroxyl complexes in solution, followed by adsorption of these complexes. The formation of a surface charge by either of these mechanisms, or even by direct adsorption of H^+ and OH^-, would result in a change of the pH of the solution.

The surface charge σ_s on a solid in water is determined by the adsorption density of potential-determining ions on the solid surface. In the case of uni-univalent salt, σ_s is given by

$$\sigma_s = F(\Gamma_{M^+} - \Gamma_{A^-}) \qquad (1)$$

where Γ_{M^+} is the adsorption density in moles/cm² of the potential-determining cation, Γ_{A^-} is that of the potential-determining anion, and F is the Faraday constant. For an oxide, M^+ and A^- are H^+ and OH^-, respectively; and for AgI, M^+ and A^- are simply Ag^+ and I^-. By means of a simple titration procedure (1, 7), the magnitude of the charge can be determined if the surface area of the solid–liquid interface is known. Figure 2 presents the results of such titration of synthetic ferric oxide (hematite) with hydrogen and hydroxyl ions in the presence of KNO_3 as supporting electrolyte (7). This figure clearly shows that the surface charge on ferric oxide reverses its sign at pH 8.6 and that it increases in absolute magnitude with increasing ionic strength and increasing concentration of potential-determining ion.

The single most important parameter that describes the electrical double layer of a solid in water is the point of zero surface charge (pzc). The pzc is expressed as the condition in the aqueous solution at which σ_s is zero and this is determined by a particular value of the activity of potential-determining ions, $(a_{M^+})_{pzc}$ or $(a_{A^-})_{pzc}$. Assuming that potential differences due to dipoles, etc., remain constant, the total double layer potential or the surface potential ψ_0 is considered to be zero at the pzc. The value of the surface potential at any activity of 1–1 valent potential-determining electrolyte is given by

$$\psi_0 = \frac{RT}{F} \ln \frac{(a_{M^+})}{(a_{M^+})_{pzc}} \qquad (2)$$

where R is the gas constant and T the temperature in °K. Table 1 presents pzc values for a number of ionic (salt-type) solids that have been investigated. Calcite, fluorite, and barite are positively charged in their saturated solution at neutral pH, whereas the others listed are negative, except for hydroxyapatite which is uncharged at this pH. Table 2 presents a tabulation of typical pzc values of a number of oxides. This table shows that the surfaces of oxides range from being acidic in nature to quite basic. Examples of how pH strongly affects adsorption at the surface of oxides will be shown later.

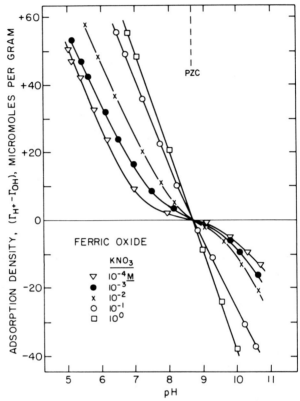

Fig. 2. The adsorption density of potential-determining ions on ferric oxide as a function of pH and ionic strength using KNO_3 as the indifferent electrolyte [after Parks and deBryun (7)].

TABLE 1

THE POINT OF ZERO CHARGE OF SOME IONIC SOLIDS

Material	pzc	Ref.
Fluorapatite, $Ca_5(PO_4)_3(F, OH)$	pH 6[a]	9
Hydroxyapatite, $Ca_5(PO_4)(OH)$	pH 7[a]	8
Calcite, $CaCO_3$	pH 9.5[a]	10
Fluorite, CaF_2	pCa 3	8
Barite (synthetic), $BaSO_4$	pBa 6.7	11
Silver iodide, AgI	pAg 5.6	1
Silver chloride, AgCl	pAg 4	1
Silver sulfide, Ag_2S	pAg 10.2	12

[a] From the hydrolysis equilibria and solubility data, the activities of the other potential-determining ions can be calculated.

The importance of the pzc is that the sign of the surface charge has a major effect on the adsorption of all other ions. For our purposes we are very interested in the adsorption of counter ions that occurs to maintain electroneutrality at the surface. In contrast to the situation in which the potential-determining ions are special for each system, any ions present in the solution can function as the counter ions. If the counter ions are adsorbed only by electrostatic attraction, they are called *indifferent*

TABLE 2
THE POINT OF ZERO CHARGE OF SOME OXIDES[a]

Material	pzc	Ref.
Quartz, SiO_2	pH 2–3.7	*13, 14*
Cassiterite, SnO_2	pH 4.5	*15*
Rutile, TiO_2	pH 6.0	*16*
Hematite (natural), Fe_2O_3	pH 4.8	*18*
Hematite (synthetic)	pH 8.6	*7*
Corundum, Al_2O_3	pH 9.0	*5*
Magnesia, MgO	pH 12	*17*

[a] These are typical results. The source of the oxide, its trace impurities, method of pretreatment, etc. cause variations in observed values.

electrolytes. As has been well established (*1*), the counter ions occur in a diffuse layer that extends from the interface out into the solution. The "thickness" of the diffuse double layer is $1/\kappa$ where κ for a symmetrical electrolyte ($z_- = z_+ = z$) is given by

$$\kappa = \left(\frac{8\pi F^2 z^2 C}{\epsilon RT}\right)^{1/2} \tag{3}$$

where ϵ is the dielectric constant of the liquid and C is the bulk concentration in mole/cm³. For a 1–1 valent electrolyte, $1/\kappa$ is 1000 Å in 10^{-5} M solutions, 100 Å in 10^{-3} M solutions, and 10 Å in 10^{-1} M solutions, for example. Actually, κ gives a measure of the center of gravity of the diffuse layer. The closest distance of approach of counter ions to the surface is one (hydrated) ionic radius away, shown as the distance δ in Fig. 1. Depending on the ionic strength of the solution, there is a considerable drop in potential between the solid surface and the plane δ. This potential drop is given by $\psi_0 - \psi_\delta$.

The charge in the diffuse double layer σ_d is given by the Gouy–Chapman relation (1) (with the Stern modification) for a symmetrical electrolyte:

$$\sigma_d = -\left(\frac{2\epsilon RT}{\pi}\right)^{1/2} (C)^{1/2} \sinh\left(\frac{zF\psi_\delta}{2RT}\right) \tag{4}$$

For systems in which the counter ions are adsorbed only by electrostatic attraction, this relationship can be used to calculate σ_s, since under these conditions $\sigma_s = -\sigma_d$. Further, if ψ_δ does not change appreciably, the adsorption density of counter ions should vary as the square root of the concentration of added electrolyte, as deBruyn (19) has found for the adsorption of dodecylammonium acetate on quartz at low concentrations. If ions adsorb only electrostatically, they cannot reverse the sign of ψ_δ but can only reduce it towards zero.

On the other hand, some ions exhibit surface activity in addition to electrostatic attraction and adsorb strongly in the Stern plane because of such phenomena as covalent bond formation, hydrophobic bonding, hydrogen bonding, solvation effects, etc., as will be discussed in detail in the next section. Because of their surface activity, such counter ions may be able to reverse the sign of ψ_δ since the charge of the layer of ions adsorbed in the Stern plane can exceed the surface charge. In such cases

$$-\sigma_s = \sigma_\delta + \sigma_d \tag{5}$$

where σ_δ is the charge due to adsorption in the Stern plane, assuming the plane of adsorbed ions coincides with the plane δ.

Electrokinetic phenomena, which involve the interrelation between mechanical and electrical effects at a moving interface, have found widespread use in colloid and surface chemistry. The two electrokinetic effects that have been most widely used are electrophoresis and streaming potential measurements. Electrokinetic results are generally expressed in terms of the ζ-potential, which is the potential at the slipping plane when liquid is forced to move relative to the solid; only those ions in the diffuse layer outside of the slipping plane are involved in the electrokinetic process. Thus, while knowledge of the ζ-potential at some single condition may be of certain value, determination of the change in ζ-potentials as solution conditions are varied is extremely useful. From these changes, modes of adsorption of various kinds of ions can be ascertained if one makes the useful assumption that the slipping plane and the Stern plane coincide (1). This approximation seems permissible because the potential differences between the plane δ and the slipping plane are small compared with the total potential differences across the double layer.

It should be pointed out that the case in which there is no ambiguity is when $\psi_\delta = 0$, for then ζ must also be 0 (20).

III. Adsorption Energies of Surfactants at Solid–Water Interfaces

In a limited number of aqueous surfactant–solid systems, adsorption appears to result from electrical interactions only and in these cases the adsorbed ions occur primarily in the diffuse layer as counter ions. For those more numerous cases where there is specific interaction with surface sites, we are interested in adsorption in the Stern plane δ.

The adsorption density of surfactant ions in the Stern plane is controlled by the (standard) free energy of adsorption ΔG°_{ads}:

$$\Gamma_\delta = 2rC \exp\left(\frac{-\Delta G^\circ_{ads}}{RT}\right) \tag{6}$$

where Γ_δ is the adsorption density in moles/cm^2, r is the effective radius of the adsorbed ion, and C is the bulk concentration in moles/cm^3. This equation is perhaps the simplest (1) for relating adsorption in the Stern plane to the bulk concentration and the adsorption free energy. The free energy of adsorption ΔG°_{ads} can be considered to be made up of a number of contributing terms (21):

$$\Delta G^\circ_{ads} = \Delta G^\circ_{elec} + \Delta G^\circ_{HM} + \Delta G^\circ_{CH_2} + \Delta G^\circ_{h} + \Delta G^\circ_{solv} + \Delta G^\circ_{chem} + \cdots \tag{7}$$

The electrostatic contribution to adsorption ΔG°_{elec} results from coulombic interaction with the surface charge in the double layer and is simply

$$\Delta G^\circ_{elec} = z_\pm F \psi_\delta \tag{8}$$

All the remaining terms give rise to specific adsorption effects, and these can be taken together as a specific adsorption free energy, ΔG°_{spec}. ΔG°_{HM} is a contribution to the adsorption of organic surfactant ions that result from the tendency of hydrocarbon chains to be expelled from water; this tendency causes the hydrocarbon chains of adsorbed surfactants to interact with each other. $\Delta G^\circ_{CH_2}$ is the hydrophobic bonding contribution that results in the interaction of the hydrocarbon chain directly with the solid surface. Although the same properties of the hydrocarbon chain give rise to $\Delta G^\circ_{CH_2}$, it is useful to distinguish between the two kinds of adsorption mechanisms. ΔG°_{h} is an adsorption free energy that results from hydrogen bond formation. $\Delta G^{\circ}_{\text{solv}}$ is the contribution of solvation effects on the polar head of the adsorbate and/or adsorbent to adsorption. $\Delta G^{\circ}_{\text{chem}}$ is the free energy of adsorption that arises from the formation of covalent bonds between surfactant ions and the adsorbent.

If adsorption results only from coulombic interaction between the surface charge and the surfactant ion, adsorption is considered to be nonspecific and

$$\Delta G^{\circ}_{\text{ads}} = \Delta G^{\circ}_{\text{elec}}$$

On the other hand, if specific adsorption does occur, the adsorption process is controlled by

$$\Delta G^{\circ}_{\text{ads}} = \Delta G^{\circ}_{\text{elec}} + \Delta G^{\circ}_{\text{spec}}$$

with ΔG_{spec} resulting from any one of a variety of causes, as given above. In the following sections of this chapter, examples will be given to illustrate how the adsorption process in various systems is controlled by both electrostatic and the various specific interaction effects.

IV. Physical Adsorption on Oxides

A. The Role of the Hydrocarbon Chain in Adsorption

A number of solid–surfactant–water systems involve adsorption of the surfactant as counter ions in the electrical double layer without chemical bond formation. Examples of such systems that have been investigated in detail include the adsorption of alkylammonium acetates or chlorides on quartz (*19, 20, 22*) and sodium alkyl sulfates or sulfonates on alumina (*23–25*). In these systems in which the hydrocarbon chain of the surfactant can play an important role, two contributions to the adsorption process have been found to be dominant, namely $\Delta G^{\circ}_{\text{elec}}$ and $\Delta G^{\circ}_{\text{HM}}$, with $\Delta G^{\circ}_{\text{chem}}$ absent. Here, the adsorption of surfactant ions at low concentrations occurs as individual counter ions, but once the adsorbed ions reach a certain critical concentration at the solid–liquid interface they begin to associate into two-dimensional patches of ions at the surface in much the same way as they associate into three-dimensional aggregates to form micelles in bulk solution. The forces responsible for this association at the surface will be the same as those operating in the bulk during micelle formation, except that the coulombic attraction for the surface adsorption sites will aid the association. The patches of associated ions at the water–solid interface have been termed "hemimicelles" (*22*). The bulk critical concentration at which the association is observed can be referred to as the hemimicelle concentration. In the case of alkylammonium ions on quartz at neutral pH, the hemimicelle concentration is about 0.01 of the bulk critical micelle concentration, (the cmc); however, a simple calculation shows that the concentration of adsorbed ions within the Stern layer approximates the cmc (26).

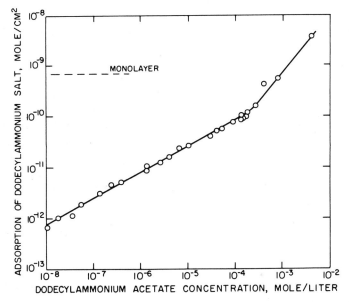

Fig. 3. Adsorption isotherm of dodecylammonium acetate on quartz (0.14 m²/g) at neutral pH [after deBruyn (19)].

Figure 3 presents the results of deBruyn for the adsorption of dodecylammonium acetate on quartz (19). Below 2×10^{-4} M the experimental isotherm of dodecylammonium ions on quartz can be described by the following empirical equation:

$$\Gamma = 8.1 \times 10^{-9} \, C^{0.5} \tag{9}$$

where Γ is the adsorption density in moles/cm² and C is the bulk concentration in moles/liter. When the adsorbed ions begin to associate above the hemimicelle concentration (indicated by the break in the curve), the adsorption density increases rapidly and is given by

$$\Gamma = 2.2 \times 10^{-6} \, C^{1.2} \tag{10}$$

In the region where dodecylammonium ions are adsorbed electrostatically, the adsorption density is proportional to the square root of the concentration, as predicted from the Gouy–Chapman relationship. When interaction of the hydrocarbon chains occurs through hydrophobic bonding between themselves (i.e., ΔG_{HM}° becomes active), the adsorption densities increase markedly and even exceed monolayer coverage.

In the adsorption of surfactants as counter ions in the double layer, an adsorption isotherm is meaningful only if the process occurs at fixed concentration of potential-determining ion in the system (provided that the adsorption of the surfactant does not shift the pzc). In the case of oxides, such as alumina or quartz, this means that the pH must be held constant. Also, to fix the magnitude of σ_s and to fix the activity coefficient of the surfactant, the ionic strength should also be held constant. A recent investigation under such controlled conditions clearly delineates adsorption phenomena in this kind of system; this involved measurement of the amount of dodecyl sulfonate adsorbed on alumina at pH 7.2 from aqueous solutions of 2×10^{-3} M ionic strength with NaCl as the supporting electrolyte (23, 25). As can be observed from Fig. 4, these carefully determined isotherms show three distinct regions, which we will call Regions 1, 2, and 3.

In Region 1, sulfonate ions adsorb on alumina as individual ions in the double layer in competition with the chloride ions used to control ionic strength. In this region, where $\Delta G_{ads}^{\circ} = \Delta G_{elec}^{\circ}$ (i.e., there is no apparent contribution from the chain), sulfonate ions will be adsorbed primarily in the diffuse layer, and because the surfactant ions are replacing chloride ions in the double layer, the sulfonate ions have no noticeable effect on the electrophoretic mobility of the alumina particles. Furthermore, since these ions are merely in exchange adsorption with chloride ions, the slope of the isotherm will be approximately unity.

Region 2 is characterized by a marked change in the slope of the adsorption isotherm. This results from the onset of association of the hydrocarbon chains of the surfactant ions adsorbed in the Stern plane, i.e., on the formation of hemimicelles. Here, $\Delta G_{ads}^{\circ} = \Delta G_{elec}^{\circ} + \Delta G_{HM}^{\circ}$. With the added adsorption potential due to the hydrophobic bonding between hydrocarbon chains, the slope of the adsorption isotherm increases markedly. The onset of Region 2 is clearly shown by the abrupt change in the electrophoretic mobility, i.e., in the ζ-potential.

In Region 3 the adsorption isotherm is characterized by having a slope less than that for Region 2. The boundary between Regions 2 and 3 is indicated by the reversal in sign of the ζ-potential, shown in Fig. 4 as

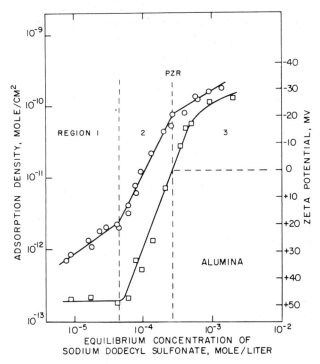

Fig. 4. The effect of the concentration of sodium dodecyl sulfonate on the surfactant adsorption density ($\bigcirc$) and the ζ-potential of alumina ($\square$) (14 m²/g) at pH 7.2 and 2×10^{-3} M ionic strength using NaCl as the indifferent electrolyte [after Wakamatsu and Fuerstenau (25)].

pzr. In Region 3, since the charge in the Stern plane exceeds that of the surface layer, the electrostatic contribution to the free energy of adsorption works against the adsorption process. However, clearly the hydrophobic bonding between hydrocarbon chains dominates the adsorption process in this region.

B. The Effect of Alkyl Chain Length on Adsorption

Since the hydrocarbon chain of a surfactant is important in adsorption processes, the number of carbon atoms in the chain should directly relate to adsorption behavior. Studies of micelle formation have shown that the cohesive free energy for the formation of micelles is -1.0 to $-1.1RT$

(about -600 cal) per mole of CH_2 groups (27, 28). If the molar cohesive free energy per mole of CH_2 group in the hydrocarbon chain is ϕ and the number of carbon atoms in the alkyl chain is N, then for hemimicelle formation (if N is about 8 or greater)

$$\Delta G^{\circ}_{HM} = N\phi \tag{11}$$

If all other specific adsorption effects are negligible, then above the hemimicelle concentration the adsorption density of surfactant ions in the Stern layer is given by

$$\Gamma_\delta = 2rC \exp\left(\frac{-zF\psi_\delta - N\phi}{RT}\right) \tag{12}$$

This equation shows that the amount of surfactant adsorbed should depend strongly on the hydrocarbon chain length, once the hemimicelle concentration is exceeded. This has been demonstrated for the adsorption of alkylammonium ions on quartz (19) and alkyl sulfonate ions on alumina (25). Figure 5 presents the results of measurement of the ζ-potential of quartz at neutral pH as a function of the concentration of alkylammonium acetates ranging from 10 to 18 carbon atoms (20). These results indicate that the ζ-potential of quartz in the presence of the various surfactants approximates that in the presence of ammonium acetate until hemimicelle formation sets in. Actually, octylammonium acetate behaves almost identically with ammonium acetate (26), indicating that the hydrocarbon chain must exceed, say, 8 carbon atoms to be sufficiently hydrophobic for hemimicelle formation to occur. After hemimicelle formation, increasing the length of the hydrocarbon chain causes a systematic increase in the amount of surfactant adsorbed.

If after association of the alkyl chains the ζ-potential is reduced to zero (20), it follows from Eq. 12 that

$$\ln C_0 = +N(\phi/RT) + \ln (\Gamma_\delta)_0 - \ln 2r \tag{13}$$

where $C = C_0$ at $\psi_\delta = 0$. If the ratio of the adsorption density at $\psi_\delta = 0$ to the factor $2r$ is relatively independent of chain length, then the logarithm of the concentration of alkyl surfactant ions at $\psi_0 = 0$ should be a linear function of the alkyl chain length, with a slope equal to $-\phi/2.3RT$. Figure 5 includes such a plot for alkylammonium ions on quartz, and from this plot ϕ is calulated to be $-1RT$ (about -0.6 kcal/mole of CH_2 groups) in agreement with values obtained from solubility data and micelle formation.

Direct measurements have been made for the adsorption of alkylsulfonates of various chain lengths on alumina at constant pH and ionic

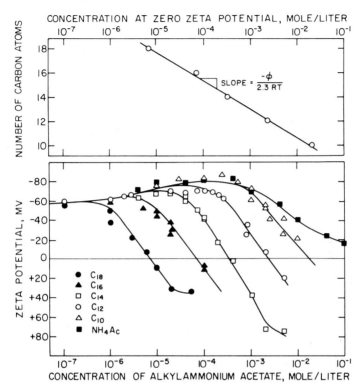

Fig. 5. The effect of hydrocarbon chain length on the ζ-potential of quartz at neutral pH in solutions of alkylammonium acetates; the variation of the concentration of alkylammonium acetate required to reverse the sign of ζ as a function of the number of carbon atoms in the chain [after Somasundaran, Healy, and Fuerstenau (20)].

strength (25). Those measurements show that before hemimicelle formation sets in, the amount of surfactant adsorbed is independent of hydrocarbon chain length, but above the hemimicelle concentration chain length has a very pronounced effect on the amount of surfactant adsorbed, causing the adsorption to increase markedly as the chain length is increased.

C. The Effect of Chain Structure on Adsorption

Whether the hydrocarbon chain possesses side chains, double bonds, or aromatic rings affects adsorption to varying degrees. Side chains would be expected to have a great effect in chemisorption systems where specific adsorbate–site interaction occurs (*29*). In physical adsorption systems the presence of a benzene ring on the surfactant is approximately equivalent to having three additional CH_2 groups in the alkyl chain. The effect of double bonds in the alkyl chain on adsorption is to reduce the hydrophobic nature of the chain and hence to reduce the tendency to form hemimicelles. Purcell and Sun (*30*) found that adsorption of soaps on rutile decreases with increasing double bonds on the chain, i.e., in the order oleate, linoleate, and linolenate.

D. The Effect of pH on the Adsorption of Sulfonates on Alumina

As discussed in the previous section, the adsorption of alkylsulfonates on alumina appears to occur through electrostatic interaction with the

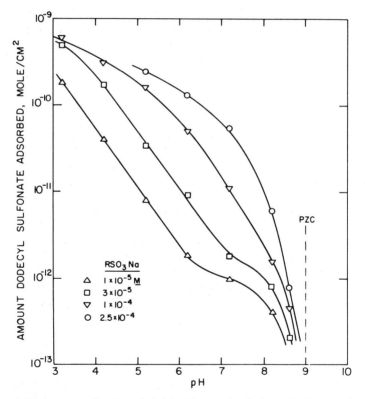

Fig. 6. The amount of sodium dodecyl sulfonate adsorbed on alumina as a function of pH at 2×10^{-3} *M* ionic strength with NaCl as the indifferent electrolyte (*31*).

surface. Thus adsorption in this system should be strongly pH-dependent. Figure 6 presents data (*31*) showing the amount of sodium dodecyl sulfonate adsorbed on alumina as a function of pH at an ionic strength of 2×10^{-3} M controlled with NaCl. These results clearly show that adsorption ceases as the pzc is approached. To increase the adsorption of

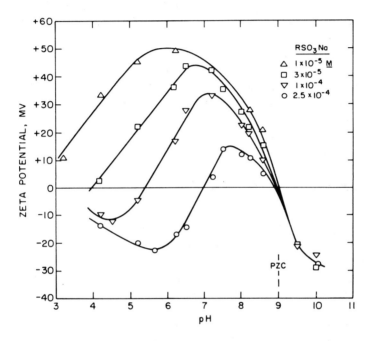

Fig. 7. The electrophoretic mobility of alumina as a function of pH at various concentrations of sodium dodecyl sulfonate at an ionic strength of 2×10^{-3} M controlled with NaCl as the indifferent electrolyte (*31*).

sulfonate the pH must be lowered and the concentration of surfactant increased.

Perhaps the way to clearly illustrate the effect of pH on the mechanism of detergent adsorption in this system is through the electrophoretic behavior of alumina particles in solutions of sodium dodecyl sulfonate. Figure 7 presents the ζ-potential of alumina as a function of pH for the same solution conditions used for determination of the adsorption isotherms given in Fig. 6. Figure 7 also shows that the electrophoretic mobility of alumina is essentially independent of the presence of the detergent above pH 9 (the pzc of alumina), indicating that sulfonate ions are not adsorbed under these conditions. As the pH is lowered the electrophoretic mobility increases until a point is reached where the pH–ζ-potential curve sharply reverses its slope. This point is the hemimicelle concentration for the pH under consideration. For example, at pH 7 the hemimicelle concentration is about 10^{-4} M sulfonate, whereas at pH 6 it is reduced to about 10^{-5} M. To cause sufficient adsorption for hemimicelle formation in dilute surfactant solutions the pH must be lowered in order to increase the surface charge density. After the ζ-potential has been reversed in sign upon lowering the pH, there is another change in the slope of the pH–mobility curves, which appears to be caused by monolayer adsorption of the detergent ions. Once this condition is reached, lowering the pH now affects the mobility primarily because of the increase in σ_s without appreciable effect on Γ_δ.

V. Chemisorption at Solid–Water Interfaces

There are several kinds of surfactant–solid systems in which chemisorption is known to occur. With such research tools as infrared spectroscopy it is possible to study the formation of chemical bonds between surfactant molecules and the solid surface. Examples of where chemisorption appears important are the adsorption of xanthates on sulfide minerals, soaps on fluorite or apatite, soaps on ferric oxide, etc. In the following sections we shall discuss briefly the chemisorption of soaps on salt-type minerals and the chemisorption of alkyl sulfates on hematite.

A. Chemisorption of Oleate on Fluorite

Evidence of chemisorption on such solids as fluorite, barite, and calcite may be shown directly by means of infrared spectroscopy (32, 33). Figure 8a presents the infrared spectra (32) of oleic acid (solid line) and calcium oleate (dotted line). The two bands in the vicinity of 3.5 μ result from C—H stretching in —CH$_3$ and —CH$_2$— groups. The 5.82 μ band is a C=O stretching mode associated with the —COOH group. In salts

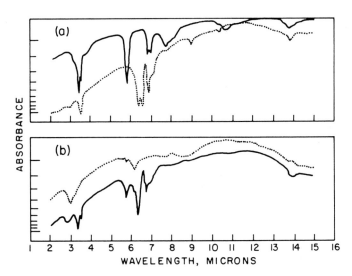

Fig. 8. (a) Infrared spectra of oleic acid (—) and calcium oleate (· · ·). (b) Infrared spectra of fluorite (· · ·) and oleic acid adsorbed on fluorite (—) [after Peck and Wadsworth (*32*)].

there are at least two C=O stretching modes, an antisymmetric mode

$$-C\overset{\displaystyle O}{\underset{\displaystyle O}{\diagup}}\quad (-) \text{ having a wave length near } 6.4\,\mu \text{ and a symmetric mode}$$

$$-C\overset{\displaystyle O}{\underset{\displaystyle O}{\diagup}}\quad (-) \text{ near } 6.9\,\mu \ (4). \text{ In the calcium oleate spectrum two bands}$$

near 6.5 μ are in evidence along with a strong band at 6.85 μ. Figure 8b presents the spectra of fluorite (dotted) and oleic acid adsorbed on fluorite (solid). The C—H stretching modes near 3.5 μ may be seen, as well as strong bands at 5.8 μ, 6.4 μ, and 6.8 μ, indicating that both free oleic acid (5.8 μ band) and a salt (6.4 μ and 6.8 μ bands) are present. Simple washing with acetone or extended washing with water removes the 5.8 μ band, leaving the calcium oleate surface bond. The physically adsorbed oleic acid is weakly held to the surface and is readily removed. The chemisorbed calcium oleate is not the same as bulk calcium oleate, as evidenced by the double band near 6.4 μ shown in Fig. 8a and the single band of Fig. 8b.

The depth of the infrared absorption bands at 5.8 μ and 6.4 μ for oleic acid adsorbed on fluorite (Fig. 8b) can be used by appropriate calibration to determine the amount of physically adsorbed oleic acid and the chemisorbed oleate, respectively. The extent of physical adsorption and chemisorption is illustrated in Fig. 9a. The amount of oleic acid chemisorbed, as

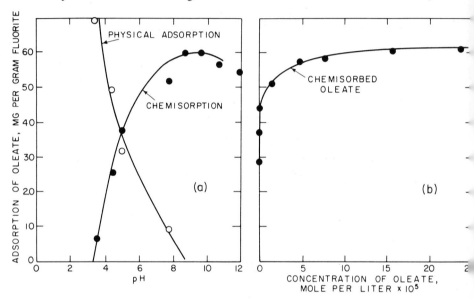

Fig. 9. (a) The amount of oleate adsorbed on fluorite (pH 8.7) at various pH values from a solution containing 9×10^{-4} M sodium oleate. The physical adsorption curve is determined from the 5.8 μ band and the chemisorption curve from the 6.4 μ band. (b) Adsorption isotherm for the chemisorption of oleate on fluorite at pH 8.7 [after Peck and Wadsworth (*32*)].

evidenced by the formation of surface calcium oleate bonds, increases with pH and passes through a maximum near pH 9. At high pH values, carbonates in solution react with fluorite to form surface and bulk calcium carbonate readily observable by infrared spectroscopy, thus competing with the calcium oleate at the surface. As the pH is lowered, the bulk concentration of oleate is reduced due to the formation of oleic acid by hydrolysis, thereby decreasing the amount of chemisorbed oleate. However, increasing amounts of molecular oleic acid adsorb physically as the pH is lowered, possibly through interaction of the hydrocarbon chains of oleic acid molecules with the little oleate that is chemisorbed. Figure 9b is an adsorption isotherm showing the amount of chemisorbed oleate (6.4 μ band) as a function of concentration at pH 8.7. This isotherm has the appearance of a typical Langmuir adsorption isotherm, showing almost complete surface coverage when there is but little residual oleate in solution. This kind of adsorption should be contrasted with the physical adsorption system of amine salts on quartz.

Somasundaran (34) has recently published the results of a study of the adsorption of oleate on calcite. His methods, however, cannot distinguish between physically adsorbed and chemisorbed oleate.

B. Chemisorption on Hematite

By means of infrared spectroscopy, Peck, Raby, and Wadsworth (35) studied the adsorption of sodium oleate and oleic acid on hematite and concluded that oleate chemisorbs on ferric oxide. Their work indicated that the solid surface is first filmed with undissociated oleic acid molecules, which react with the solid to form ferric oleate surface bonds, displacing surface hydroxyls and physically adsorbed water molecules. Using infrared data they postulated the following reactions between oleic acid and hematite surface sites:

$$M—OH + HOl \rightarrow M—OH \cdots HOl$$
$$M—OH \cdots HOl \rightarrow M—Ol + H_2O$$

where HOl represents oleic acid, M—OH mineral surface sites with chemisorbed hydroxyls, and M—Ol mineral surface sites with chemisorbed oleate. This particular study also indicated that maximum chemisorption occurs near the pzc of their hematite.

In a recent paper, Shergold and Mellgren (18) presented the results of an investigation of the adsorption of sodium dodecyl sulfate on natural hematite. The natural mineral used in their study was found by electrophoresis to have a pzc at pH 4.8. Figure 10 presents the ζ-potential of hematite as a function of pH at various sodium dodecyl sulfate concentrations at an ionic strength of $6 \times 10^{-3} M$. This figure indicates strong chemisorption of dodecyl sulfate ions by hematite.

In contrast, Fig. 7 clearly shows that once the pzc of alumina is reached, dodecyl sulfonate ions have little or no effect on the ζ-potential of alumina, whereas in the case of iron oxide in the presence of sodium dodecyl sulfate, the pH must be raised nearly four pH units above the pzc before dodecyl sulfate ions cease to have any effect on the ζ-potential of hematite. Figure 11 presents the adsorption of dodecyl sulfate ions on hematite as a function of pH for various concentrations of sodium sulfate at an ionic strength of 6×10^{-3} M. These adsorption isotherms clearly show that dodecyl sulfate ions strongly adsorb at pH values above the pzc; in fact, at four

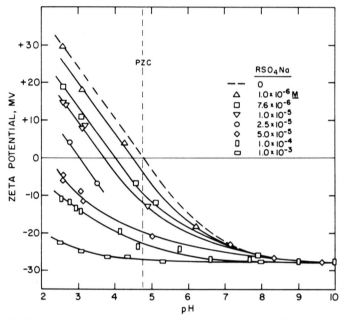

Fig. 10. The ζ-potential of natural hematite as a function of pH at various concentrations of sodium dodecyl sulfate at an ionic strength of 6×10^{-3} M [after Shergold and Mellgren (18)].

pH units above the pzc, the adsorption density of the surfactant is greater than 10^{-11} mole/cm. This chemisorption behavior should be contrasted with the adsorption isotherms presented in Fig. 6 for the alumina–sulfonate system, a system in which adsorption is clearly electrostatic in nature. In the case of the adsorption of alkyl sulfates on hematite, the adsorption is clearly controlled by $\Delta G^{\circ}_{\text{chem}}$ and $\Delta G^{\circ}_{\text{HM}}$, with relatively small effects due to $\Delta G^{\circ}_{\text{elec}}$.

In a very detailed study of the hematite–dodecyl amine system, Joy and Watson (36) found analogous behavior to the alumina–sulfonate system. Namely, dodecylammonium ions adsorb essentially through physical interaction with the surface with chemisorption absent.

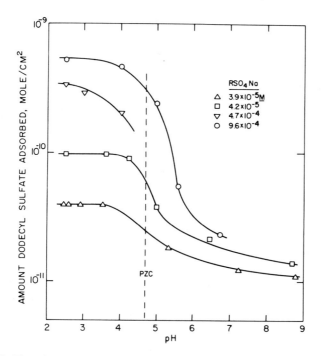

Fig. 11. The adsorption of dodecyl sulfate on hematite as a function of pH at a constant ionic strength of 6×10^{-3} M [after Shergold and Mellgren (*18*)].

These results show that to delineate whether adsorption in any solid–surfactant system is physical or chemical in nature, a number of experiments must be performed to determine the extent of adsorption in the regions where the surface charge is reversed in sign.

VI. Activation Phenomena in Surfactant Adsorption

In the following paragraphs it will be seen that inorganic ions (other than potential-determining) added to the system may either inhibit or assist the adsorption of the surfactant. Any inorganic ions that can compete with surfactant ions for sites in the double layer can inhibit the adsorption of the surfactant. It has been shown that the adsorption of ionic surfactants on positively charged alumina is markedly reduced by the addition of NaCl or Na_2SO_4. For example, at pH 6 the adsorption of dodecyl sulfate on alumina (*37*) is markedly reduced by both Cl$^-$ and

SO_4^{2-}, but the effect of SO_4^{2-} is about 500 times that of Cl^-. This greater effect of SO_4^{2-} over Cl^- in depressing surfactant adsorption results from the specific adsorption potential of SO_4^{2-} on alumina, thereby causing SO_4^{2-} to preferentially displace the alkyl sulfate ions from the positive alumina surface. On the other hand, it is possible to find conditions under which the same inorganic salt can promote surfactant adsorption. Specifically, at pH 6 alumina does not adsorb dodecylammonium chloride since

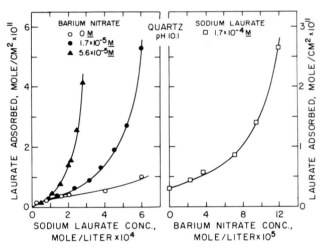

Fig. 12. The effect of barium nitrate on the adsorption of sodium laurate on quartz from aqueous solutions at pH 10.1. Barium ions function as an activator in this system [after Gaudin and Chang (38)].

the organic cations do not interact with the positively charged solid; but by adding sufficient SO_4^{2-} ions to the system the charge in the Stern plane can be reversed and the solid can now adsorb the organic cations. In such systems sulfate ions are said to function as an activator; they serve as a link between a surface and a surfactant with similar charges.

Gaudin and Chang (38) measured the adsorption of barium ions and laurate ions on quartz at pH 10.1. Figure 12 shows limited adsorption of laurate anions on negatively charged quartz in the absence of Ba^{2+}; however, when Ba^{2+} is present even in small amounts the adsorption of the carboxylate ions increases markedly. Figure 12 further shows that at a fixed sodium laurate concentration increasing the concentration of barium nitrate in solution causes the uptake of laurate by quartz to increase. In this case adsorption appears to result from Ba^{2+} specifically adsorbing in the Stern plane and thereby serving as a link between laurate and the quartz surface.

More recently it has been shown that marked adsorption of anionic surfactants on quartz appears to take place when the activator begins to hydrolyze (39). Furthermore, the very extensive studies of coagulation phenomena that have been carried out during the past decade clearly show that hydrolyzing cations have very pronounced effects at extremely low concentrations on the stability of colloidal suspension. This has been attributed to the specific adsorption of multicharged, polymeric hydrolysis products in the Stern layer. The mechanism for activation in such systems is for the hydrolyzing cation to adsorb very strongly onto the solid surface, perhaps reversing the ζ-potential in the process and thereby permitting adsorption of the anionic surfactant adjacent to the positively charged Stern layer. A guide for conditions under which a given inorganic cation might function as an activator is the determination of its hydroxide precipitation pH. At a pH slightly less than this, adsorption should be a maximum.

VII. Adsorption of Surfactants on Silver Iodide

The classical material used by colloid chemists for the investigation of colloidal phenomena has been silver iodide, whose potential-determining ions are clearly Ag^+ and I^-, with the pzc occurring at pAg 5.6. Only quite recently has it become established that AgI is a partially hydrophobic solid (40–42). Billett and Ottewill (42) observed advancing contact angles as large as 47° on AgI in water. Thus there is a strong possibility for surface-active agents having a hydrophobic bonding contribution to the adsorption free energy.

In the case of quartz, where hydrophobic bonding with the surface itself is absent, octylammonium acetate appears to behave identically with ammonium acetate in adsorption phenomena. On the other hand, at pI 5 (pAg 11), Pravdic and Mirnik (43) found that C-8 and C-6 amines reverse the ζ-potential of negatively charged AgI. This can be interpreted in terms of the hydrophobic bonding phenomena mentioned above, where even the squeezing out of a 6-carbon hydrocarbon chain from water onto a hydrophobic solid surface results in a considerable decrease in free energy, although some chemisorption could also be occurring in the case of the amine adsorption (44). Other indications of hydrophobic bonding on AgI are given in the ζ-log C curves of anionic surfactants on positively charged AgI. Ottewill and Watanabe (45) found that the concentration region over which ζ is most markedly affected by the added sodium dodecyl sulfate or sulfonate begins in dilute solutions and occurs over three orders of magnitude change in surfactant concentration. In contrast, the concentration range over which ζ is affected when hemimicelles

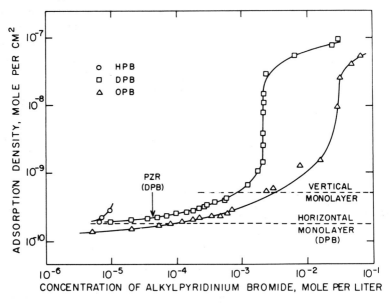

Fig. 13. The adsorption of octyl (OPB), dodecyl (DPB), and hexadecyl (HPB) pyridinium bromides at pI 4 on silver iodine with a specific surface 0.7 m²/g. The point of ζ-reversal is also shown [after Jaycock, Ottewill, and Rastogi (46)].

abruptly form is very small, occurring in as little as a two- or three-fold change in surfactant concentration (20, 22).

Adsorption isotherms (46) have been determined for the uptake of octylpyridinium bromide, dodecylpyridinium bromide, and hexadecyl-pyridinium bromide by AgI from aqueous solutions at pI 4; these results are summarized in Fig. 13. Several interesting adsorption phenomena can be observed in this figure. First, even at the lowest concentrations tested, a complete monolayer of horizontally lying ions is formed, indicating strong adsorption indeed. Perhaps a chemical bond is formed between the pyridinium group and AgI in the surface. As can be seen, there is no particular change in the adsorption isotherm where the ζ-potential reverses. The vertical rise in the isotherms is most striking in that the uptake of the pyridinium surfactant appears to reach the equivalent of several hundred flat-lying monolayers. Recent electronmicrographs have indicated that the actual formation of a new chemical compound may be occurring at the surface, resulting in the nucleation of small crystals at the surface (44).

Thus it now appears that the adsorption of surfactants at the surface of what colloid chemists have long used as their classic material may be one of the most complicated systems with which one can work, since a complex combination of electrical, hydrophobic, chemical, and other effects may be occurring in the AgI system.

VIII. Adsorption on Carbon

It has long been established that carbon (in particular graphite) is hydrophobic. Adsorption of any solute at the surface of a hydrophobic solid takes place by displacement of the solvent molecules (47). Thus, if an organic surfactant is adsorbed at the surface of carbon in an aqueous medium, water molecules are displaced as the adsorption process proceeds. Energies involved here are clearly those associated with hydrophobic bonding.

A colloidal carbon (graphite) with a very homogeneous surface is Graphon, which is produced by a high-temperature heat treatment procedure. Graphon is essentially free of surface oxygenated sites and hence carries essentially no electrical charge on its surface. However, carbons with oxygenated surface sites will be charged in water because of ionization of these polar groups. Thus, nonheat-treated carbons will have different adsorption properties because of the electrical charge that might result from surface ionization and because of the strong interaction of polar water molecules with the oxygenated surface sites. Day, Greenwood, and Parfitt (48), who studied the effect of heat treatment of carbon on the adsorption of surfactants from aqueous solution, have found that increasing the heat-treatment temperature removed increasing amounts of surface oxygen and hydrogen and that the adsorption of the surfactant increased correspondingly. They also measured the adsorption of sodium dodecyl sulfate on Graphon; Fig. 14 presents their adsorption isotherm for this system.

In this case, the adsorption process is controlled by hydrophobic bonding phenomena, with the hydrocarbon chains from the surfactant displacing water from the Graphon surface. It should be pointed out that in this system the ΔG°_{elec} term will always act against the adsorption process since the only electrical effects will be those due to the charge on the heads of the adsorbed surfactant ions (in the case of the nonionic surfactant, this term would be absent). The isotherm in Fig. 14 shows two distinct steps. Initially, the sharp rise in the isotherm results from adsorption with the chains lying flat, and the first inflection corresponds to a close-packed monolayer of flat-lying surfactant ions, occupying about

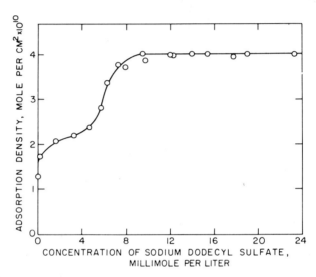

Fig. 14. The amount of sodium dodecyl sulfate adsorbed on Graphon with a specific surface of 79 m²/g [after Day, Greenwood, and Parfitt (48)].

70 Å² per ion. At higher concentrations the adsorbed ions begin to stand up, orienting the polar group toward the aqueous phase. The second plateau corresponds to a vertically oriented monolayer, occupying about 40 Å² per ion. In the first step of this kind of adsorption isotherm (where the hydrocarbon chains interact with the solid) the primary contribution to the adsorption free energy is $\Delta G^{\circ}_{CH_2}$, whereas in the second step (where the hydrocarbon chains interact with each other) ΔG°_{HM} is acting in addition to $\Delta G^{\circ}_{CH_2}$. The ultimate adsorption is reached at about the cmc of sodium dodecyl sulfate; because of the orientation of the layer of ionized polar groups being toward the aqueous solution, there is no mechanism by which multilayers can build up.

Since the adsorption of any organic surfactant at the surface of carbon depends on the magnitude of the hydrophobic bonding free energy, the amount of surfactant adsorbed should increase directly with increased chain length, in accordance with the well-known Traube's rule. Thus adsorption can readily be expressed in terms of what has been called the "reduced concentration." This is an activity ratio—the ratio of the concentration of solute to its solubility limit or to its cmc. Numerous examples of adsorption isotherms as a function of reduced concentration have been presented—for example, by Hansen and Craig (49).

IX. Adsorption on Noble Metals

Reactive metals are coated with oxide films, and adsorption at their surfaces is thus analogous to adsorption on oxides; for example, adsorption on aluminum should be similar to that on alumina and should be highly pH dependent.

Mercury is the classic material that has been very widely studied in electrical double layer research; therefore, much is known about its double layer and adsorption of ions at its surface. Mercury is hydrophobic (50), and the results of Smolders (51) clearly show that surfactants strongly adsorb at the mercury surface through hydrophobic bonding. Some recent work (52) on gold will be used to illustrate surfactant adsorption phenomena on noble metals. Although there is considerable controversy over whether gold is hydrophobic or hydrophilic (50), clean gold does appear to be a hydrophobic solid. Like most metals, gold carries a negative charge in water, with the ζ-potential about -50 mV (52). Benton and Sparks (52) have measured the amount of alkyl pyridinium bromides of various alkyl chain lengths (C-12, C-14, and C-16) adsorbed from water onto gold and also the adsorption of sodium alkyl sulfates. Figure 15 presents the adsorption isotherms of these various surfactants in the form of the amount of surfactant adsorbed as a function of the reduced concentration, where the reduced concentration is given in terms of the ratio of the solution concentration C to the critical micelle concentration C_m. Because the data can be described in terms of a reduced concentration, hydrophobic bonding appears to be the basis for adsorption in this system. Although Benton and Sparks interpreted the shape of their curves in terms of adsorption in two monomolecular layers, the adsorption seems somewhat more analogous to the mechanism of adsorption on Graphon or on a latex; namely, adsorption first occurs in a horizontal orientation and then in a vertical orientation. For the cationic surfactants, Benton and Sparks (52) found that the ζ-potential is reversed at $C/C_m = 0.012-0.016$. Although the effect is small, the value of C/C_m required for ζ-potential reversal increases slightly in the order of C-16, C-14, and C-12.

Figure 15 shows that both cationic and anionic surfactants adsorb on gold, although the total uptake of the anionic surfactants is about half

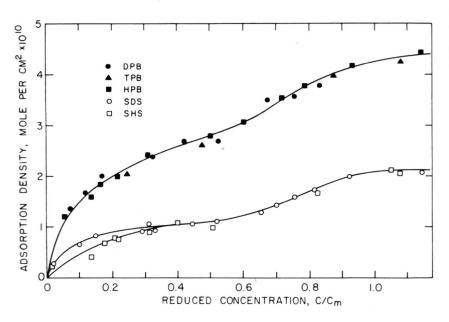

Fig. 15. The adsorption of dodecylpyridinium bromide (DPB), tetradecylpyridinium bromide (TPB), hexadecylpyridinium bromide (HPB), sodium dodecyl sulfate (SDS), and sodium hexadecylsulfate (SHS) on gold of 0.3 m²/g specific surface area. The isotherms are presented in terms of reduced concentrations C/C_m [after Benton and Sparks (52)].

that of the cationic surfactants. Because the ζ-potential reversal occurs at such low values of C/C_m, the initial attractive electrostatic contribution to adsorption in the case of cationic surfactants and the initial repulsive electrostatic contribution in the case of anionic surfactants would not seem to have much effect on the adsorption at high concentrations. Considerably more work remains to be done before these various phenomena can be clarified, but we see that hydrophobic bonding results in adsorption of both cationic and anionic surfactants at the surface of gold.

X. Adsorption on Carboxylated Polystyrene Latices

Latices are becoming important materials for colloidal studies because they can be made monodispersed in size, are spherical in shape, and can be prepared with ionizable surface groups. These materials are of interest for surfactant adsorption research because they provide a controlled combination of electrostatic and hydrophobic effects.

Latices can be prepared by emulsion polymerization of styrene using sodium dodecanoate as the emulsifying agent and H_2O_2 as the initiator; the sodium dodecanoate is subsequently removed by dialysis (53, 54). Potentiometric, electrophoretic, and infrared studies confirmed the presence of carboxyl (—COOH) groups on the surface. Thus the surface charge of this material can be varied by ionization of the surface carboxyls through variation of the pH.

Figure 16 presents very detailed isotherms for the adsorption of alkyl trimethylammonium halides (C-8, C-10, and C-12) on polystyrene latices. The dashed line marks the concentration at which the ζ-potential is reversed (pzr). In each of the cases there is initially a relatively sharp increase in adsorption as the concentration of surfactant is increased. The first break in the curve occurs at the same adsorption density for each surfactant, i.e., at about 8×10^{-12} mole/cm² or about 2600 Å² per adsorbed ion. To achieve this adsorption density, however, the equilibrium concentration in solution must be increased as the chain length of the surfactant decreases. Above the reversal of the ζ-potential the slope of the isotherms first decreases and then has a second rise, followed finally by a leveling off of the isotherm at the cmc. All of these isotherms can be normalized by a reduced concentration plot (log Γ versus log C/C_m). These results suggest that the first adsorption region corresponds to a decrease in free energy due to hydrophobic bonding plus electrostatic interaction with the charged carboxyl groups. In this region the chains must be lying down. Between the pzr and the second inflection the adsorption probably continues with the surfactant ions lying down, but here ΔG°_{elec} acts against adsorption because of the reversal of ζ. At the second inflection the adsorbed surfactant ions begin to align vertically, causing increased adsorption, with the final saturation in adsorption occurring when the surface is covered with a monolayer of vertically oriented adsorbed ions. Those covering carboxyl groups would be

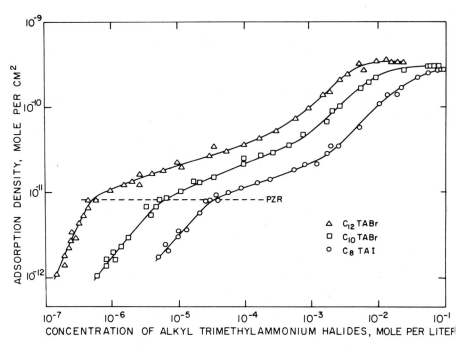

Fig. 16. The adsorption of alkyl trimethylammonium halides on a carboxylated polystyrene latex with specific surface of 60 m²/g. The point of ζ-reversal is also shown [after Connor (*55*)].

oriented with their polar head toward the solid, but with the majority being oriented with their heads toward the aqueous phase; thus, adsorption will be limited to a monolayer.

REFERENCES

1. Overbeek, J. Th. G., in *Colloid Science*, (H. R. Kruyt, ed.), Vol. I, Elsevier, Amsterdam, 1952, Ch. 4 and 5.
2. Wood, L. A., *J. Amer. Chem. Soc.* **68**, 437 (1946).
3. Verwey, E. J. W., *Colloid Chem.* **7**, 47 (1950).
4. Hair, M., *Infrared Spectroscopy in Surface Chemistry*, Dekker, New York, 1967, Ch. 4.
5. Yopps, J. A. and Fuerstenau, D. W., *J. Colloid Sci.* **19**, 61 (1964).
6. Healy, T. W. and Fuerstenau, D. W., *J. Colloid Sci.* **20**, 376 (1965).
7. Parks, G. A. and deBruyn, P. L., *J. Phys. Chem.* **66**, 967 (1962).
8. Parks, G. A., *Advan. Chem. Ser.* **67**, 121 (1967).
9. Somasundaran, P., *J. Colloid Interface Sci.* **27**, 659 (1968).
10. Somasundaran, P. and Agar, G. E., *J. Colloid Interface Sci.* **24**, 433 (1967).

11. Buchanan, A. S. and Heymann, E., *Proc. Roy. Soc. A* **195**, 150 (1948).
12. Freyberger, W. L. and deBruyn, P. L., *J. Phys. Chem.* **63**, 1475 (1957).
13. Gaudin, A. M. and Fuerstenau, D. W., *AIME Trans.* **202**, 66 (1955).
14. Iwasaki, I., Cooke, S. R. B., and Choi, H. S., *AIME Trans.* **220**, 394 (1961).
15. Johansen, P. G. and Buchanan, A. S., *Austral. J. Chem.* **10**, 3980 (1957).
16. Berube, Y. G. and deBruyn, P. L., *J. Colloid Interface Sci.* **27**, 305 (1968).
17. Robinson, M., Pask, J. A., and Fuerstenau, D. W., *J. Amer. Ceram. Soc.* **47**, 516 (1964).
18. Shergold, H. L. and Mellgren, O., *Trans. Inst. Min. Met.* (*London*) **78**, C121 (1969).
19. deBruyn, P. L., *Trans. AIME* **202**, 291 (1955).
20. Somasundaran, P., Healy, T. W., and Fuerstenau, D. W., *J. Phys. Chem.* **68**, 3562 (1964).
21. Haydon, D. A. and Taylor, F. H., *Proc. 3rd Int. Cong. of Surface Activity* Vol. 1, 1960, p. 157.
22. Gaudin, A. M. and Fuerstenau, D. W., *Trans. AIME* **202**, 958 (1955).
23. Tamamushi, B. and Tamaki, K., *Proc. 2nd Int. Cong. of Surface Activity* (*London*) Vol. 3, 1957, p. 449.
24. Somasundaran, P. and Fuerstenau, D. W., *J. Phys. Chem.* **70**, 90 (1966).
25. Wakamatsu, T. and Fuerstenau, D. W., *Advan. Chem. Ser.* **79**, 161 (1968).
26. Fuerstenau, D. W., *J. Phys. Chem.* **60**, 981 (1956).
27. Stigter, D. and Overbeek, J. Th. G., *Proc. 2nd Int. Cong. of Surface Activity* (*London*) Vol. 1, 1957, p. 311.
28. Shinoda, K., et al., *Colloidal Surfactants*, Academic, New York, 1963.
29. Schulman, J. H. and Smith, T. D., *Recent Developments in Mineral Dressing*, Inst. Min. Met. London, 1953, p. 393.
30. Purcell, G. and Sun, S. C., *Trans. AIME* **226**, 6 (1963).
31. Fuerstenau, D. W. and Wakamatsu, T., unpublished results.
32. Aplan, F. F. and Fuerstenau, D. W., in *Froth Flotation—50th Anniv. Vol.*, AIME, New York, 170 (1962).
33. Peck, A. S. and Wadsworth, M. E., *Proc. 7th Int. Mineral Processing Cong.*, Gordon and Breach, New York, 1965, p. 259.
34. Somasundaran, P., *J. Colloid Interface Sci.* **31**, 557 (1969).
35. Peck, A. S., Raby, L. H., and Wadsworth, M. E., *Trans. AIME* **238**, 301 (1966).
36. Joy, A. S. and Watson, D., *Trans. IMM* **73**, 323 (1964).
37. Modi, H. J. and Fuerstenau, D. W., *Trans. AIME* **217**, 381 (1960).
38. Gaudin, A. M. and Chang, M. C., *Trans. AIME* **193**, 193 (1952).
39. Fuerstenau, M. C., Martin, C. C., and Bhappu, R. B., *Trans. AIME* **226**, 449 (1963).
40. Hall, P. G. and Tomkins, F. C., *Trans. Faraday Soc.* **58**, 1734 (1962).
41. Tcheurekdjian, N., Zettlemoyer, A. C., and Chessick, J. J., *J. Phys. Chem.* **68**, 773 (1964).
42. Billett, D. F. and Ottewill, R. H., *Wetting*, S.C.I. Monograph No. 25, 1967, p. 253.
43. Pravdic, V. and Mirnik, M., *Croatia Chim. Acta* **32**, 1 (1960).
44. Ottewill, R. H., Private communication.
45. Ottewill, R. H. and Watanabe, A., *Kolloid Z.* **170**, 132 (1960).
46. Jaycock, M. J., Ottewill, R. H., and Rastogi, M. C., *Proc. 3rd Int. Cong. of Surface Activity*, Vol. 2, 1960, p. 283.
47. Skewis, J. D. and Zettlemoyer, A. C., *Proc. 3rd Int. Cong. of Surface Activity*, Vol. 2, 1960, p. 401.

48. Day, R. E., Greenwood, F. G., and Parfitt, G. D., *Proc. 4th Int. Cong. on Surface Activity*, Vol. II, 1967, p. 1005.
49. Hansen, R. S. and Craig, R. P., *J. Phys. Chem.* **58**, 211 (1954).
50. Fowkes, F. M., *I&EC*, **56**: 12, 40 (1964).
51. Smolders, C. A., *Rec. Trav. Chim.* **80**, 651, 699 (1961).
52. Benton, D. P. and Sparks, B. D., *Trans. Faraday Soc.* **62**, 3244 (1966).
53. Ottewill, R. H. and Shaw, J. N., *Kolloid Z. Z. Polym.* **218**, 161 (1967); **218**, 34 (1967).
54. Shaw, J. N., *J. Polym. Sci.* **27C**, 237 (1969).
55. Connor, P., Doctoral Dissertation, University of Bristol, 1967.

5

Adsorption of Proteins and Lipids to Nonbiological Surfaces

J. L. BRASH

Stanford Research Institute
Menlo Park, California

AND

D. J. LYMAN

University of Utah
Salt Lake City, Utah

I. Introduction

The adsorption of proteins and lipids to nonbiological surfaces is important within the context of this book in at least two respects. First, such interactions can provide model systems for the study of adsorption in physiological systems where both reactants are biological in nature—for example, the interaction of cells with proteins and lipids in blood and tissue, and the formation of cell membranes by formation of lipid–protein "surfaces." Second, with the increasing use of prosthetic materials in the body, it is essential to know how nonbiological surfaces will interact with biological media. Thus, potential damage to blood proteins, lipids, and formed elements (which can in some instances be regarded as the equivalent of lipid micelles) coming in contact with prosthetic surfaces is of vital importance in the field of cardiovascular surgery. Similar interactions are also of importance for prostheses in other biological environments, e.g., in the synovial fluid of joints, in brain tissue where aneurysms may be treated by reinforcement with polymeric materials, in tissues where surgical adhesives may be used, and in numerous other situations.

The properties of proteins and lipids in solution have been covered in other chapters. The manner in which these properties can be altered by contact with a surface is the major concern of this chapter. Among the various possibilities is that the surface–protein or surface–lipid complex is intrinsically different from the unbound protein or lipid. This might be the case if the protein or lipid adsorbs to the surface by binding at a specific site, thus preventing the "normal" reactions of this site or exposing a predominance of other sites to the contacting medium. Alternatively, the protein may change its conformation (e.g. by partial unraveling of the polypeptide chains), thereby changing its reactivity. Also to be considered are the possible changes in proteins and lipids which have been in contact with a surface and subsequently released or desorbed back into the medium. The various forms and degrees of denaturation, e.g., partial uncoiling, breaking of intrachain bonds, and exposing or masking of active sites, must also be considered in this context.

This chapter will attempt to summarize current knowledge of such protein and lipid adsorption to nonbiological interfaces. Since this area of work is still at an early stage of development, it would seem to be most useful to attempt a critical review of recent work. The review will not be exhaustive but hopefully the papers covered will represent all current points of view. Reflecting the authors' own interest, the discussion will deal primarily with solid–liquid interfaces and with the proteins of blood.

II. Proteins at Fluid Interfaces

This area has been reviewed extensively in the past, notably by Bull (*1*), by Cheesman and Davies (*2*), and by Fraser (*3*). Liquid interfaces have historically been an attractive subject of study for surface chemists, due in large measure to the fact that the so-called film pressure can be readily varied by increasing the quantity of protein in a constant area of film, or by changing in controlled fashion the area occupied by a given quantity of protein. The flexibility thus conferred on these systems makes them particularly suited for theoretical studies.

One of the first practical uses of protein layers on water was in determining protein molecular weight. In the low surface pressure region, i.e., below about 1 dyne/cm, protein films can be shown to behave like ideal gases in two dimensions. Thus, by analogy with ideal gases, film pressure is inversely related to area and a plot of pressure × area against pressure is linear (1). The latter plot sometimes exhibits a minimum, indicating finite forces of attraction between molecules at the surface. When the plot is linear over its entire range, it can be shown that the intercept is equal to nRT, where n is the number of moles of protein, R is the gas constant, and T is the absolute temperature. The molecular weight is therefore $24.6 \times 10^2/(FA)_{F \to 0}$, where F is the film pressure in dynes/cm and A is the area in m²/mg of protein. Use of this procedure leads to values of molecular weight in good agreement with those obtained by other methods. Thus Bull (4) deduced the molecular weight of ovalbumin to be 44,000, and Guastalla (5) obtained 40,000; these compare with a value of 45,000 obtained by osmometry. In cases where results are different from those obtained by techniques that measure molecular weight in the bulk solution (e.g., osmometry), it is possible to make deductions as to changes that might be occurring in the interface. Thus, a molecular weight of 17,000 for β-lactoglobulin at the air–water interface (6) contrasts with approximately double this value obtained by conventional methods (7). This difference is taken as an indication of dissociation into halves at the interface. Other examples of this type of behavior have been reported; it would appear in many cases to be pH-dependent (8). The interpretation of such results is not by any means certain, and, indeed, some authors contend that more work is needed on precise characterization of mono-layers before they can validly be used for determination of molecular weights.

Protein films at the air–water interface have also been used to form layers of molecular dimensions at solid surfaces by Langmuir–Blodgett transfer. The technique was first described for salts of long-chain fatty acids, such as palmitic and stearic acids (9), and later extended to proteins (10). The method consists simply of dipping a glass or metal slide into (or withdrawing from) a liquid with a film spread at its air interface. The number of molecular thicknesses deposited is directly related to the number of passages through the interface. The experiment permits determination of the thickness of a monomolecular protein film by deposition of a known unimolecular or multimolecular film on the solid surface.

The structure of protein films at the air–water interface has been the subject of much study and speculation (*1*). It is generally agreed that the films are oriented more or less with the polar groups directed toward the aqueous phase and the nonpolar side groups of the peptide chain directed toward the air. The extent to which the chains are uncoiled and extended is still controversial and would seem to depend in large measure on the film pressure. As might be expected, the degree of uncoiling appears to decrease as the pressure increases. Films at pressures below 1 dyne/cm exhibit a high degree of uncoiling since presumably there is sufficient free area at the surface for maximum spreading. (These are the so-called gaseous films referred to above.) At high film pressures the film may consist entirely of globular molecules (*11*), and at intermediate pressures there may exist a mixture of uncoiled and globular species (*12*).

Such conclusions are to a large extent based on calculations of the film thickness, which assume that protein density is the same as in the crystal and that the film is of uniform thickness over its entire area. It is probable that protein density in the film is less than in the crystal, so that if a true density were used, greater thicknesses would be deduced. The assumption of uniformity is not subject to confirmation or denial but would seem to be a simplifying one at best.

Denaturation induced exclusively by spreading at the air–water interface is not always equivalent to total destruction of secondary and tertiary structure. Thus, spread films of ovalbumin can be further expanded by ultraviolet irradiation (*13*), and heat-denatured proteins show larger areas than native proteins on spreading (*14*). These observations might indicate that in the interface more of the helical structure and intra chain bonds remain intact than under the influence of heat or light.

More recently, Baier and Zobel (*15*) have presented electron microscope evidence that monolayers of the structural protein myosin in a highly spread condition (1.9 m²/mg, surface pressure less than 1 dyne/cm) contain arrays of filamentous aggregates 100–200 Å wide, as well as an overlaying, more diffuse layer. This type of evidence suggests retention of some structure in that system. Much infrared evidence has also been produced for both proteins and polypeptides to show that α-helical as well as extended chain structures can exist at the air–water interface (*16–19*).

Change of biological activity of proteins has also been observed at surfaces. Data are available indicating retention of some activity in layers of spread enzymes (20) and antigenic proteins (21, 22). It is tempting to assume that such layers are not fully unfolded and extended, and this may well be the case. However, reversibility of the unfolding process (23) or retention of intact active sites would also explain such results.

Another probe for changes in structure at the air–water interface is the interaction of dissolved protein with the layer. Thus protein–protein interaction may occur between a surface molecule and a solution molecule of the same protein due to the alteration in structure at the surface. These "monolayer penetration" phenomena are most validly studied as changes in area of the film at constant pressure (24–26) since under these conditions the adsorbed film is not undergoing continuous pressure-induced change as well. Using such "isobaric" measurements it has been shown that expansion of both insulin and albumin films occurs when albumin is injected into the sub-phase. The expansion rate is greater with the insulin film, indicating a specific interaction in that case. Support for these results was obtained from force–area isotherms of mixtures. Arnold and Pak have also discussed some of the biological implications of these interactions (24).

III. Lipids at Fluid Interfaces

The interfacial behavior of lipids results from their amphipathic character. By reason of their ionic or polar character, they exhibit some affinity for water. The nonpolar hydrocarbon portions, however, are more attracted to the air at an air–water interface and to the oil at an oil–water interface. Thus, lipids possessing polar and/or ionic groups will form an adsorbed, oriented layer at the interface between water and air or oil. This adsorption is manifested as a lowering of the interfacial tension. The limiting area of such layers varies from about 20 Å^2 per molecule for simple lipids like stearic acid, to 40 Å^2 for cholesterol, to about 100 Å^2 for lecithin (27). When the interface becomes saturated, further lipid is accommodated by formation of micelles (at the critical micelle concentration), which present a polar or ionic exterior to the water. At even higher concentration, liquid crystals can be formed consisting of lipid bilayer sheets and interspersed water. Bangham has given a detailed discussion of these phenomena (28).

As with protein films, more recent interest has centered in "penetration" phenomena especially with proteins, as discussed in Sect. IV. Mixed lipid layers have recently been discussed by Demel et al. (*29, 30*). Renewed impetus for such studies stems from the recent availability of synthetic phospholipids of precisely characterized structure. For a series of phospholipid monolayers in which the chain length of the acid residues was varied, it was found that the area per molecule increases as the acid chain length decreases. Similarly, the area per molecule increases with increasing unsaturation in the acid residues. The hypothesis was advanced that organisms can work to synthesize phospholipids of identical physical properties even though the chemical composition may vary. In this way the surface physical properties of the cell membranes are preserved so that membrane viability is not compromised.

Studies of the so-called "condensing" effect of cholesterol on phospholipid monolayers were made by the same authors. In this effect, the area per phospholipid molecule in the mixed layer is smaller than the area per molecule in the respective pure monolayers. Various intermolecular interactions are postulated to account for this behavior, such as van der Waals interactions and structural alterations in the water layers adjacent to the interface. These interactions in turn depend to a large extent on the fatty acid structure, e.g., chain length and degree of unsaturation. There is some suggestion that such interactions may also occur in the formation of biological membranes.

In contrast to Demel's findings, which apply only to unsaturated phospholipids, Rosenberg (*31*) reported a similar condensing effect with the fully saturated di-palmitoyl lecithin.

IV. Mixed Protein–Lipid Adsorption at Fluid Interfaces

The principal inducement to study of such systems is their close analogy to the membranes of normal cells. They have therefore been investigated as working models for cell membranes from the point of view of both membrane structure and function.

The structure of cell membranes has been discussed adequately elsewhere (*32, 33*). The electron microscope has revealed that cell membranes are about 80 Å thick and consist of two similar bounding surfaces with a central region of lower stain (OsO_4) absorbing capacity (*34*). Based on considerations of chemical composition (approximately equal weights of protein and lipid) and on the general plausibility of the model, it is widely accepted that the darkly staining regions are the polar ends of an oriented bimolecular lipid sheet with protein in close association, and the light region is the inward-facing hydrocarbon chains of the lipid bilayer. This is the well-known "paucimolecular" model of Davson and Danielli (35).

Classical surface chemistry has made significant contributions to the development of this model, and studies of mixed lipid–protein films at the air–water and oil–water interfaces have been of particular value. Fraser (3) has reviewed the important work in this field to 1957. Though some work has been done on the interaction of protein layers with lipid in the subphase (36), most investigations have been concerned with protein adsorption to lipid layers, which is more relevant to the formation of biological membranes. Pioneering work in the latter area was done by Schulman and coworkers (37, 38) who studied changes in the surface pressure at constant area (and vice versa) of films of cephalin, cholesterol, and cardiolipin upon addition of various proteins (e.g., albumin, hemoglobin) to the subsolution. Interactions were strong when the ionogenic groups of the interacting species were oppositely charged, but were weak between neutral or similarly charged species.

More quantitative studies were later reported by Eley and Hedge (39–41) on the interactions of various proteins with monolayers of stearic acid, cholesterol, lecithin, and cephalin at the air–water interface. The pressure-concentration isotherms showed two breaks, which were taken to correspond to the formation of two separate layers of protein. From the quantities adsorbed, they deduced that the first layer consisted of unfolded protein, which did not "penetrate" the lipid layer as discrete molecular entities but rather was adsorbed by polar interactions between the peptide groups of the protein and the polar ends of the lipid. The nonpolar side chains of the unfolded protein were thought to penetrate the layer. The second protein layer contained much more material per unit area and was judged to be a monomolecular layer of native globular protein. These conclusions support the membrane model of Davson and Danielli.

Most recently, Colaccico et al. (42, 43) have examined other lipid–protein systems including the complex glycosphingolipids and proteins present in naturally occurring lipoproteins (e.g., the apoprotein of high density serum lipoprotein). Initial work confirmed the order of reactivity found by earlier workers. N-Palmitoyl dihydroceramide lactoside was between cholesterol (the most reactive lipid) and cephalin in the magnitude of its film pressure increase on addition of γ-globulin to the subphase. Colaccico (43) presents evidence for specific lipid–protein interactions, to be distinguished from the penetration (i.e., film pressure increase) that lipid films experience on reaction with most proteins. An example is the complexing of the antilactoside antibody with the lactose residue of the glycosphingolipid. A model of naturally occurring lipoprotein complexes somewhat different from that of Davson and Danielli is presented, which envisages a mosaic structure with "packages" of lipid surrounded by protein. Structuring of water in the mosaic is closely connected with its stability.

In connection with the structure of lipoproteins, it is of interest to note recent work on the line widths of proton nmr signals in serum lipoproteins (44). When serum albumin was added to lysolecithin, a marked broadening of the methylene proton signal was observed, whereas the methyl protons of the choline residue were unaffected. This behavior was taken as an indication of lipid–protein binding via the hydrocarbon chain of the lipid, i.e., hydrophobic bonding. The spectra of serum lipoproteins, on the other hand, do not exhibit line broadening, indicating neither polar nor apolar binding in significant amounts. These authors propose a micellar structure for the serum lipoproteins.

Adsorption of proteins to lipids at the oil–water interface has also been studied as a model for cell membrane formation. This area has also been reviewed extensively by Fraser (3). Behavior similar to that at the air–water interface is often observed. Study of the flocculation of lipid-stabilized oil emulsions by protein showed that the interactions are largely ionic (45). Later work along the same lines has been summarized by Schulman and Fraser (46), who made extensive investigations of the adsorption of the enzymes trypsin and catalase to lipid-stabilized emulsions. In this work, changes in enzyme activity as well as the quantities adsorbed were measured. The results are summarized in Fig. 1, which shows the adsorption of catalase at various lipid surfaces. Behavior varied considerably from surface to surface. On anthracene–C_{21}–sulfonate surfaces, 10 $\mu g/cm^2$ of catalase was adsorbed, and enzyme activity in the outer part of the multilayer was retained and could be recovered on desorption. On lecithin surfaces, no adsorption could be detected and enzyme activity was not altered. Most of these adsorptions could be explained again by an ionic mechanism, and supporting evidence was obtained from measurements of particle electrophoretic mobility.

Electrophoresis is a technique much used to study protein adsorption and will be referred to again in Sect. V. One interesting study using this method is that of Douglas and Shaw (47) who measured changes in mobility as a function of pH for liquid hydrocarbon droplets dispersed in water and bearing adsorbed layers of protein, lipid, polysaccharide, or nucleic acid. Relationships between surface structure and mobility were

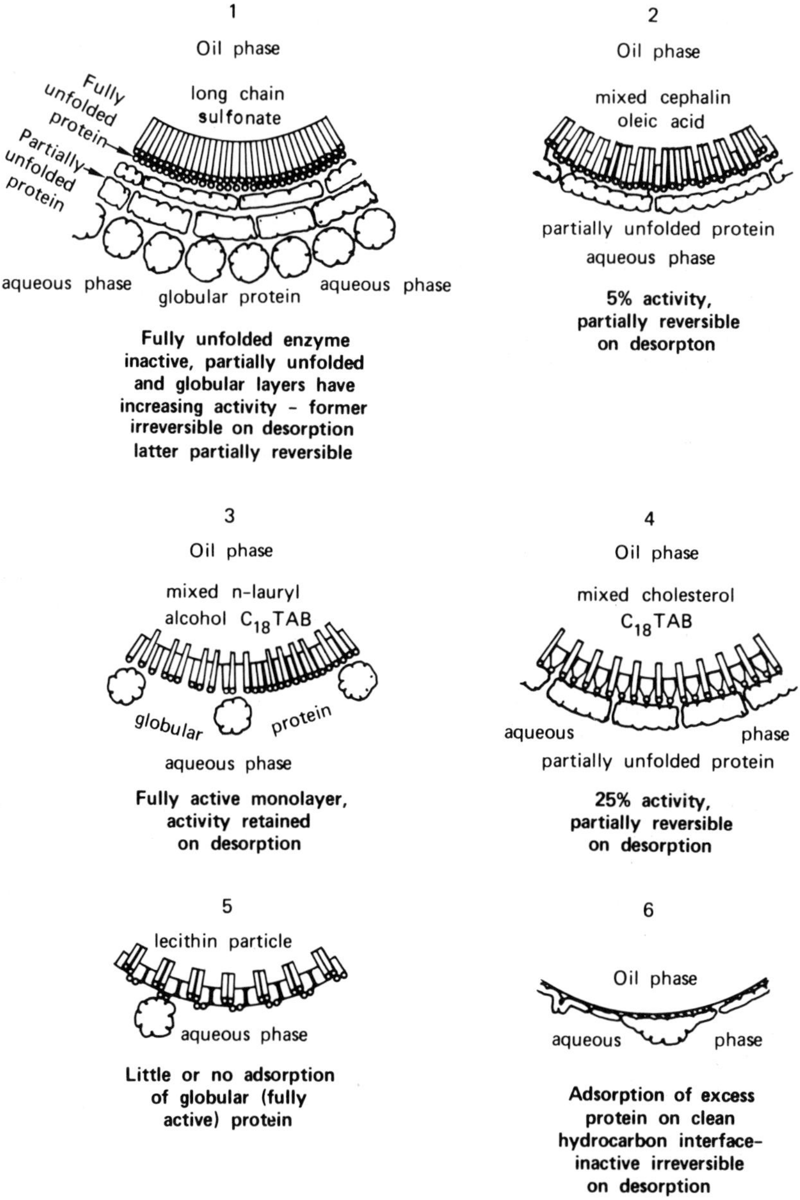

Fig. 1. Adsorption of catalase at various lipid-stabilized oil–water interfaces. [Reprinted from Ref. (46), p. 69, by courtesy of *Ver. Kolloid Ges.*]

considered. Mobility measurements revealed the same isoelectric point for adsorbed and dissolved fibrinogen or ribonuclease and the same general shape for the mobility–pH curves. Mobilities at a given pH, however, were two to three times greater for the adsorbed proteins, which would indicate perhaps some surface orientation of molecules with the ionogenic groups pointing away from the surface. The opposite effect was noted by Bull (48) for albumin adsorbed on glass particles. Reduced mobility, and therefore adsorption via ionic groups, was indicated for that system, as discussed below.

Lipid coated droplets gave mobility–pH curves to be expected on the basis of their ionogenic groups, although adsorbed lecithin and cephalin did not show the same isoelectric points as their phosphatidyl choline and phosphatidyl ethanolamine groups would predict them to have.

V. Adsorption of Proteins at Solid–Liquid Interfaces

The adsorption of proteins at solid–liquid interfaces is important not only in the handling of solutions *in vitro* (e.g., electrophoresis, chromatography, simple adsorption to the containing vessel) but also in *in vivo* situations where foreign materials are in contact with physiological fluids, such as blood. The adsorption of proteins is of particular importance in relation to surface-induced blood coagulation, which is in turn one of the crucial unsolved problems in the use of artificial organs, such as vascular prostheses, heart valves, and artificial hearts and kidneys.

Since aqueous systems are overwhelmingly the most important systems when dealing with proteins, the discussion in this section will be limited to water as solvent.

Protein adsorption at solid–solution interfaces is of course amenable to the general theoretical treatments developed for solid–solution systems. These in turn have developed from corresponding treatments for gas-solid adsorption. The best known and most successful equations in describing the adsorption isotherms found in practice are the Freundlich equation

$$\frac{x}{m} = kC^h \tag{1}$$

and the Langmuir equation

$$\frac{x}{m} = \frac{aC}{1 + bC} \tag{2}$$

where x is the quantity of solute adsorbed at the equilibrium concentration C by weight m of the solid, and h, a, and b are experimental constants.

The Freundlich equation was originally empirical but can be derived from the Gibbs adsorption equation for the case of dilute solutions (*127*). The Langmuir equation also fits a large amount of experimental data, usually over a more extended concentration range, since it allows for approach to a limiting value of adsorption (a/b) at higher concentrations. This equation can be derived from kinetic considerations (*127*) by assuming a steady-state surface concentration at which the rate of condensation

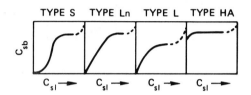

Fig. 2. Fundamental types of isotherm for adsorption from solution. C_{sb} = equilibrium concentration of adsorbed solute in substrate; C_{sl} = equilibrium concentration of solute in external solution. [Reprinted from Ref. (*49*), p. 457, by courtesy of Butterworths Scientific Publications.]

(adsorption) is equal to the rate of evaporation (desorption) from the surface. The rate of condensation is assumed to be proportional to the remaining unoccupied surface area, whereas the rate of evaporation is given by an Arrhenius expression involving an energy barrier.

It should be noted that in the case of solutions, adsorption of solute involves a corresponding desorption or displacement of solvent from the surface, so that in effect a competition exists between the two components for surface sites. In gas adsorption, on the other hand, only a single adsorbate component is present. Consequently, the Langmuir and Freundlich equations break down at high concentrations of solute. In the case of proteins, however, we are normally dealing with relatively dilute solutions of the order of several milligrams per 100 ml, and in this region, as we shall see, the Freundlich or Langmuir isotherms are quite often obeyed. A discussion and classification of isotherm shapes has been given by Giles and MacEwan (*49*) who have distinguished the four types shown in Fig. 2. In type L (Langmuir), by far the commonest type encountered in practice, the flat portion usually corresponds to monolayer formation.

A. Adsorption to Glass Surfaces

Glass has been the most extensively studied solid substrate for protein adsorption partly because of the realization that very dilute protein solutions could behave anomalously and exhibit unexplained changes in concentration when prepared in glass vessels (50). The most extensive contributions to the subject have been made by Bull and co-workers in a series of papers dating from 1938 (51). A study was made, for example,

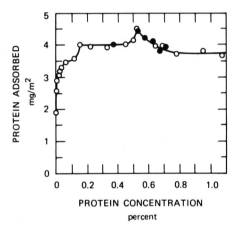

Fig. 3. Plot of milligrams of protein adsorbed per square meter of Pyrex glass at pH 4.66 as a function of the protein concentration expressed in percent. Closed circles indicate adsorption from solution of glass initially exposed to 0.416% protein. [Reprinted from Ref. (52), p. 466, by courtesy of *Biochim. Biophys. Acta.*]

of the adsorption of bovine serum albumin to powdered Pyrex glass, using the simple technique of measuring the decrease in solution concentration of the protein (52). The isotherm at pH 4.66 was of the Langmuir type with a plateau region beginning at solution concentrations above 0.2% (see Fig. 3). The plateau of adsorption corresponded to about 0.5 μg/cm² and was assumed to be associated with the formation of a complete monolayer. Adsorption was also dependent on pH with a "flat" maximum occurring at pH 4–5. The adsorption was not completely reversible and desorption showed evidence of hysteresis and a dependence on pH which mirrored that shown by adsorption. The calculated thickness of the monolayer in the optimum pH range of 4–5 was about 30 Å, decreasing to about 10 Å at pH below 4. These two thicknesses correspond to the dimensions of the native protein in solution and of a monolayer of protein at the air–water interface, respectively.

Substantially similar behavior was found for the adsorption of egg albumin to glass powder (53) (see Fig. 4) with a limiting surface concentration of about 0.4 μg/cm². Complete desorption of these films was again difficult and about 30% of the adsorbed material remained firmly attached to the surface through repeated washings with buffer solution or

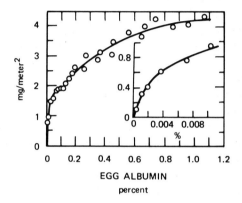

EGG ALBUMIN
percent

Fig. 4. Adsorption isotherm of egg albumin on Pyrex glass powder at 30°C, pH 4.6 in 0.05 ionic strength sodium acetate buffer. Insert graph has expanded scales to accommodate low protein concentrations. [Reprinted from Ref. (53), p. 103, by courtesy of *Arch. Biochem. Biophys.*]

water. It was further noted that these residual films adsorbed additional protein when immersed in fresh solutions. This additional amount was approximately equal to the quantity desorbed from the original layer, thus indicating that the desorption was not followed by subsequent spreading of the residual protein to occupy the total area.

The mechanism of adsorption and the mode of binding of these proteins are not by any means clear from Bull's experiments. Maximal adsorption near the isoelectric point would seem to suggest a predominantly nonionic mechanism, and the partial reversibility in both systems indicates different types or degrees of binding, with some molecules more firmly bound than others. This difference may depend on the existence of different types of surface site or on the precise part of the protein that collides with the surface.

More recent studies of Bull and co-workers on glass surfaces were concerned with the electrophoresis of adsorbed proteins (48, 54, 55) and with the enzyme activity of adsorbed ribonuclease (*56*). Both of these topics are related to the structure of the protein at the surface and consequently to the mechanism of adsorption.

These studies indicated that the mobilities of dissolved and adsorbed proteins (bovine serum albumin and egg albumin) were substantially equal under conditions where the mobility of the adsorbed protein would be expected to be greater by a factor of about 3 if the radii were assumed equal. It was concluded that the unexpectedly low mobilities of the adsorbed protein are due to its reduced charge density under given conditions of pH and ionic strength, suggesting that the ionizable groups of the proteins are in contact with the glass, leaving a relatively uncharged surface exposed to the solution. Since glass is negatively charged at the pH of these experiments, it is expected that the positively charged groups of the protein are involved in the binding. This suggestion of an ionic adsorption mechanism is somewhat at odds with the conclusion from the solution depletion experiments (*52*) discussed above, which tended to show from the maximal adsorption near the isoelectric point that adsorption is not predominantly ionic and might therefore involve any of the other possible processes, such as hydrogen-bonding and dispersion forces.

In this connection, recent work by Messing (*57*) has shown that the initial (first 20 min) adsorption of various proteins (including ribonuclease, cytochrome *c*, chymotrypsin, and pepsin) to porous glass surfaces increases as the isoelectric point increases. Since proteins of higher isoelectric point have a higher density of positive charge (mainly protonated amine groups), Messing deduced that the initial fast adsorption involves formation of ionic bonds between the amine groups and the dissociated silanol SiO^- groups on the glass surface. Subsequent adsorption was inversely related to the molecular weight of the protein, suggesting diffusional control of entry into pores. Desorption as in Bull's experiments proved difficult, and only a combination of acid and urea was able to elute the protein completely. This was taken as an indication that both hydrogen bonding and ionic bonding are involved in the adsorption of these proteins to glass. It would seem that a study of the thermochemistry and temperature dependence of these reactions, from which values of heats of adsorption and estimates of bond strength could be obtained, would be of immense value in settling the question of the adsorption mechanism. From the available evidence, this mechanism certainly seems complex.

Again, Chatoraj and Bull (*54*) studied the electrophoretic mobility of adsorbed protein as a function of the solution concentration used in the adsorption experiment. They found that the mobilities of human serum albumin and egg albumin adsorbed on glass, Nujol, and solid paraffin wax particles exhibit two concentration maxima at pH values on the acid side of the isoelectric point (see Fig. 5). The maxima correspond to points of inflection in the adsorption isotherms and are attributed to structural changes in the adsorbed layer. The first maximum corresponds to the completion of coverage by a "spread" layer when the solution

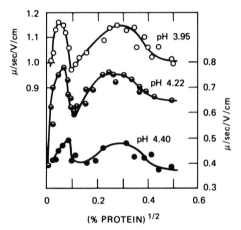

Fig. 5. Mobilities of Pyrex glass particles as functions of the square root of percentage of bovine serum albumin concentration. Left scale pH 3.95; right scale pH 4.22 and 4.40. Sodium acetate buffers ionic strength 0.05 at 25°C. [Reprinted from Ref. (*54*), p. 5129, by courtesy of *J. Amer. Chem. Soc.*]

concentration is still relatively small ($< 0.01\%$). In this layer the protein molecules are to some extent uncoiled and there is increased exposure of ionogenic groups to the solution. As the concentration increases, more protein is adsorbed on a given area, resulting in partial compression of protein molecules and a reversion to native form. Charged sites are thus buried, and electrophoretic mobility reduced. The second maximum, at about 0.1% concentration, is postulated to correspond to completion of the compression process and formation of a compact monolayer of globular protein molecules. The charge density of a completely covered surface would probably be the same whether the protein is spread or globular, an assumption borne out by the very similar mobilities at the two maxima. The decrease in mobility beyond the second maximum is tentatively attributed to a reorganization of the molecules in such a way as to conceal charged sites within the layer. The change involved is likened to the transition from a crystalline to an amorphous layer. Interestingly, it was found that the lower concentration maximum (but not the higher one) is exhibited in the reverse process of desorption, so evidently "crystallization" of the layer is not reversible.

It should be pointed out that the initial work on electrophoresis (*48*) was carried out at a fixed concentration of 1%, which is well beyond the second maximum and presumably in the range of the "amorphous" native layer.

TABLE 1

ISOELECTRIC POINTS OF ADSORBED RIBONUCLEASE[a]

Surface	Buffer	pI
Nujol	Michaelis	7.8
Nujol	Michaelis	7.95
Paraffin	Michaelis	>9.10
Paraffin	Tris	6.00
Paraffin	Acetate	5.22
Glass	Acetate	4.85
Dowex 50	Acetate	4.85
Dowex 2	Tris	8.54
Free solution	—	8.95

[a] Ionic strength = 0.05; data taken from (*55*).

As another approach to obtaining information on the mechanism of these interactions, Bull and Barnett (*55*) studied changes in the isoelectric points of proteins upon adsorption. Results for ribonuclease adsorbed on ion exchange resins (both cationic and anionic) as well as on glass and neutral paraffin surfaces are shown in Table 1. It is seen that in all systems studied by these authors, pI is shifted from its value of 8.95 in free solution.

The direction of the shift (except in the single case of paraffin particles in Michaelis buffer) is toward a lower value of pI. Such a shift would be expected if the interactions are predominantly ionic and result in a disproportionate exposure of negative charge to the solution. Adsorption to glass or to a cation-exchange surface such as Dowex 50 should, and in fact does, lead to this behavior. However, shifts of a similar direction and magnitude are observed for paraffin surfaces in acetate and tris buffers and, for these neutral surfaces, ionic interactions would not be expected. The confusion is further compounded if one observes that when the buffer anion is of the Michaelis type, i.e., for the paraffin system, pI is shifted in the opposite direction. These perplexing results might lead one to infer that the proteins are denatured to varying extents at some surfaces, thereby leading to apparent inconsistencies in surface charge distribution. At any rate, no consistent picture of ribonuclease adsorption emerges from these pI studies.

A final observation on this system by Barnett and Bull (55) is that ribonuclease enzymatic activity is not impaired by adsorption to glass, Dowex 50, or Dowex 2, and that the optimum pH for activity is the same for both the free and adsorbed enzyme. This result suggests that the active site of ribonuclease is not involved in binding to these various surfaces and seems also to rule out any extensive denaturation as a concomitant of adsorption.

Other studies of the adsorption of ribonuclease on glass were carried out by Shapira (58) and by Hummel and Anderson (50). These authors were primarily concerned with unpredictable results in assays for ribonuclease, which were apparently caused by adsorption to the surfaces of glass apparatus. Shapira found that this was indeed the case and that enzyme "lost" in a succession of assays could be quantitatively recovered by elution with electrolyte solutions of minimum ionic strength, but not by water. These results indicate an ionic binding mechanism and, in agreement with Bull, suggest that surface denaturation is not sufficient to impair enzyme activity seriously.

The more systematic and quantitative study of Hummel and Anderson (50) revealed that adsorption of ribonuclease to glass is very strong, the surface becoming saturated at concentrations as low as 0.5 mg%. As shown in Figs. 6 and 7, adsorption is a maximum at a pH of about 8 and decreases monotonically with increasing ionic strength. Ionic binding between the protein and glass is again indicated. The apparent contradiction that adsorption does not increase steadily with decreasing pH is explained by the additional finding, shown in Fig. 8, that the adsorption isotherm reaches a lower plateau at the lower pH values. The ionic mechanism is not necessarily being replaced by some other mode of adsorption. It is more likely that the protein molecules themselves are changing shape just prior to, or instantaneously upon, adsorption. The isotherm forms certainly suggest monolayers at both pH values, and the calculated areas per molecule indicate that at pH 7.5 the layer is a compact array of native globular molecules, whereas at pH 5.5 the molecules are probably to some extent spread and therefore denatured. The behavior

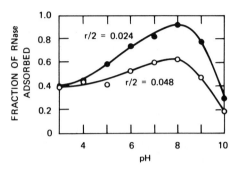

Fig. 6. The influence of pH upon the adsorption of RNase (19.8 μg/ml) on glass (40 mg/ml) at ionic strengths of 0.024 (●) and 0.048 (○). The buffers and their pH were as follows: formate, pH 3.0 and 4.0; acetate, pH 5.0; cacodylate, pH 6.0 and 7.0; Tris, pH 8.0 and 9.0; glycine, pH 10.0. [Reprinted from Ref. (*50*), p. 445, by courtesy of *Arch. Biochem. Biophys.*]

at lower pH could also be due to a decrease in the number of negative surface sites, but the data of Chatoraj and Bull (*54*) indicate that this effect is too small to explain quantitatively the decrease in adsorption.

Interestingly, it was also found by Hummel and Anderson that adsorption is inhibited to some extent by pyrophosphate, which is known to complex with the active site of ribonuclease. This implies either that the

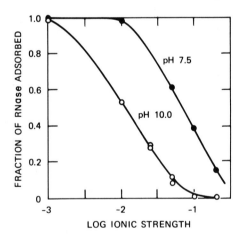

Fig. 7. The adsorption of RNase on glass as a function of log ionic strength of Tris, pH 7.5 (●) and glycine, pH 10.0 (○). The same concentrations of RNase and powdered glass were used as in Fig. 6. [Reprinted from Ref. (*50*), p. 445, by courtesy of *Arch. Biochem. Biophys.*]

pyrophosphate is preferentially adsorbed or that the active site of ribonuclease is involved in its adsorption. If the latter is true, then the enzyme–glass complex should not exhibit activity, a conclusion which has not been confirmed. Barnett and Bull (55) and Shapira (58) were able to recover enzyme activity by elution, but the enzyme–glass combination was not studied.

In other enzyme–glass surface studies, Hartman, Bateman, and Edelhoch (59) investigated the differences in electrophoretic mobility between

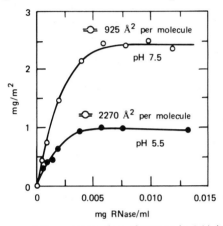

Fig. 8. The influence of the concentration of RNase in 0.01 ionic strength buffer (acetate pH 5.5 ●; Tris, pH 7.5 ○) on the RNase adsorption (expressed as mg/M²) on powdered glass (40 mg/ml). [Reprinted from Ref. (50), p. 446, by courtesy of *Arch. Biochem. Biophys.*]

dissolved and adsorbed trypsin. Such differences were attributed by these authors to a surface-catalyzed lytic reaction, which produces "split-products" that are enzymatically inactive and of different electrokinetic properties.

Reed and Rossall (60) investigated the adsorption of radio-iodinated human serum albumin to glass surfaces. Their study is noteworthy for the novelty of the experimental technique. Measurements were carried out in a Geiger–Muller counter of the liquid annulus type, and an increase in the counting rate could be correlated with adsorption on the glass walls of the counter. The Freundlich isotherm was followed and adsorption was maximal near the isoelectric point. These authors cast some doubt on the applicability to uniodinated proteins of conclusions drawn from studies with iodinated proteins. Obviously, the introduction of iodine in the position *ortho* to the phenolic hydroxyl group of tyrosine residues (the most probable iodination reaction) could change the properties in general and surface behavior in particular. This factor will be alluded to in more detail in the discussion of adsorption to polymer surfaces using radio-labeled proteins of other than biosynthetic origin.

In addition to the techniques already discussed, i.e., solution depletion, electrokinetic property changes, radiolabeling, and changes in enzyme activity where applicable, electron microscopy has been used to study proteins adsorbed on glass. The dimensions of most proteins fall well within the limitations of resolving power of modern microscopes (5–10 Å), and the technique would seem to be an obvious one for observing such morphological alterations as may occur when proteins are in contact with surfaces. Much work has been done on determination of the shape of isolated molecules, notably by Hall (61) and by Valentine (62). These authors studied proteins whose solutions had been sprayed in fine droplet form onto the surface of microscope sample grids. Providing the droplets are small enough and solvent evaporation is fast enough, the molecules should be deposited rather than adsorbed and their morphology should not be influenced by solid–solution interfacial forces.

Hall used the metal shadow/replication technique to increase electron scattering power and therefore contrast. The well-known triple dumbbell form of fibrinogen was deduced from such studies (63).

Valentine adapted the technique of negative staining or negative contrast as an aid to visualization. In this technique the molecules are embedded in a stain material of high electron opacity, such as phosphotungstic acid, which surrounds but does not penetrate them. Micrographs then show the molecules as light objects in a dark background, the converse of the metal shadow/replication technique. From studies using this method, Valentine (62) concluded that many protein molecules, which had been deduced from sedimentation measurements to be cigar-shaped ellipsoids with high length-to-width ratios, were in fact spherically symmetrical and in some cases, such as γ-globulin, actually spherical. It is still a matter of controversy whether the 70-Å diameter sphere or the 235×44-Å ellipsoid (64) is the correct shape for γ-globulin. Valentine contends that disagreement between his direct observations and the deductions from ultracentrifuge data may be taken as an indication of the inapplicability of classical hydrodynamic theory to species of molecular dimensions.

It should be noted that the negative contrast technique is not applicable to proteins such as serum albumin, which are internally hydrated and therefore susceptible to penetration by the stain.

Both methods of improving image contrast have been used by the authors to study the adsorption of plasma proteins to glass and plastic surfaces (*65*). The negative contrast work with plastic surfaces will be discussed in Sect. V.B.1. The most suitable glass-type substrate for studying adsorption by metal shadow/replication procedures was found to be polished silica, which gave micrographs whose homogeneity was limited only by the grain of the shadowing metal (usually platinum). Figure 9 shows a typical replica of a clean silica surface on which poly-

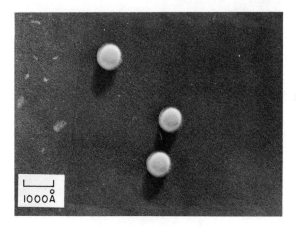

Fig. 9. Polystyrene latex on polished silica. Iridium shadowed.

styrene latex spheres were deposited as a calibration marker and focusing aid. Figure 10 shows a typical micrograph of γ-globulin adsorbed on silica from a 16 mg/100 ml solution. At lower concentrations, occasional rod forms were observed (Fig. 11) and although these were not reproducible, it is tempting to identify them as bundles of uncoiled peptide chains resulting from surface denaturation. The more usual pebble grain surface at the higher concentration is interpreted as a compact monolayer of native molecules. Individual molecules are not easily distinguishable, but the "units" of the layer appear as spherical segments with an average diameter of about 100 Å. The depth of the layer was estimated by double deposition of marker spheres, before and after protein adsorption. Measurement of the shadow length differences of the two groups of spheres yielded a thickness of 240 Å. A monolayer is thus indicated if we assume the dimensions of Oncley et al. (*64*) or a triple layer if Valentine's dimensions are correct. The uniform character of the layer elements would certainly indicate retention of native form. Studies in the pH range 6.4 to 7.4 failed to reveal any effect of pH on the appearance of the layer.

The mechanism of adsorption suggested by these studies is that the first few molecules reaching the surface may be distorted by interfacial forces, causing uncoiling to either extended chain or helical form. The

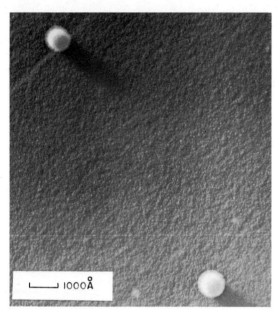

Fig. 10. γ-Globulin adsorbed on polished silica. 0.16 mg/ml, 10-min exposure.

arrival of subsequent molecules causes the rapid filling of the surface, and further denaturation is prevented by lack of free area.

In summarizing our knowledge of the interactions at the glass–protein solution interface, it may be said that there exists a high affinity between glass and proteins and that the surface very quickly becomes saturated. However, the interaction would not seem to be seriously damaging, especially at pH values near neutral or within 2 units of the isoelectric point. At more acid pH values, there is some evidence of formation of thin layers which are suggestive of denatured protein. No evidence exists of a dynamic equilibrium where proteins in the layer would be continuously exchanged with those in solution and released in denatured form. The observation that enzymes can be desorbed without loss of activity militates against this possibility.

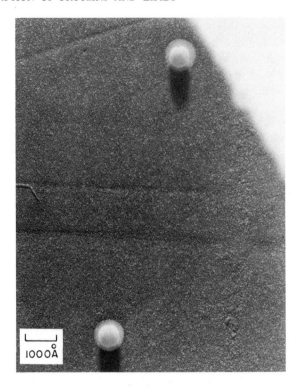

Fig. 11. γ-Globulin adsorbed on polished silica. 0.006 mg/ml, 10-sec exposure.

B. Adsorption to Polymer Surfaces

Prior to about 1960 there is little published work on this class of substrate, but a recent upsurge of interest in the possibility of finding a synthetic material compatible with blood has focused attention on the importance of protein adsorption in this connection. Consequently, a number of investigators have turned their attention to this problem. Since polymer films represent a group of materials of widely divergent surface properties ranging from completely inert and hydrophobic to electrically charged, hydrophilic, or chemically reactive, a further subcategorization is necessary for a rational coherent discussion.

1. Hydrophobic, Neutral Surfaces

These are simple materials, which are not penetrated or even wetted by water and possess no permanent net electric charge either actual or potential (excluding the ζ-potential possessed by all materials in contact with solutions and due to the associated electrical double layer). They are typically represented as a class by such materials as polyethylene and Teflon. Since such materials are electrically neutral, ionic mechanisms of adsorption would not be expected. However, if the proteins are highly charged, as in certain ranges of pH, there may be increasing intermolecular repulsive effects as the surface sites are occupied.

One of the detailed earlier investigations was a study of the adsorption of human γ-globulin to polystyrene latex particles (66) by solution

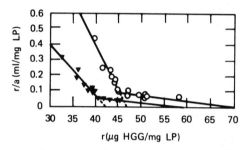

Fig. 12. Adsorption of human γ-globulin to polystyrene latex particles, 8020 Å diameter ○, pH 9.01; ▽, 3.98. [Reprinted from Ref. (66), p. 339, by courtesy of *J. Immunol.*]

depletion measurements using finely divided polymer (0.8μ and 0.2μ diameter spheres). The data followed the Langmuir isotherm as shown for the 0.8μ diameter particles in Fig. 12. This figure is based on a modified version of the Langmuir equation:

$$r/a = KN - Kr$$

which yields linear plots of r/a against r. In this expression, r is the amount of protein adsorbed in μg/mg of polystyrene, a is the equilibrium solution concentration in μg/ml, N is the maximum possible adsorbable amount of protein, and K is a constant. The intercept at $r/a = 0$ yields N and at $r = 0$ yields K.

The interesting feature of the results is the existence of two lines for a given set of conditions. These are taken to be representative of two different processes. The line of greater slope is interpreted as corresponding to formation of an adsorbed monolayer, and values of N for this line are 0.61 μg/cm^2 at pH 3.98 and 0.68 at pH 9.01. These values lead to layer thicknesses of 46 and 50 Å, respectively, whereas the diameter of the γ-globulin ellipsoid is given as 44 Å (64). The layer is therefore very reasonably interpreted as an array of γ-globulin molecules lying flat on the surface in a side-on disposition. The transition point between the lines is taken to be the point of completion of the monolayer.

The second line is open to two interpretations. First, it could represent the formation of a second protein layer. From the relative values of N for the two lines it can be shown that such a second layer does not contain as much protein as the first. If it covers the total area it is probably a spread layer and, if not, it is simply an incomplete second native layer. Alternately, the second line may represent a reorganization of molecules in the native monolayer in such a way as to accommodate more protein per unit area of surface. This reorganization could occur, for example, if additional molecules are adsorbed "on-end" on the free interstitial area between flat-lying molecules. Such an interpretation, though perhaps overdrawn, is again consistent with the extra amount of protein adsorbed and with the available interstitial area. Adsorption is greater at pH 9.01 than at pH 3.98, and at the more alkaline pH the protein probably carries a lower net charge, resulting in less self-repulsion of protein in the layer.

It is interesting to note that these authors were studying adsorbed γ-globulin from the standpoint of its use in the diagnosis of rheumatoid arthritis. In this technique, latex particles with adsorbed γ-globulin are agglutinated by serum from rheumatoid arthritis patients. The mechanism is unknown but the reaction is certainly due to interaction between serum macroglobulins and adsorbed γ-globulin, resulting in destabilization of the suspension either by a surface reaction or by cross-linking to form aggregates.

A more recent study of γ-globulin (IgG) adsorption to polystyrene microspheres by Kochwa et al. (67) emphasized elucidation of configurational changes in protein at the surface. Potentiometric titration was used to detect the exposure of acid-titrable groups; the acquired immunogenicity of the protein–polystyrene complex was also studied. The former approach is similar to that of Chatoraj and Bull (54) in their studies of electrophoretic mobility, inasmuch as it attempts to correlate differences in surface charge with configurational changes in the protein.

Several types of polystyrene were studied, all with particle diameters in the region of 0.2 μ. These materials, known as Tytrons (Monsanto Chemical Co.), are made in negatively and positively charged, as well as neutral, compositions by inclusion of comonomers containing carboxyl and amine groups, respectively. Comparison with the type of polystyrene studied by Oreskes and Singer (66) is therefore of doubtful validity, though the neutral material may be similar.

As with many other similar systems, adsorption was found to follow a Langmuir isotherm with leveling off occurring beyond a certain solution concentration. Data on this aspect of the interaction are most complete for the carboxylated surface, although similar trends are apparent for the other materials. For the carboxylated material, the saturation surface concentration was about 0.23 $\mu g/cm^2$, corresponding to 10,900 $Å^2$/molecule, in excellent agreement with the calculated value of 10,300 $Å^2$ for a monolayer of close-packed molecules side-on to the surface. The values for the other materials are not greatly different, except for those containing amine functions. These latter materials exhibit saturation surface concentrations corresponding to as much as 26,000 $Å^2$/molecule.

The greatest number of titrable groups per molecule formed during adsorption occurred with the neutral material, an observation taken to indicate maximum unfolding at this type of surface. However, the smaller number observed for the carboxylated surface could be due as much to complexing of ionic groups by the surface as to a lesser degree of unfolding. The fact that no titrable groups were formed on adsorption to aminated materials is more difficult to explain in terms of surface configuration and was simply taken by the authors to be a consequence of reduced adsorption on these materials.

For the carboxylated materials, the number of titrable groups per molecule was inversely related to the surface concentration of adsorbed protein, suggesting simply that as the surface fills up the positively charged sites on the protein become less available for reaction with acid. The pH dependence of adsorption, showing a maximum number of titrable groups at pH 6.8 and none at pH 8, was taken as an indication that amine and histidine groups are exposed by the unfolding that accompanies adsorption.

Stronger evidence of surface configurational change came from immunogenicity considerations. IgG molecules bound to the surface acquire immunogenic character akin to that of heat-denatured material. Thus they react readily with serum containing rheumatoid factor and possess enhanced ability to generate antibodies in rabbits. As with the number of titrable groups, immunogenicity decreased with increase in surface concentration.

Kochwa et al. (67) convey the distinct impression of having tacitly accepted the occurrence of surface dimensional denaturation. It seems to the authors that their results do not unequivocally show this, but are merely not inconsistent with such an occurrence. In fact, their data suggest, as do those of Oreskes and Singer (66), that at low surface concentrations there may well be some chain unfolding, but at saturation surface concentrations the layer is composed of globular ellipsoidal molecules.

Perhaps the most comprehensive study of neutral hydrophobic polymer surfaces has been carried out in the authors' laboratory (*68*). We have made a fairly detailed study of the adsorption of albumin, γ-globulin, and fibrinogen to polyethylene, polystyrene, polydimethylsiloxane, and a perfluorinated ethylene/propylene copolymer. The objective was to correlate protein adsorption with the varying tendency of these surfaces to

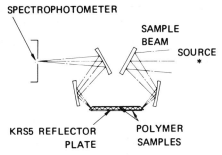

Fig. 13. Schematic of optical arrangement for infrared reflection spectroscopy.

coagulate or thrombose blood in contact with them (*69*). In selecting an experimental approach we were acutely aware of the fact that most previous studies involved depletion methods, which are indirect in the sense that the solution and not the surface is studied. It is always possible, despite the most careful blank experiments, that such indirect methods may produce results due to effects other than that of the surface. In addition, most previous studies have used finely divided powders to achieve high surface-to-volume ratios and particle size may influence behavior.

A technique was sought, therefore, that would directly examine the surface and could be applied to plane surfaces in film form. Infrared internal reflection spectroscopy, developed by Fahrenfort (*70*) and reviewed recently by Harrick (*71*), proved to have these various attributes. Figure 13 shows the principle of the method. The beam of radiation is made to enter a prism of high refractive index at normal incidence so that it strikes the back surface at greater than the critical angle. Under these circumstances the beam is totally reflected. It can then be made to undergo a series of such total reflections along the length of the prism, finally emerging normal to the exit plane at the other end. If a material of low refractive index is pressed against the prism, the beam is strictly speaking not totally reflected but penetrates into the optically rare material to some extent. Thus the beam effectively samples this material at each reflection and, if the emerging beam is then scanned, we observe an infrared spectrum corresponding to that of the material and which is very similar to its transmission infrared spectrum. Figure 14 shows such a spectrum. The solid line is the reflection spectrum of a clean polyethylene surface and the broken line is the same surface covered with a layer of serum albumin.

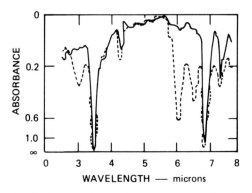

Fig. 14. Infrared reflection spectrum of human serum albumin on polyethylene: (—) untreated polyethylene: (– – –) protein treated polyethylene. [Reprinted from Ref. (*68*), p. 180, by courtesy of *J. Biomed. Mater. Res.*]

The layer was deposited by drying an aqueous solution and should represent an average surface concentration of 10 μg/cm². We can readily see the N-H stretching vibration around 3 μ and the amide I and amide II bands at 6.1 and 6.5 μ respectively.

This spectrum was obtained using a 1-mm thick KRS5 prism with entrance and exit face angles of 45°. The arrangement gives 50 theoretical reflections, and for the average organic material, the penetration is on the order of about 1 μ. Therefore, as the spectrum shows, the beam penetrates the polyethylene substrate as well as the protein. An advantage actually occurs here since one of the most crucial factors in determining the intensity of the spectrum is the degree of contact between the prism and the sample, i.e., the fraction of the area of the sample in absolute contact with the prism. The absorbance of a suitable band in the polymer spectrum can be used as a measure of this area of contact, so that the absorbance of the protein band can be made relative to a standardized contact area.

The technique can thus be made reasonably quantitative by measurement of the ratio of the protein to the polymer absorbance. In this case, we used the strong amide I band at 6.1 μ and the methylene bending vibration at 6.9 μ. For all the protein–polymer systems we have studied, this ratio of absorbances has been found to be a linear function of protein concentration at the surface.

Figure 15 shows that adsorbed layers can also be detected and indicates the adsorption of fibrinogen from a 20 mg% solution to the surface of a

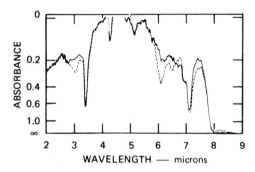

Fig. 15. Infrared reflection spectrum of human fibrinogen on Silastic: (—) untreated Silastic: (– – –) protein treated Silastic. [Reprinted from Ref. (68), p. 181, by courtesy of J. Biomed. Mater. Res.]

medical grade Silastic (polydimethylsiloxane). Again the amide A, amide I and amide II bands are clearly in evidence.

For the various protein–polymer systems studied, this technique was used to measure the quantity of protein adsorbed as a function of solution concentration at pH 7.4. Figure 16 shows the Langmuir-type isotherms obtained for polyethylene. Each protein has its own distinct saturation surface concentration, which appears to increase with protein molecular weight. If these saturation values are treated as monolayers, we can calculate the dimensions of the layers, and these are shown in Table 2. Table 3 shows the layer dimensions calculated from the data of Oncley et al. (64) on the dimensions of native protein molecules in solution. The measured layer thicknesses correspond closely to the diameters of the proteins, whereas the area data correspond more to a close-packed layer adsorbed on end. Attempts to distinguish between these configurations may well not be justified by the limited reproducibility of the data. However, it does seem clear that the adsorbed material does not consist of a single layer of uncoiled protein, since this would have a thickness of

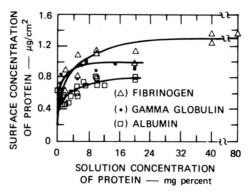

Fig. 16. Adsorption of plasma proteins to polyethylene at 37°C. Δ, Fibrinogen; ●, γ-globulin; $\square$, albumin. [Reprinted from Ref. (68), p. 183, by courtesy of *J. Biomed. Mater. Res.*]

TABLE 2

EXPERIMENTAL DIMENSIONS OF PROTEIN LAYERS ADSORBED ON POLYMER SURFACES[a]

Protein	Layer thickness, Å	Average area per molecule, Å²
On polystyrene		
Albumin	44	2300
γ-Globulin	54	3800
Fibrinogen	130	4000
On polyethylene		
Albumin	62	1400
γ-Globulin	77	2660
Fibrinogen	96	5340
On Silastic		
Albumin	120	720
γ-Globulin	138	1500
Fibrinogen	120	4200
On Teflon FEP		
Albumin	62	1440
γ-Globulin	0	0
Fibrinogen	108	4760

[a] Reprinted from Ref. (68) by courtesy of the *J. Biomed. Mater. Res.*

about 10 Å. We concluded, therefore, that in general, plasma proteins are not dimensionally denatured by adsorption to this type of surface.

Again within experimental error, the adsorption of each protein appears to be similar on all of these surfaces, the γ–globulin–Teflon system (showing no adsorption) being a possible exception. We thus infer that in a mixture, a monolayer would be adsorbed and that its composition would be the same as that of the solution. It would then be expected that from blood or plasma, albumin would be the most abundant protein in the adsorbed layer. Contrarily, Vroman and Adams (72) found that on tantalum and silicon surfaces fibrinogen was preferentially adsorbed from

TABLE 3
DIMENSIONAL DATA FOR PLASMA PROTEINS[a]

Protein	Diameter, Å	Projected area, end-on, Å²	Length, Å	Projected area, side-on, Å²
Albumin	40	1700	115	4,600
γ-Globulin	44	2000	235	10,300
Fibrinogen	65	4200	475	13,000–30,000

[a] Reprinted from Ref. (68) by courtesy of the J. Biomed. Mater. Res.

plasma. Further discussion follows on the relationship between protein adsorption and surface-induced blood coagulation or thrombus formation.

It should also be noted that our conclusions regarding mechanism are similar to those already discussed for glass surfaces; namely, that the proteins are physically but irreversibly adsorbed as close-packed monolayers in the native state. There would appear to be little difference, at least at neutral pH, between the negatively charged glass surface and the neutral polymer surfaces.

Some of these same systems were also studied under conditions of flow when, in addition to adsorption, corresponding changes in optical rotation were measured as a means of following denaturation (73). For polyethylene it was found that when flow was turbulent, the initial rate of adsorption was lower, but the plateau value of surface concentration corresponded to a multilayer. In the laminar flow region, adsorption was indistinguishable from that occurring under static conditions. Denaturation occurred in both flow regions, but the rate in turbulent flow was about twice that in laminar flow. It seems more probable that the flow per se is responsible for denaturation rather than that the surface layer is in a state of dynamic equilibrium with a continuous turnover of molecules undergoing denaturation in the act of desorption.

Additional studies of these systems were made using the electron microscope (74) to determine the behavior of individual molecules, as only
average behavior was obtained from the infrared measurements. This
information is important from the standpoint of activation of clotting due
to the enzyme amplifier character of that process (75). Initial experiments
using metal shadowing showed migration of metal on these low energy,
nonwettable surfaces. A modified negative-staining technique proved

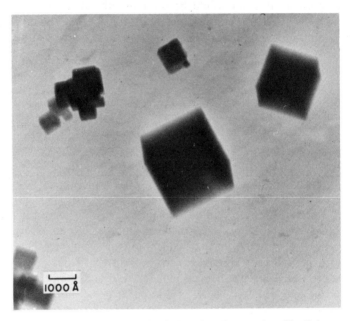

Fig. 17. Teflon FEP negatively stained with phosphotungstic acid. Cubes are magnesium oxide markers.

applicable to the problem. This technique involved coating the polymer/
adsorbed protein surface with a layer of polyvinyl alcohol (PVA). The
dry PVA was then stripped from the polymer (taking along any adsorbed
protein) and the exposed side vacuum-coated with carbon. The PVA was
then dissolved away in hot water, and the exposed side of the carbon film
(bearing the protein) stained by spraying with 2% phosphotungstic acid
solution.

A typical selection of micrographs for γ-globulin adsorbed on Teflon
FEP is shown in Figs. 17–20. In negative staining, the area around the
molecule adsorbs the stain so that the outline of the protein is revealed.

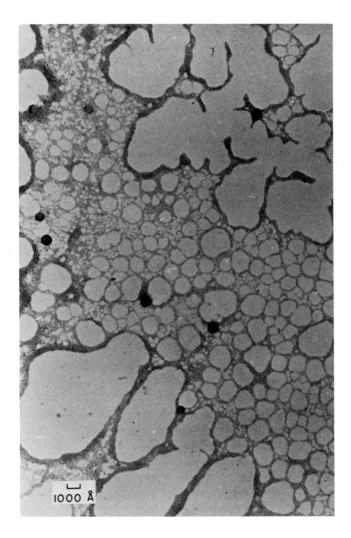

Fig. 18. γ-Globulin (0.16 mg/ml) adsorbed on Teflon FEP negatively stained with phosphotungstic acid.

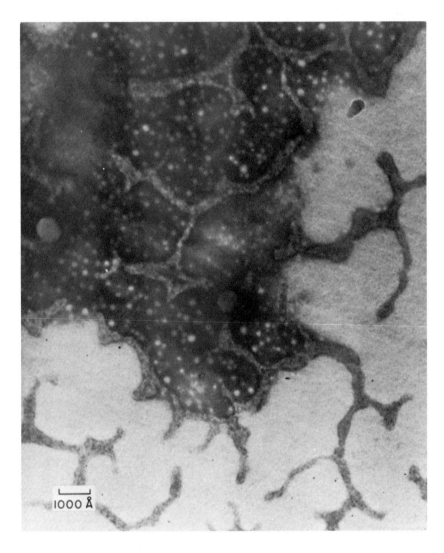

Fig. 19. γ-Globulin (0.16 mg/ml) adsorbed on Teflon FEP negatively stained with phosphotungstic acid.

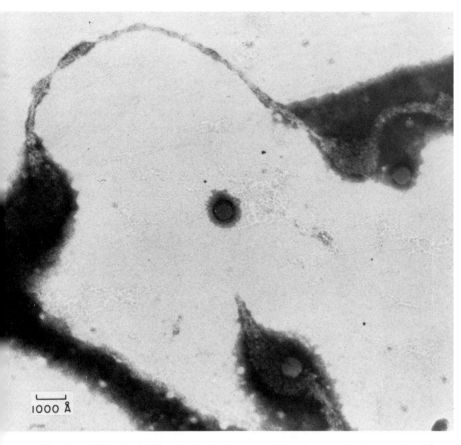

Fig. 20. γ-Globulin (0.16 mg/ml) adsorbed on Teflon FEP negatively stained with phosphotungstic acid.

Thus, dark areas represent stain with protein in the immediate vicinity. Blank experiments in which distilled water was substituted for the protein solution show no structure except for the magnesium oxide crystals used as markers (see Fig. 17). Therefore the structures seen in Figs. 18 to 20 are indeed attributable to protein.

The adsorption pattern of γ-globulin on Teflon FEP resembles a lace-like meshwork with only a fraction of the total surface actually covered by protein (the protein molecules are represented by the small dots within the lace). This would explain the lack of adsorption observed in the infrared results. Within the lace the dots are consistent with compact monolayer formation, and individual molecules are in the form of spheres. From the average center-to-center distance, these spheres can be deduced to be about 70 Å in diameter.

An alternate interpretation, prompted by the larger dots in the areas of excess stain that accumulated adjacent to the lace (see Fig. 19), is that the entire surface is covered by a loose or diffuse layer of protein and that the lace represents the beginning of a second layer. These larger dots (about 200 Å in diameter) indicate perhaps a squashed form of the native protein. Thus the observations of Valentine (62) on γ-globulin sprayed on carbon (assumed to be in the native form) indicate spheres about one-fourth to one-half as big as those in the adjacent areas. This interpretation is also consistent with considerably less than compact monolayer coverage and again would be in line with the infrared results.

Similar results were obtained for polyethylene and polydimethylsiloxane except that the "lace density" was considerably higher, again in agreement with the results of the infrared studies. There would appear to be a basic disagreement between the electron microscope and infrared results as to distribution on the surface. The infrared data, in fact, give no information as to distribution, though the average surface concentrations are suggestive of a uniform monolayer. On the other hand, surface replication procedures could introduce several artifacts in the image finally produced in the electron microscope. Whichever distribution is correct, it can at least be said that there is no compelling evidence of surface denaturation, either extensive or in isolated areas.

Other studies in our laboratory (76) have been concerned with changes in the activity of enzymes adsorbed on polymer surfaces. Acid phosphatase was shown both by infrared and depletion techniques to adsorb to polyethylene and polystyrene, but no activity could be eluted from these materials. Either lack of desorption or enzyme inactivation could have been the cause. Slurries of powdered polymer with adsorbed protein had no activity, however, indicating either denaturation or blocking of the active site by the surface or by adjacent molecules in the adsorbed layer.

In this same context of adsorption-activation of enzyme systems, the adsorption of prothrombin was also studied. The data shown in Table 4 indicate fairly consistent behavior among the various surfaces. The area per molecule lies between the values of 4000 and 1100 Å² expected for undenatured prothrombin adsorbed lengthwise and endwise, respectively. Thus, prothrombin adsorption is similar to that of the more abundant plasma proteins. Since the adsorption was not reversible, no attempt was made to determine the presence of thrombin in eluted material.

Other work on thrombin adsorption has been described recently by Waugh and Baughman (77). Using a depletion technique based on a clotting time assay, these authors determined that the average concentration of thrombin adsorbed at equilibrium to surfaces (such as polymethylmethacrylate, polyethylene, and polypropylene) was 0.077 μg/cm^2 corresponding to surface coverage of about 20%. It is not clear why thrombin should behave differently from other proteins in this respect.

Leininger et al. (78, 79) have used both ζ-potential and radiolabeled proteins to study adsorption to polymer films. For example, when

TABLE 4

ADSORPTION OF PROTHROMBIN TO SEVERAL POLYMER SURFACES
(2-hour immersions in 20 mg% solution at 37°C)

Surface	Surface concentration of prothrombin, μg/cm^2 (average of 4 expts.)	Area per molecule corresponding to 69,000 MW and $\rho = 1.3$
Polystyrene	0.545	2110
Polyethylene	0.474	2430
Silastic	0.636	1810
Teflon FEP	0.553	2080

polystyrene is immersed in fibrinogen solutions, the ζ-potential is shown to change to more positive values, and the change is attributed to protein adsorption. A surface to which heparin (a mucopolysaccharide containing sulfate and sulfonate groups) had been electrostatically bonded exhibited a smaller change in ζ-potential, indicating less protein adsorption in that case. The results of these studies are not interpretable in terms of quantities sorbed per unit area.

Studies with radiolabeled proteins (79), as shown in Table 5, indicated that both heparinized and unheparinized surfaces adsorb very large quantities of protein from solutions of physiological concentration. Of the proteins studied, albumin gave the smallest surface concentration, 3 μg/cm^2, corresponding to about three molecular layers. Other proteins, including Hageman factor—the plasma protein normally associated with surface activation of the intrinsic clotting system (80)—were adsorbed in considerably larger amounts. Thrombin adsorption on a polypropylenestyrene grafted surface was 0.62 μg/cm^2, approximately equivalent to a monolayer. The same material treated with heparin adsorbed 4.3 μg/cm^2 of the thrombin, suggesting that the anticlotting character of heparinized materials might be due in some measure to their ability to remove large amounts of thrombin from the blood.

The multilayers found by these authors are in contrast to the more usual monolayer formation in other solid–solution systems, but it should be kept in mind that the solution concentrations are also high and that the behavior of these proteins may also be modified by the diisopropyl fluorophosphate (DFP) residue introduced in the labeling reaction. DFP is known to acylate a specific serine site on certain enzymes (81).

TABLE 5

ADSORPTION OF PLASMA PROTEINS ON SURFACES[a]

Protein solution (at physiological concentration)	Proteins adsorbed, mg $\times$ 10^3/cm^2		
	Styrene-grafted polypropylene ($\zeta = -16$ mV)	Quaternized polypropylene ($\zeta = +9$ mV)	Heparinized polypropylene ($\zeta = -14$ mV)
Plasma	48.3	7.5	8.3
Albumin	3.0	6.6	2.8
γ-Globulin	19.3	22.4	21.4
Hageman factor	4.5	53.8	14.8
Thrombin	0.62	2.5	4.3

[a] Reprinted from Ref. (79), p. 248, by courtesy of the *J. Biomed. Mater. Res.*

The complexed protein is therefore modified in that a reactive site is blocked and that it possesses an added phosphate residue; its surface behavior thus may be different from that of the native protein.

Vroman (72, 82–87) et al. have made extensive studies of protein adsorption using the ellipsometer to measure film thickness. These studies have dealt with supported hydrophobic surfaces, such as ferric stearate, lignoceric acid, and pure silicon. Much of Vroman's work, however, is concerned with hydrophilic surfaces, and we will reserve discussion for the section dealing with such materials.

2. Hydrophilic Polymer Surfaces

It might be expected that proteins, which themselves are heavily hydrated, would interact less strongly with hydrophilic than with hydrophobic surfaces since if two species are sufficiently hydrated, there is effectively no contact or interface between them. An adsorbing molecule would have to displace water from the surface, but in the case of hydrophilic materials the water is more strongly held, usually by hydrogen bonding. The displacing molecule would therefore have to possess very strong affinity for the surface (on the order of 10 kcal/mole or more).

Apart from glass, already discussed, little work has been done on hydrophilic polymer surfaces. One drawback to quantitative study is that such materials are usually porous to some degree, and the area available for adsorption will vary depending on the size of the molecule. In any case, it will be different from the geometric area and usually unknown.

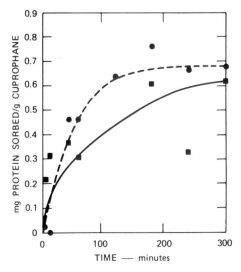

Fig. 21. Sorption of albumin (– – – ●) and γ-globulin (— ■) by cuprophane as a function of time.

Most of the recent research discussed here is of a qualitative character and to a large extent concerned with the relationships between blood clotting and adsorption.

Our own recent efforts have included studies of protein interaction with cellulose in some of its various forms (*88*). The infrared reflection technique can be used to study changes in the amide II protein band at 6.5 μ. Results for cuprophane dialysis membrane (a regenerated cellulose formed by the cuprammonium process) are shown in Figs. 21 and 22.

Sorption of albumin as a function of time is somewhat more rapid than sorption of γ-globulin, although the equilibrium quantities of both proteins are about the same. As a function of concentration (Fig. 22), there is little variation in the quantity of γ-globulin sorbed at solution concentrations between 5 and 45 mg%. On the other hand, sorption of albumin shows a continuous increase in the same range, with no apparent leveling. Thus it would appear that albumin interacts somewhat more strongly than γ-globulin. Similar experiments with fibrinogen indicated zero or negligible sorption of that protein. Thus the sorption of proteins would appear to decrease with increasing molecular weight and eventually to cease for proteins of sufficiently high molecular weight.

In comparing these systems with hydrophobic surfaces, three points should be made. The first, already referred to, is the dependence of

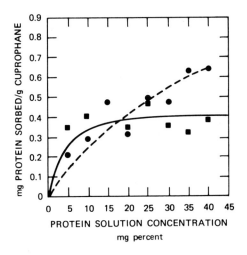

Fig. 22. Sorption of albumin (– – – ●) and γ-globulin (— ■) by cuprophane as a function of solution concentration.

adsorption on protein molecular weight. Hydrophobic surfaces adsorb these three proteins in identical fashion. Second, we can calculate from the data in Figs. 21 and 22 that the equilibrium sorption of γ-globulin corresponds to about 0.5 mg/g cuprophane. If we assume that the area available for sorption is identical to the geometric area of the sample, we can further deduce that equilibrium sorption corresponds to about 0.4 μg/cm^2. By contrast, a typical value of surface concentration on a hydrophobic surface would be about 1.0 μg/cm^2. It is difficult to assess the validity of such a comparison, but if the geometric area of cuprophane is close to the area available for adsorption, then sorption would appear to be less for cuprophane than for a typical hydrophobic surface. Our estimate of 0.5 μg/cm^2 is somewhat higher than the 0.08 μg/cm^2 reported by Salzman et al. (89). However, since the estimate of surface area is questionable, it seems remarkable that the values agree as well.

The third point of comparison concerns the reversibility of sorption. For hydrophobic surfaces, adsorption is generally irreversible (68). On the other hand, albumin and γ-globulin are completely desorbed from cuprophane by immersion in water. These differences all suggest, as anticipated, that proteins interact less with hydrophilic than with hydrophobic materials. Other studies with cellulosics of different pore size (88) suggest that sorption may be a pore penetration process.

Preliminary studies of changes in enzyme activity at cellulose surfaces (76) have shown that thrombin activity may be somewhat reduced but that prothrombin is not converted to thrombin.

Since the discovery by Gott et al. (90) that surfaces coated with heparin exhibit a high degree of thromboresistance, much work has been done both to improve the original heparinized material and to elucidate the modified blood–surface interactions (including protein adsorption) that confer this nonclotting character. These surfaces may be considered appropriately as hydrophilic polymers.

The work of Falb et al. (79) on the adsorption of DF^{32}P-labeled proteins has already been mentioned. More recently, Salzman et al. (89) have reported on heparinized cellulose surfaces, prepared using ethylene imine as the intermediate providing positively charged centers for ionic binding. Using radio-iodinated proteins, they found that fairly low surface concentrations resulted even from solutions of relatively high concentration. For example, a 60 mg% albumin solution deposited 0.099 μg/cm^2 on a heparinized surface and 0.08 μg/cm^2 on an unmodified cellulose control. The authors, nonetheless, felt that these surfaces might owe their thromboresistance to the layer of protein that develops in contact with blood. In what way this layer is different from that formed on most other surfaces that are not thromboresistant (52, 68, 79) is not clear.

It was also found by these authors that surface layers of albumin are uniquely effective in reducing the adherence of blood platelets from heparinized plasma. Fibrinogen and γ-globulin layers, on the other hand, caused an increase in platelet adherence. This result is supported by recent work of Zucker and Vroman (91). Also Packham et al. (92) have shown that platelets in suspension have little tendency to adhere to glass surfaces treated with albumin; with fibrinogen treatment the platelets adhere and with γ-globulin treatment platelets adhere, release constituents, and form aggregates. Work in our laboratory (93) has shown that platelet adherence from fresh whole blood is virtually eliminated when surfaces are coated with plasma or individual protein fractions of whatever identity, including fibrinogen, albumin, hemoglobin, and ribonuclease. These varying results may well be due to the method of applying protein to the surface and/or to the technique of exposing the surface to the platelet-containing medium.

The adsorption of coagulation factor proteins from citrated plasma to various types of hydrophilic polymer surface, including heparinized surfaces, was recently reported by Oja et al. (94). Using what is basically a depletion technique (in which adsorption was taken as equivalent to loss of factor activity in the plasma), various forms of cellulose (including neutral, anionic, cationic, and heparinized, as well as ion-exchange resins of both polarities) were studied. With cationic and heparinized cellulose, significant adsorption of factors II and IX was observed, whereas unmodified cellulose was without effect. This result suggests that factors II and IX are amphoteric proteins capable of complexing with ions of either charge. On the other hand, no significant adsorption of factors was observed with various other anionic materials.

In our discussion so far, we have stressed the fact that proteins are more or less universally adsorbed to all surfaces. In the work of Oja et al. we seem to have examples of surfaces that do not adsorb protein or that are selective in their adsorption behavior. These experiments were conducted in plasma, however, and there is a strong possibility that the various surfaces become quickly saturated with the more abundant plasma proteins. Albumin, for example, present at about 6g/100 ml, is an order of magnitude higher in concentration than fibrinogen (0.3 g/100 ml) and several orders higher than most of the other clotting factors. We could speculate that if all proteins other than the clotting factors were absent from the medium the results would be very different and that large depletions of most factors would be observed.

Nonetheless, it must be acknowledged that the proteins of blood plasma are peculiar in exhibiting selective adsorption. The standard method of removing fibrinogen is to treat with the clay mineral, bentonite. Barium stearate, on the other hand, selectively adsorbs factors XI and VIII according to Vroman (85). These are definitely not concentration-dependent effects and selectivity presumably arises from differences in charge density or distribution.

Recent work of Vroman (72, 86) has been concerned with surfaces containing positively charged ammonium groups. These surfaces, formed by tridodecylmethylammonium chloride (TDMAC) and a polymer of 2-hydroxy-3-methacryloyloxypropyltrimethylammonium chloride (GMAC), were originally designed to complex with heparin and thereby hopefully acquire a degree of nonthrombogenicity (*95*). Using ellipsometry it was shown that GMAC adsorbed a film of 100–150 Å thickness from plasma. The film contained a high proportion of fibrinogen, but globulins and albumin were also present. Heparin-coated GMAC did not adsorb protein from plasma. This result is in contrast to the finding of Falb et al. (*79*) that heparinized surfaces adsorbed as much protein as did the corresponding nonheparinized surfaces. TDMAC surfaces showed similar but less pronounced effects.

C. MISCELLANEOUS SOLID SURFACES

Vroman has conducted extensive studies on the adsorption of blood proteins to a variety of surfaces in an effort to unravel the mechanism of surface activation of blood clotting. Most of this work has been concerned with tantalum and silicon surfaces treated in different ways, to render them wettable or nonwettable. For example, a wettable tantalum surface was produced by oxidation with nitric acid, and a nonwettable tantalum surface by adsorption of lipids such as ferric stearate or lignoceric acid. Wettable silicon was also produced by nitric acid treatment and nonwettable by hydrofluoric acid treatment to remove oxide. All these surfaces are optically suitable for study by ellipsometry.

Initial studies (*82, 85*) were concerned with demonstrating that fibrinogen could be converted to fibrin by adsorbed thrombin. Little difference in its ability either to adsorb thrombin or to be converted at the surface was observed among the various types of surface. Films of thrombin 25–60 Å thick (probably monolayers) were deposited at equilibrium, the exact amount depending on the type of thrombin (see Fig. 23). These thrombin layers, in turn, adsorbed fibrinogen to a depth of 30–55 Å. The quantities of thrombin are about the same as those observed by Falb et al. (*79*) on a styrene-grafted polypropylene surface, which should be similar to the nonwettable surfaces of Vroman. They are considerably greater, however, than the quantities found by Waugh and Baughman (*77*) on polystyrene and other nonwettable polymers.

In a more detailed study of wettable Ta and Si surfaces (*96*), Vroman concluded that a layer of 30–50 Å thickness was adsorbed from plasma and that a subsequent partial desorption occurred spontaneously when coagulation factors XII and XI were present. This finding is in conformity with the notion (*80*) that wettable surfaces initiate blood clotting by adsorption of factor XII, which in turn adsorbs factor XI. The complex is then desorbed in activated form and triggers a series of enzyme–substrate reactions, leading to the final fibrin clot. This subject is dealt with fully in Chapter 10.

Later work on thrombin adsorption (96) showed that ellagic acid, which is a known plasma activator and probably complexes with the polar and charged groups of proteins, caused additional thrombin to be adsorbed. The same was also true for the adsorption of factor XII.

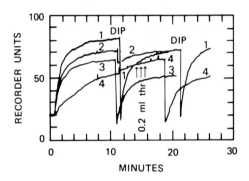

Fig. 23. Adsorption of thrombin topical; recorder tracings. Curve 1: TaW slide, 1 ml 0.4% thrombin added to cuvette at 0 min. Dipping causes second adsorption curve. Curve 2: TaN slide, same treatment. Dipping causes instantaneous increment. Curve 3: as curve 2, but liquid in cuvette covered with plastic film. Instantaneous increment prevented. Curve 4: 0.2-ml amounts of thrombin added at 0, 11, and 14 to 16 min. Upon dipping, new adsorption slope reflects total thrombin concentration. [Reprinted from Ref. (85), p. 476, by courtesy of *Thromb. Diath. Haemorrh.*]

This was taken as an indication that hydrophobic van der Waals bonding is involved in forming the second layer, since the first layer would be oriented with hydrophobic residues facing the solution. The free ends of the ellagic acid-complexed protein molecules would also be hydrophobic.

Pursuing the desorption of protein supposedly under the influence of factors XII and XI, later studies (again on wettable Ta and Si surfaces) were concerned with the possibility that desorption could be caused by the enzyme plasmin rather than or as well as by the release of the factor XII-XI complex (87). Using antihuman fibrinogen serum as a probe, it was shown that fibrinogen is indeed a major component of the protein film adsorbed from plasma and that this surface fibrinogen could be removed by a variety of plasminogen activators. The results suggested that processes involving both factors XII and XI and fibrinogen–plasmin probably occur simultaneously in some kind of concerted process.

Similar to the hydrophobic tantalum surfaces of Vroman is the surface formed by barium stearate deposition on chromium (97). This surface was also studied by optical reflection techniques, and values of refractive index, thickness and surface concentration were obtained for adsorbed films of bovine serum albumin. The thickness of the adsorbed layer was somewhat concentration-dependent. Using 500 mg% solutions, values ranged from 34 Å for distilled water, to about 19 Å for isoelectric acetate buffer, to 15 Å for dilute HCl of pH 3. The difference between distilled water and buffer is attributed to variation in elution behavior in media of different ionic strength rather than to any basic change in the interfacial phenomena. The reduced thickness at the lower pH is similar to that observed by other authors (52) for protein at glass surfaces and is explained as resulting from electrostatic repulsion between adjacent charged molecules. Since refractive index is also reduced in acid solutions and since further uptake of protein results from re-exposure to isoelectric solutions, erosion of the film rather than spreading is envisaged as pH decreases.

Again, those authors emphasize that the different thicknesses are all consistent with monolayers, which can be formed from variously disposed native albumin molecules. No distortion or spreading need be invoked to explain the results.

Values of surface concentration for isoelectric solutions ranged from 0.18–0.45 μg/cm^2, depending on ionic strength. These compare with 0.13–0.45 found by Bull (52) for Pyrex surfaces and to 0.5–1.6 for the hydrophobic polymer surfaces studied in our laboratory (68). These various values are remarkably similar, considering the variations in technique, and may be taken as an indication of the similarity of protein behavior on a variety of surfaces. Thus, due to their combination of amphoteric and hydrophobic character, proteins may be able to show apparently similar behavior on surfaces that are charged in either sense or not at all.

Some work has also been carried out on clay mineral surfaces such as kaolinite (98), bentonite (99), and others (100). The impetus for such studies arises in large measure from the fact that adsorption of proteins in soil is probably intimately involved in soil biochemistry, influencing such factors as protein metabolism by soil bacteria and enzymatic activity.

Kaolinite is a simple clay mineral having the empirical formula $Al_2O_3 2SiO_2 2H_2O$ and has a net surface charge that depends on the pH of the contacting solution. McLaren found (98) that the equilibrium surface concentrations are maximal at the isoelectric point for several proteins, including pepsin, ovalbumin, lactoglobulin, trypsin, and chymotrypsin.

Such behavior is similar to that already discussed for glass. A Langmuir isotherm was followed for the lysozyme–kaolinite system with a slower approach to the plateau at higher pH. The available area of the kaolinite and the area required to form a lysozyme monolayer are in good agreement if the protein is assumed to retain its native globular shape. Elution of lysozyme from kaolinite (as from glass) could only be accomplished by an increase of pH to around 11, or by increasing ionic strength at neutral pH. The eluted protein showed no reduction in enzymatic activity, again suggesting that it is not unfolded. (Reports by others (*101*) have shown that enzymes in the bound state on clays are probably inactive and are thus presumably adsorbed at the active site.) The studies outlined above indicate that adsorption to kaolinite involves ion-exchange with surface metal ions, as well as simple physical adsorption, and they again emphasize the multifaceted nature of possible protein–surface interactions.

Bentonite, with a modified crystal lattice type in which iron atoms are capable of replacing aluminum, is basically similar to kaolinite. As would be expected, therefore, Armstrong and Chesters (*99*) found that lysozyme adsorption to this substrate shows many characteristics similar to those of the kaolinite system, e.g. maximal adsorption at the isoelectric point and significant desorption only at alkaline pH.

Chiang and Chao (*100*) showed that for casein and gelatin adsorption on bentonite is related to cation exchange capacity. (McLaren (*98*), on the other hand, concluded that the relationship between these quantities was only qualitative.) Moreover, these proteins could apparently displace monovalent cations easier than divalent cations and the latter more easily than trivalent cations. Adsorption was thus interpreted as a cation exchange reaction.

Metal surfaces, including nichrome, rhodium, gold, copper, and chromium, were found by Mathot and Rothen (*102*) to adsorb varying amounts of γ-globulin, ranging from 0 on copper to about 80 Å layers on gold. These amounts probably represent various stages of monolayer formation.

A particularly detailed study was made of adsorption to chromium surfaces with and without an electric current flowing between the surface as anode and a platinum wire as cathode. The former process is known as electroadsorption. Layer thicknesses were measured by ellipsometry. The simple adsorption of γ-globulin from a 5 mg % solution to chromium showed three phases as a function of time. First and almost instantaneously, a 30 Å layer was adsorbed, followed by an additional 45 Å at 180 sec. Beyond this time, a continuous slower increase occurred (see

Fig. 24). These increases were proportional to the square root of time and adsorption was therefore postulated to occur by diffusion of protein through a highly structured water layer of high viscosity and low permeability adjacent to the metal surface (*103*). The change of slope at 75 Å was seen as corresponding to completion of a monolayer, and the continuation of thickness build-up beyond this value seemed to imply multilayer formation, in sharp contrast to the behavior of most other surfaces already discussed.

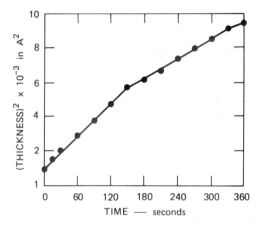

Fig. 24. Adsorption of γ-globulins on a chromium-plated glass slide. [Reprinted from Ref. (*102*), p. 53, by courtesy of *J. Colloid Interface Sci.*]

Under the influence of a current (300 μA for 30 sec), multilayers of about five molecules in thickness are quite definitely formed, and evidence is adduced that γ-globulin is preferentially adsorbed from serum under those conditions. The forces between layers are of both van der Waals and hydrogen-bonding types, and the γ-globulin is shown not to be denatured since deliberately denatured material did not form multilayers. The object of this study was to better define optimum conditions for use of the so-called immunoelectroadsorption method for quantitative study of immunological reactions (*104*). In this method a layer is adsorbed from a carrier solution which may or may not contain antigen, then a second layer is adsorbed from an immune serum. The thickness of the second layer is greater when antigen is present in the first, thus demonstrating that the antigen possesses reactivity in the surface-adsorbed state.

Other works illustrating the adsorption of protein on protein and aimed at obtaining a better understanding of enzymatic reactions at surfaces are those of Rothen (*105*, *106*) and Trurnit (*107*, *108*). In one investigation, Rothen showed that protein–protein interaction occurs across a thin (600 Å) plastic membrane. Thus if layers of either egg albumin or bovine serum albumin were coated on a glass slide and then covered by a protective membrane of formvar, it was found that trypsin (100 mg% solution) could traverse the membrane and hydrolyze the albumin. The greater the number of protein layers on the slide, the faster was the transfer of enzyme, and it was postulated that some long-range interaction (undefined), rather than simple permeation of the membrane under a concentration gradient, is responsible.

Trurnit (*107*, *108*) studied the interfacial enzyme reaction between adsorbed serum albumin and chymotrypsin and found that both adsorption of enzyme and removal of substrate occur simultaneously. Temperature dependence of the rate yielded an Arrhenius activation energy of 8–9 kcal or about half the value for the same reaction in solution, probably reflecting the influence of interfacial forces. It is interesting to note that Trurnit tacitly assumes that the albumin transferred from the air–water interface is surface-denatured, but that so-called "S-layers" formed by adsorption at the solid–liquid interface contain native protein.

VI. Adsorption of Lipids at Solid–Liquid Interfaces

The behavior of lipids at solid–liquid interfaces, as well as at fluid interfaces, is of biological interest because of the model it provides for behavior at cell surfaces. Phospholipids are of particular interest in this connection. Examination of the literature reveals that little work has been done which is concerned directly with lipid adsorption to solids in a biological context. Such adsorption studies as have been done would seem to be a concomitant of investigations of adsorption chromatography, a commonly used technique for separation of the various lipid classes. Again, many studies of fatty acid adsorption to metals have resulted from the interest in these acids as lubricants and corrosion inhibitors. These two areas have been concerned with lipids dissolved in organic solvents and are thus of lesser relevance to biological systems in which lipids exist in an aqueous milieu and therefore are probably present in the form of micelles or other molecular aggregates (see Chapter 2). Nonetheless, some discussion of these areas is pertinent here, since it is possible that interfacial forces may cause micelles to rearrange to independent lipid molecules. Adsorption from simple organic solutions would then provide a model for such behavior.

A. HYDROPHOBIC POLYMER SURFACES

In our laboratory we have investigated the adsorption of colloidal phospholipid to hydrophobic polymer surfaces and found that adsorption increases with the surface free energy of the polymer (*109*). Of the four surfaces studied, polyethylene and polystyrene adsorbed about 1 μg/cm² from 200 mg% lecithin suspensions, whereas silicone rubber and Teflon surfaces did not adsorb any detectable amounts. Adsorption of lecithin to polyethylene bearing adsorbed albumin monolayers (1 μg/cm²) averaged 2 μg/cm², indicating perhaps some ionic interaction in addition to the hydrophobic bonding, which probably accounts for adsorption to the polymers. This result is significant with regard to blood coagulation, since it contrasts with the lack of adherence of platelets to proteinated surfaces. Phospholipids are a major component of all cell membranes, and indeed their micelles can substitute for platelets *in vitro* with respect to clot-promoting activity (*110*). The difference in adsorption to protein surfaces could result either from the fact that the platelet phospholipid surface is completely masked by the outer layer of mucopolysaccharide or that the orientation of phospholipid in the two particles is different. It is of interest to note that a 2μg/cm² deposited layer of lecithin adsorbs 1 μg/cm² of albumin, so the 2:1 ratio of lecithin:albumin would appear to be significant. Furthermore, if we use the micelle model of Robinson (*111*), i.e., discs of about 900 Å diameter and 70 Å thickness, a monolayer of close-packed flat-lying discs would result in a surface concentration of about 0.6 μg/cm² and close-packed, edge-on discs would give about 7 μg/cm². The layers formed in our systems would appear to be somewhere between those extremes.

Other studies of polymer surfaces are those of Halbert et al. (*112*). These studies were primarily concerned with changes in the composition of blood plasma after exposure to various surfaces. Among other variables, changes in lipid and lipoprotein composition were determined. It was concluded that most neutral surfaces are relatively inert to plasma, but certain heparinized surfaces denature the lipoproteins liberating cholesterol, which is then adsorbed into the surface. The active part of the surface is not the heparin but the underlying positively charged quaternary ammonium salt.

B. Chromatographic Substrates

An excellent and very comprehensive review of lipid chromatography on silica gel was published by Wren in 1960 (*113*). Silica gel is the most commonly used adsorbent and is used in chromatographic literature synonymously with silicic acid and silica. It is basically a glass-like substrate used in the form of a highly hydrated powder. Like glass, its fundamental adsorptive property stems from the ability of adsorbate molecules to form hydrogen bonds with the surface silanol groups. This quality is reflected in the so-called "eluotropic series" of solvents in order of their eluting power:

methanol > ethanol > 1-propanol > acetone > methyl acetate >
ethyl acetate > ether > dichloromethane > benzene > toluene >
1,1-dichloroethane > 1,1,2,2-tetrachloroethane > chloroform >
trichloroethylene > carbon tetrachloride > cyclohexane > ligroin

This series should be related to the strength of hydrogen bonding possible between adsorbent and solvent, but will obviously be influenced in a given separation by competing interactions between particular lipids and adsorbent. The order of elution of the various lipid classes is shown in Table 6. Higher homologs of a series are generally eluted before lower homologs, and olefinic double bonds tend to promote stronger adsorption. With proper selection of conditions, separations of high resolution can be achieved. Even better resolution is claimed for a silica substrate impregnated with silver nitrate (*114*). Little information is available on kinetics, energetics, or conformational details of the adsorptions involved.

A more recent discussion of lipid chromatography, which includes helpful information on the nomenclature of polar lipids, is that of Rouser et al. (*115*). Procedures using alumina are not recommended for polar lipids, since adsorption can result in chemical changes; for example, lecithin is converted to lysolecithin on alumina. The mechanism of this transformation is not discussed, but it presumably occurs via preferential adsorption at the acyl group with subsequent deacylation during elution. Other adsorbent materials are discussed, including silicic acid, silicate–silicic acid mixtures and diethylaminoethyl-cellulose. Most of these materials separate by a physical adsorption process and lipids are eluted unchanged. Some ion-exchange processes may also be involved.

Lipid separation and analysis by gas–liquid chromatography can also be effected, although the lipids are usually first broken down to simpler moieties in a preliminary step (*116*). This is fundamentally a gas–liquid partition process and is not within the scope of a chapter on adsorption. Green has recently reported on the adsorption of lecithin to glass (*117*) using ¹⁴C-labeled lecithin. The results showed wide variation in the quantities adsorbed from different organic solvents. Adsorption was

TABLE 6

ORDER OF ELUTION OF LIPIDS FROM SILICIC ACID[a]

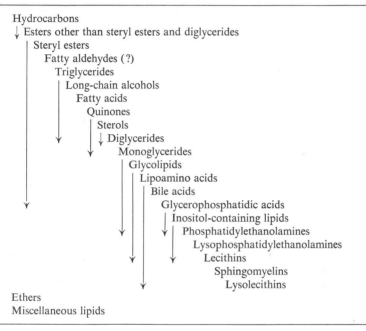

Hydrocarbons
Esters other than steryl esters and diglycerides
Steryl esters
Fatty aldehydes (?)
Triglycerides
Long-chain alcohols
Fatty acids
Quinones
Sterols
Diglycerides
Monoglycerides
Glycolipids
Lipoamino acids
Bile acids
Glycerophosphatidic acids
Inositol-containing lipids
Phosphatidylethanolamines
Lysophosphatidylethanolamines
Lecithins
Sphingomyelins
Lysolecithins
Ethers
Miscellaneous lipids

[a] Reprinted from Ref. (*113*) by courtesy of the *J. Chromatog.*

greatest from dioxane, toluene, and benzene; smaller amounts were adsorbed from hexane, carbon tetrachloride and ethanol, and none from methanol. The solvents that allow the greatest adsorption are those with the lower dielectric constants. Treatment of the glass with dimethyldichlorosilane (rendering its surface more hydrophobic) reduced adsorption from benzene to almost zero, suggesting a possible ionic mechanism with glass.

Zisman has investigated the adsorption of intermediate chain-length fatty acids and alcohols on silica as well as on various metal and metal oxide surfaces (*118*). These studies focused attention on such molecules as so-called autophobic liquids, i.e., liquids that do not spread on their own adsorbed monolayers. The layers are therefore deduced to be oriented with the hydrocarbon part of the molecule away from the substrate. It was concluded that these alcohols and acids (e.g., nonanoic acid) are adsorbed to silica and alumina via hydrogen bonding with a tightly held water layer.

Kipling and Wright investigated the adsorption of stearic acid to silica (119) and found that it adopts an orientation with its major axis parallel to the surface. Reflecting its condition in cyclohexane solution, the acid is probably in dimeric form at the surface. Similar results were obtained for stearic and lauric acids on carbon surfaces (120). In these systems, equilibrium surface concentrations are dependent on the solvent as well as on the type of carbon surface. Carbons that have been heated to high temperature acquire complete monolayers consisting of acid dimers oriented parallel to the surface. This paper emphasizes the dangers of assuming that a plateau in the isotherm or the applicability of the Langmuir equation necessarily corresponds to compact monolayer formation.

C. Metal and Metal Oxide Surfaces

Kipling and Wright (119) also considered the adsorption of stearic acid to alumina and titania. On these substrates the acid adopts a perpendicular orientation with bonding to the surface presumably involving the carboxyl group. The adsorbed quantities are less than close-packed monolayers and this was attributed to bonding at specific surface sites either by hydrogen bonding or chemisorption (aluminum stearate could be detected in desorbed material).

Other studies of stearic acid on metals (121) indicated monolayer adsorption with perpendicular orientation, and again adsorption was thought to be at specific sites on the metal lattice. It was also shown (122) that stearic acid adsorbs to glass and iron with the same packing density corresponding to the perpendicular orientation. This is the same orientation as adopted at the air–water interface near the film collapse pressure. Studies of exchange of radioactive with unlabeled material showed that about 60% of the stearic acid is weakly held and the remaining 40% may be chemisorbed.

De Boer et al. (123) also found that lauric acid is adsorbed in monolayers (corresponding to about 0.13 $\mu g/cm^2$) to aluminum hydroxide and oxide surfaces when the materials are heated to remove water and collapse pores. These data correspond to an area of about 27 $Å^2$/molecule and can be accounted for by assuming that the acid is adsorbed on the octahedral plane of the cubic oxygen lattice and that each acid molecule is associated with four oxygen atoms in that plane. Assumption of the perpendicular orientation is again required.

Smith and co-workers (124-126) studied n-nonadecanoic acid adsorption to freshly machined metal surfaces out of contact with air. They found that monolayer adsorption occurs on metals more electropositive than silver and that this behavior is associated with formation of metal soaps at the surface. The energy of electron emission from the freshly formed surfaces is sufficient to activate the soap-forming reaction. The adsorbed layer was also shown to be in dynamic equilibrium with a solution of the acid; desorption and replenishment is a continuous process.

The adsorption of fatty acids, esters, and alcohols on metals is more extensively reviewed by Kipling (127).

REFERENCES

1. Bull, H. B., Adv. Prot. Chem. 3, 95 (1947).
2. Cheesman, D. F. and Davies, J. T., Adv. Prot. Chem. 9, 439 (1954).
3. Fraser, M. J., J. Pharm. Pharmacol. 9, 497 (1957).
4. Bull, H. B., J. Amer. Chem. Soc. 208, 1078 (1945).
5. Guastalla, J., Compt. Rend. 208, 1078 (1939).
6. Bull, H. B., J. Amer. Chem. Soc. 68, 745 (1946).
7. Block, R. G. and Weiss, K. W., Arch. Biochem. Biophys. 55, 315 (1955).
8. Haurowitz, F., Boucher, P., Dicks, M. and Therriault, D., Arch. Biochem. Biophys. 59, 52 (1955).
9. Blodgett, K. B., J. Amer. Chem. Soc. 57, 1007 (1935).
10. Langmuir, I., Schaefer, V. J. and Wrinch, D. M., Science 85, 76 (1937).
11. Ray, B. R. and Augenstine, L. G., J. Phys. Chem. 60, 1193 (1956).
12. Augenstine, L. G., Ghiron, C. A. and Nims, L. F., J. Phys. Chem. 62, 1231 (1958).
13. Kaplan, J. G. and Fraser, M. J., Biochim. Biophys. Acta 9, 585 (1952).
14. Kaplan, J. G. and Fraser, M. J., Nature 171, 559 (1953).
15. Baier, R. E. and Zobel, C. R., Nature 212, 351 (1966).
16. Malcolm, B. R., Nature 195, 901 (1962).
17. Malcolm, B. R., Polymer 7, 595 (1966).
18. Loeb, G. I. and Baier, R. E., J. Colloid Interface Sci. 27, 38 (1968).
19. Loeb, G. I., J. Colloid Interface Sci. 31, 572 (1969).
20. Kaplan, J. G. and Fraser, M. J., J. Biol. Chem. 210, 57 (1954).
21. Rothen, A., J. Biol. Chem. 168, 75 (1947).
22. Rothen, A., J. Amer. Chem. Soc. 70, 2732 (1948).
23. Nelson, C. A. and Hummel, J. P., J. Biol. Chem. 237, 1567 (1962).
24. Arnold, J. D. and Pak, C. Y., J. Amer. Oil Chem. Soc. 45, 128 (1968).
25. Arnold, J. D. and Pak, C. Y., J. Colloid Sci. 17, 348 (1962).
26. Pak, C. Y. and Arnold, J. D., J. Colloid Sci. 16, 513 (1961).

27. Adam, N. K., *The Physics and Chemistry of Surfaces*, Oxford Univ. Press, England, 1941.
28. Bangham, A. D., *Advan. Lipid Res.* **1**, 65 (1963).
29. Demel, R. A., van Deenen, L. L. M. and Pethica, B. A., *Biochim. Biophys. Acta* **135**, 11 (1967).
30. Demel, R. A., *J. Amer. Oil Chem. Soc.* **45**, 305 (1968).
31. Rosenberg, M. D., *Protoplasma* **63**, 168 (1967).
32. Elbers, P. F., in *Recent Progress in Surface Science*, Vol. II, J. F. Danielli, K. G. A. Pankhurst, and A. C. Riddiford (eds.), Academic, New York, 1964.
33. Stein, W. D., *Movement of Molecules Across Cell Membranes*, Academic, New York, 1967.
34. Robertson, J. D., *Cellular Membranes in Development*, M. Locke (ed.), Academic, New York, 1964.
35. Davson, H. and Danielli, J. F., *The Permeability of Natural Membranes*, 2nd ed., Cambridge Univ. Press, Cambridge, England, 1952.
36. Schulman, J. H., *Trans. Faraday Soc.* **33**, 1116 (1937).
37. Doty, P. and Schulman, J. H., *Discuss. Faraday Soc.* **6**, 21 (1949).
38. Matalon, R. and Schulman, J. H., *Discuss. Faraday Soc.* **6**, 27 (1949).
39. Eley, D. D. and Hedge, D. G., *Discuss. Faraday Soc.* **21**, 221 (1956).
40. Eley, D. D. and Hedge, D. G., *J. Colloid Sci.* **11**, 445 (1956).
41. Eley, D. D. and Hedge, D. G., *J. Colloid Sci.* **12**, 419 (1957).
42. Colaccico, G., Rapport, M. M. and Shapiro, D., *J. Colloid Interface Sci.* **25**, 5 (1967).
43. Colaccico, G., *J. Colloid Interface Sci.* **29**, 345 (1969).
44. Steim, J. M., Edner, O. J. and Bargoot, F. G., *Science* **162**, 909 (1968).
45. Elkes, J. J., Frazer, A. C., Schulman, J. H. and Stewart, H. C., *Proc. Roy. Soc.* **A184**, 102 (1945).
46. Schulman, J. H. and Fraser, M. J., *Ver. Kolloid Ges.* **18**, 68 (1958).
47. Douglas, H. W. and Shaw, D. J., *Trans. Faraday Soc.* **53**, 512 (1957).
48. Bull, H. B., *J. Amer. Chem. Soc.* **80**, 1901 (1958).
49. Giles, C. H. and MacEwan, T. H., *Proc. 2nd Int. Congr. Surface Activity*, Vol. 3, p. 457, London, 1957.
50. Hummel, J. P. and Anderson, B. S., *Arch. Biochem. Biophys.* **112**, 443 (1965).
51. Neurath, H. and Bull, H. B., *Chem. Rev.* **23**, 391 (1938).
52. Bull, H. B., *Biochim. Biophys. Acta* **19**, 464 (1956).
53. Bull, H. B., *Arch. Biochem. Biophys.* **68**, 102 (1957).
54. Chatoraj, D. K. and Bull, H. B., *J. Amer. Chem. Soc.* **81**, 5128 (1959).
55. Barnett, L. B. and Bull, H. B., *J. Amer. Chem. Soc.* **81**, 5133 (1959).
56. Barnett, L. B. and Bull, H. B., *Biochim. Biophys. Acta* **36**, 244 (1959).
57. Messing, R. A., *J. Amer. Chem. Soc.* **91**, 2370 (1969).
58. Shapira, R., *Biochem. Biophys. Res. Commun.* **1**, 236 (1959).
59. Hartman, R. S., Bateman, J. B. and Edelhoch, H. E., *J. Amer. Chem. Soc.* **75**, 5748 (1953).
60. Reed, G. W. and Rossall, R. E., *Radioisotopes Sci. Res., Proc. Int. Conf., Paris*, 1957, Vol. 2, pp. 502–517. See *Chem. Abstr.* **54**, 655c (1958).
61. Hall, C. E., *J. Biophys. Biochem. Cytol.* **2**, 625 (1956).
62. Valentine, R. C., *Nature* **184**, 1838 (1959).
63. Hall, C. E. and Slayter, H. S., *J. Biophys. Biochem. Cytol.* **5**, 11 (1959).

64. Oncley, J. L., Scatchard, G. and Brown, A., *J. Phys. Colloid Chem.* **51**, 134 (1947).
65. Lyman, D. J., Brash, J. L., Chaikin, S. W., Klein, K. G. and Carini, M., *Trans. Amer. Soc. Artif. Int. Organs* **14**, 250 (1968).
66. Oreskes, J. and Singer, J. M., *J. Immunol.* **86**, 338 (1961).
67. Kochwa, S., Brownell, M., Rosenfield, R. E. and Wasserman, L. R., *J. Immunol.* **99**, 981 (1967).
68. Brash, J. L. and Lyman, D. J., *J. Biomed. Mater. Res.* **3**, 175 (1969).
69. Lyman, D. J., Muir, W. M. and Lee, I. J., *Trans. Amer. Soc. Artif. Int. Organs* **11**, 301 (1965).
70. Fahrenfort, J., *Spectrochim. Acta* **17**, 698 (1961).
71. Harrick, N. J., *Internal Reflection Spectroscopy*, Wiley, New York, 1967.
72. Vroman, L. and Adams, A. L., *J. Biomed. Mater. Res.* **3**, 43 (1969).
73. Jirgensons, B., *Arch. Biochem. Biophys.* **39**, 261 (1952).
74. Brash, J. L. and Niemeyer, I. C., unpublished work.
75. Hemker, H. C., Hemker, P. W. and Leoliger, E. A., *Thromb. Diath. Haemorrh.* **13**, 155 (1964).
76. Brash, J. L., Fritzinger, B. K. and Lyman, D. J., unpublished work.
77. Waugh, D. F. and Baughman, D. J., *J. Biomed. Mater. Res.* **3**, 145 (1969).
78. Leininger, R. I., Cooper, C. W., Falb, R. D. and Grode, G. A., *Science* **152**, 1625 (1966).
79. Falb, R. D., Takahashi, M. T., Grode, G. A. and Leininger, R. I., *J. Biomed. Mater. Res.* **1**, 239 (1967).
80. Hardisty, R. M. and Margolis, J., *Brit. J. Haematol.* **5**, 203 (1959).
81. Balls, A. K. and Jansen, E. F., *Advan. Enzymol.* **13**, 321 (1952).
82. Vroman, L., *Nature* **196**, 476 (1962).
83. Vroman, L., Nat. Bureau Standards, Misc. Publication No. 256, p. 335 (1963).
84. Vroman, L. and Lukosevicius, A., *Nature* **204**, 701 (1964).
85. Vroman, L., *Thromb. Diath. Haemorrh.* **10**, 455 (1964).
86. Vroman, L. and Adams, A. L., *J. Colloid Interface Sci.* **31**, 188 (1969).
87. Vroman, L. and Adams, A. L., *Thromb. Diath. Haemorrh.* **18**, 522 (1967).
88. Brash, J. L., Fritzinger, B. K., Klein, K. G., Loo, B. H., Lyman, D. J. and Niemeyer I. C., Annual Report No. 3 to National Heart Institute, Sept. 1969.
89. Salzman, E. W., Merrill, E. W., Binder, A., Wolf, C. F. W., Ashford, T. P. and Austen, G. W., *J. Biomed. Mater. Res.* **3**, 69 (1969).
90. Gott, V. L., Whiffen, J. D. and Dutton, R. C., *Science* **142**, 1297 (1963).
91. Zucker, M. B. and Vroman, L., *Proc. Soc. Exp. Biol. Med.* **131**, 318 (1969).
92. Packham, M. A., Evans, G., Glynn, M. F. and Mustard, J. F., *J. Lab. Clin. Med.* **73**, 686 (1969).
93. Klein, K. G., Lyman, D. J., Brash, J. L. and Fritzinger, B. K., to be published.
94. Oja, P. D., Holmes, G. W., Perkins, H. A. and Love, J., *J. Biomed. Mater. Res.* **3**, 165 (1969).
95. Grode, G. A., Anderson, S. J., Grotta, H. M. and Falb, R. D., *Trans. Amer. Soc. Artif. Intern. Organs* **15**, 1 (1969).
96. Vroman, L., in *Blood Clotting Enzymology*, W. H. Seegers (ed.), Academic, New York, 1967.
97. Bateman, J. B. and Adams, E. D., *J. Phys. Chem.* **61**, 1039 (1957).
98. McLaren, A. D., *J. Phys. Chem.* **58**, 129 (1954).
99. Armstrong, D. E. and Chesters, G., *Soil Sci.* **98**, 39 (1964).

232

100. Chiang, C. M. and Chao, C. H., *T'u Jang Hsueh Pao* **12**, 411 (1964). See *Chem. Abstr.* **65**, 1429a (1966).

101. Mortland, M. M. and Gieseking, J. E., *Proc. Soil Sci. Soc. Amer.* **16**, 10 (1952).

102. Mathot, C. and Rothen, A., *J. Colloid Interface Sci.* **31**, 51 (1969).

103. Davies, J. T., in *Recent Progress in Surface Science*, Vol. 2, J. F. Danielli, K. G. A. Pankhurst, and A. C. Riddiford (eds.), Academic, New York, 1964.

104. Mathot, C., Rothen, A. and Casals, J., *Nature* **202**, 1181 (1964).

105. Rothen, A., *J. Phys. Chem.* **63**, 1929 (1959).

106. Rothen, A., *J. Colloid Sci.* **17**, 124 (1962).

107. Trurnit, H. J., *Arch. Biochem. Biophys.* **47**, 251 (1953).

108. Trurnit, H. J., *Arch. Biochem. Biophys.* **51**, 176 (1954).

109. Brash, J. L. and Lyman, D. J., unpublished work.

110. Wallach, D. F. H., Maurice, P. A., Steele, B. B. and Surgenor, D. M., *J. Biol. Chem.* **234**, 2829 (1959).

111. Robinson, N., *Trans. Faraday Soc.* **56**, 1260 (1960).

112. Halbert, S. P., Anken, M. and Ushakoff, A. E., Proceedings Artificial Heart Program Conference, Washington, D.C., 1969.

113. Wren, J. J., *J. Chromatog.* **4**, 173 (1960).

114. deVries, B., *Chem. Ind. (London)* 1049 (1962).

115. Rouser, G., Kritchevsky, G., Galli, C. and Heller, D., *J. Amer. Oil Chem. Soc.* **42**, 215 (1965).

116. Kuksis, A., Stachnyk, O. and Holub, B. J., *J. Lipid Res.* **10**, 660 (1969).

117. Green, F. A., *J. Lipid Res.* **10**, 710 (1969).

118. Hare, E. F. and Zisman, W. A., *J. Phys. Chem.* **59**, 335 (1955).

119. Kipling, J. J. and Wright, E. H. M., *J. Chem. Soc.* 3535 (1964).

120. Kipling, J. J. and Wright, E. H. M., *J. Chem. Soc.* 855 (1962).

121. Timmons, C. O. and Zisman, W. A., *J. Phys. Chem.* **69**, 984 (1965).

122. Timmons, C. O., Patterson, R. L. and Lockhart, L. B., Jr., *J. Colloid Interface Sci.* **26**, 120 (1968).

123. de Boer, J. M., Houben, G. M. M., Lippens, B. C., Meigs, W. H. and Walrave, W. K. A., *J. Catal.* **1**, 1 (1962).

124. Smith, H. A. and Allen, K. A., *J. Phys. Chem.* **58**, 499 (1954).

125. Smith, H. A. and McGill, R. M., *J. Phys. Chem.* **61**, 1025 (1957).

126. Smith, H. A. and Fort, T., Jr., *J. Phys. Chem.* **62**, 519 (1958).

127. Kipling, J. J., *Adsorption from Solutions of Non-Electrolytes*, Academic, New York, 1965.

6

Bilayer Lipid Membranes: An Experimental Model for Biological Membranes

H. T. TIEN

Department of Biophysics
Michigan State University
East Lansing, Michigan

I. Introduction

This chapter is concerned with a type of interfacial lipid membrane of bimolecular (bilayer) thickness separating two aqueous solutions. As a result of ultrathinness (<100 Å), this type of membrane appears to be "black" when viewed by reflected light; hence, membranes of this type are also frequently referred to as "black" lipid membranes (BLM).

The formation of BLM in aqueous media is of recent origin (1960). Nevertheless, the resemblance of BLM to the postulated model of biological membranes (to be elaborated in the next paragraph) has aroused considerable interest. To date, more than one hundred papers on BLM have been published. A symposium in May, 1967, devoted an entire session to BLM and was published subsequently (1). In the past two years the subject has been reviewed by Castleden (2), Henn and Thompson (3), Rothfield and Finkelstein (4), Bangham (5), Kajiyama (6), and by Tien and Diana (7). Details contained in the previous reviews will only be covered briefly here. The purposes of this paper are twofold: (1) to present an account of the BLM work from its inception with emphasis on experimental principles and main findings and (2) to summarize in detail recent contributions that have not been previously reviewed or published. The contents of this chapter are organized as follows: The first part, after a brief consideration of the BLM and its relationship to the biological membrane, will cover basic aspects of the experimental techniques and the intrinsic properties of BLM; the effects of modifying agents (or modifiers) on the properties of BLM will be considered next; and the last part will be devoted to a discussion of BLM as an experimental model for various types of biological membranes.

A. Natural Membranes and BLM

It is difficult to define precisely what is meant by natural or biological membranes. The popular view held by the majority of investigators is that membranes are a larger part of cell structure. Membranes not only separate the individual cells but they also constitute a great part of its internal structure. It is because of this widespread recognition of the membranous nature of cells and the crucial role of membranes associated with the living state that an increasing amount of research is being devoted to membrane studies.

From a functional point of view, biological membranes are believed to be involved in a large number of vital processes. Some of the important functions served by the membranes are listed in Table 1. Details of

TABLE 1

SOME IMPORTANT FUNCTIONS OF BIOLOGICAL MEMBRANES

(a) Permeability barrier of ions and molecules
(b) Ion accumulation or active transport
(c) Conduction of nervous impulse
(d) Conversion of light into chemical energy
(e) Conversion of light into electrical energy
(f) Oxidative and photosynthetic phosphorylation
(g) Site of immunological reactions
(h) Protein synthesis

various aspects of membrane functions may be found in a recent monograph (8).

In contrast to membrane functions, the basic structure of all biological membranes appears to be unique and possesses a common design consisting of a bimolecular lipid leaflet covered on both sides by a layer of protein or other nonlipid material (9). Deduction of the bimolecular leaflet model was based upon early studies made with the Langmuir monolayer technique and the classical permeability measurements (10).

Historically, the idea of the biological membrane comprising of lipids originated from Overton (11). He found that lipids or lipid-like materials readily diffused across plant plasma membranes. In 1925 Gorter and Grendel extracted the lipids from red blood cells and spread them on a Langmuir trough (12). From an estimate of the interfacial area for the intact red cells used in the experiment and the measured area occupied by the lipid monolayer (which was about twice that of the red blood cells used) Gorter and Grendel made the historic suggestion that red blood cells "are covered by a layer of fatty substance that is two molecules

thick" (*13*). The reader interested in pursuing the development of biological membrane structure from Gorter–Grendel's suggestion up to 1965
is referred to a collection of classical papers selected by Branton and
Park (*14*).

Although our understanding of the structure of biological membranes
has become increasingly sophisticated, due mainly to the use of the
electron microscope, x-ray diffraction techniques, and model systems, it
can be said without much exaggeration that the simple bilayer concept of
Gorter and Grendel has dominated thinking concerning the biological
membrane structure for nearly half a century. The bimolecular leaflet
model has been extended to practically all types of biological membranes
that have been studied. These membranes include the plasma membrane
of erythrocyte, the nerve membrane of axon, the cristae membrane of
mitochondrion, the thylakoid membrane of chloroplast, and the outer
segment sac membrane of retinal rod. Schematic representation of these
basic units of life processes are depicted in Fig. 1.

Historically, the complexity of biological membranes has compelled
investigators to study model systems. Until recently these have largely
consisted of precipitated copper ferrocyanide, gelatine, thick "oil"
layers, collodion, cellophane, ion exchange materials of various kinds,
and lipid-soaked filter papers (*7*). In addition, monolayers at the air/water
and oil/water interfaces have been also employed as model systems (*15*).
The obvious shortcomings of these models are either that they lack the
required thickness of cell membranes or approximate poorly the known
properties of natural membranes and their environment. Further, the
chemical composition of any adequate model should be similar to the
composition of biological membranes, but it is perhaps less obvious that
the model must also possess similar dimensions to its natural counterpart,
i.e., a thickness of about 100 Å or less. This has been the least frequently
met criterion in previous models. From the viewpoint of colloid and
interfacial chemistry, biological membranes may be considered as the
limiting case of a three-phase system, i.e., a lipid or lipoprotein phase
separating two aqueous phases (intracellular–fluid/cell membrane/
extracellular–fluid). As revealed by electron microscopy, this cell membrane phase for a number of biological membranes has a thickness of the
order of a bimolecular lipid leaflet (*9, 16, 17*).

It has been evident for many years that if the bimolecular lipid layer
were indeed the major structural component of biological membranes
knowledge concerning the properties and the formation of such a structure
in vitro would be of both experimental and theoretical significance. It was

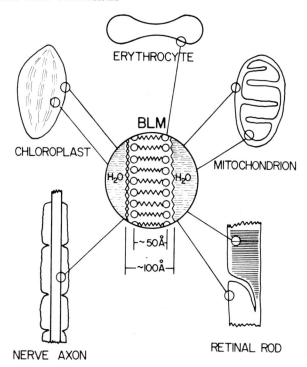

ERYTHROCYTE

BLM

CHLOROPLAST

MITOCHONDRION

H_2O H_2O

$\vdash\sim 50\text{Å}\dashv$

$\vdash\!-\sim 100\text{Å}\!-\dashv$

NERVE AXON

RETINAL ROD

Fig. 1. Diagrammatic representation of the fundamental units of life as they are generally visualized under the electron microscope. The center diagram represents the molecular structure of the five basic types of biological membranes based upon the bimolecular leaflet concept.

also apparent that a detailed physical chemical description of natural membranes would be best approached by studies of simpler, well-defined models owing to the great complexity of living systems.

B. The Formation of BLM

The search for a better membrane model has led to the discovery of a method for the formation of BLM in aqueous media (*18*). It is perhaps of some interest to relate the train of thought leading to the method of BLM formation.

In the late 1950s while Rudin et al. (*19*) were investigating the ion specificity of lipid monolayers and the Langmuir–Blodgett multilayers, two inspiring publications appeared which exerted great influence on the thinking of the early workers and altered the course of their research. The first was a reprinting of Boys' classic book on soap bubbles, which was based on his original lectures given some seventy years earlier (*20*). The second was a volume dedicated to N. K. Adam (*21*) in which A. S. C. Lawrence recounted succinctly the highlights in the development of monolayers, soap films, and colloid chemistry. Lawrence's account in particular, brought into focus the relationship among the various topics. Most significant was the mention of Newton's observation (described in

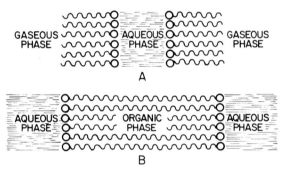

Fig. 2. Orientation of long-chain hydrocarbon molecule containing polar group at the interface between air and water, and between organic solvent and water. Circles denote polar groups and wavy lines represent the hydrocarbon chain of the molecule. (A) Ordinary soap film in air; (B) molecular film in aqueous solution. Diagrams illustrate the limiting structures—the so-called secondary black films. These films are probably only 2 molecules thick with the polar groups oriented either toward each other as in (A) or against as in (B).

detail in Boys' book) of the so-called black soap films. Thus the orientation of molecules in monolayers at the air/water surface as deduced by Langmuir, the thickness of black soap films measured by several investigators, since Newton's time, and the bimolecular leaflet model for the plasma membrane suggested by Gorter and Grendel which was then being confirmed by numerous electron microscopic investigations (*22*), provided the background information for a new experimental approach. It was simply decided to attempt to form a black film in aqueous solution by extending the methods used in black soap film work (*19*). The formation of an underwater black film turned out to be easier than anticipated. Figure 2 illustrates the limiting structure of the two types of black films. According to the terminology of thin film optics, a thin film is a layer with parallel faces whose thickness is of the order of the wavelength of light. A "black" or an *ultrathin* film is about 0.01 of the wavelength of light. Thus BLM properly belongs to the category of ultrathin films (see Sect. II.E for further discussion).

The current status of work on the BLM is summarized in Table 2, where a comparison of known properties of BLM with those of natural

TABLE 2

COMPARISON OF SOME PHYSICAL CHARACTERISTICS OF BIMOLECULAR LIPID MEMBRANES (BLM) WITH NATURAL MEMBRANES

Property	Natural membranes	BLM[a]
Thickness (Å)		
Electron microscopy	40–130	60–90
X-ray diffraction	40–85	—
Optical methods	—	40–80
Capacitance (using assumed dielectric constant)	30–150	40–130
Potential difference (mV) (resting)	10–88	0–140
Resistance (Ω-cm^2)	10^2–10^5	10^3–10^9
Breakdown voltage (mV)	100	100–550
Capacitance (μF/cm^2)	0.5–1.3	0.4–1.3
"Excitability"	Observed	Observed
Excitation by light	Observed	Observed
Refractive index	About 1.6	1.37–1.66
Interfacial tension (ergs/cm^2)	0.03–3.0	0.2–6.0
Water permeability (10^{-4} cm/sec)	0.25–400	8–50

[a] Taken from a review article on BLM with minor changes and additions (7).

membranes is presented. From an inspection of Table 2, it may be seen that the BLM possess certain dimensional, electrical, permeability, and "excitability" characteristics which closely resemble those of biological membranes.

Thus, viewed in the light of our current knowledge of membrane structure and function, the BLM system represents the closest approach to the biological membrane. In later sections the usefulness of the BLM system for further advances in the understanding of the role of biological membranes in permeation, ion selectivity and energy transduction will be discussed.

II. General Considerations

A. Basic Experimental Techniques of Formation

Conceptually, the formation of a BLM in aqueous solution is extremely simple. It involves the creation of two coexisting aqueous–solution/lipid–solution interfaces or a *biface*. Under favorable circumstances thinning of the lipid solution takes place spontaneously and leads to a BLM.

BLMs are generally formed by one of three basic methods. In the original method the membrane is generated literally by brush-painting the BLM-forming solution across a small hole (1–2 mm diameter) in a polyethylene or Teflon cup (5–10 ml capacity) immersed in a salt solution. This simple method is now known as the brush technique (*23*). The second method involves passing a Teflon loop or frame through an oil/water interface (or vice versa). It is useful for studying the formation characteristics and thickness and avoids the transfer of the BLM-forming solution through air. The third method, much less used than the other two, consists of attaching a hypodermic needle to a small syringe filled with aqueous solution, dipping the needle into BLM-forming solution, placing the needle in a beaker containing the same or different aqueous solution, and "blowing a bubble." With this method a spherical BLM of large surface area may be formed under optimal conditions (*24, 25*). For a detailed description of various BLM techniques the reader is referred to the literature (*2–7*) and to (*26*) on techniques of surface chemistry and physics.

B. Composition of BLM-Forming Solutions

BLMs can be formed from a variety of materials; these include brain lipids, proteolipids, phospholipids (e.g., lecithin), *Escherichia coli* lipids, chloroplast extracts, surfactants, and oxidized cholesterol. A liquid hydrocarbon (hexane to hexadecane) is generally required as solvent.

Depending upon the BLM-forming solution used, the membrane can be generated in a wide range of salt concentrations, pH values, and temperatures (5–50°C). That a BLM can be formed from a variety of materials and under quite different conditions suggests that the phenomenon is of general occurrence in nature. This will be elaborated on in a later section (pp. 280–282).

C. FORMATION CHARACTERISTICS AND BLM STRUCTURE

The formation characteristics of lipid membranes are quite similar to those of soap films in air, with two notable exceptions. First, the thinning by evaporation is evidently not possible for lipid membranes in aqueous solution (Fig. 3). Instead, diffusion of solvent and/or solute must play an

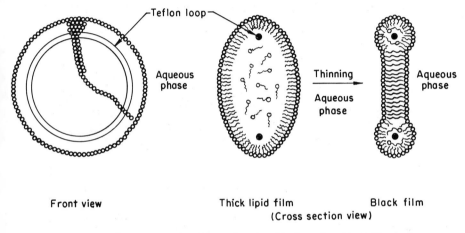

Teflon loop

Aqueous phase

Thinning

Aqueous phase

Aqueous phase

Aqueous phase

Front view   Thick lipid film   Black film
(Cross section view)

Fig. 3. Schematic representation illustrating the thinning of a thin lipid membrane to a lipid bilayer (7).

important role in the thinning processes. Second, the black spots are seen frequently to appear at portions of the membrane other than the edge. Once formed, these black spots grow at a fairly constant rate. When the membrane is sufficiently thick (0.1–1 μ) the structure of the membrane is pictured as being a sandwich consisting of an organic phase between two adsorbed monolayers of lipid molecules. At a distance greater than 1000 Å (about the thickness of a silvery film), the attraction of the lipid monolayers at the two interfaces due to the van der Waals forces is small; the two interfaces are therefore essentially independent of each other. The main process of thinning takes place because of the presence of Plateau–Gibbs borders (see p. 268). However, as the thinning proceeds it seems probable that (for mobile lipid films at least) chance contacts of the hydrocarbon chains of the adsorbed monolayers situated at the opposite interfaces may occur. Although the analogy of behavior between soap films in air and lipid membrane in aqueous solution has been previously invoked, it should be stressed that such an analogy could be entirely misleading. Although both of these systems possess two interfaces (see Fig. 2), the lipid membrane system consists of two condensed bulk phases and the energetics and factors governing the stability of the two systems can be drastically different, in spite of the superficial similarities of optical behavior.

Before discussing the structure of BLM, it is relevant to briefly consider the behavior of lipids (and amphipathic molecules) at an air/water or oil/water interface. The structure of lipids is unique in that one part of the

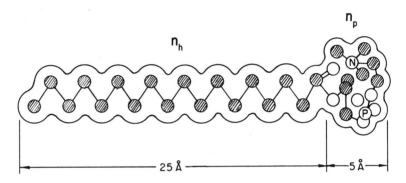

Fig. 4. A schematic drawing of a lecithin molecule with compact polar group at the membrane-water interface. The hydrocarbon chain is about 25 Å. n_h and n_p denote, respectively, the refractive indices of hydrocarbon layer and polar region of the molecule.

molecule is water–soluble (hydrophilic) and the other part is oil–soluble (hydrophobic). For example, a lecithin molecule (phosphatidyl choline, Fig. 4) when introduced at an oil-water interface will spontaneously orient itself with the long-chain hydrocarbon portion immersed in the oil phase. This orientation of lipid molecules at the interface leads to a reduction of interfacial free energy and is of fundamental importance in the formation of molecular aggregates and biological membranes. In Fig. 5 the structural orientations of some important biocompounds at an oil/water interface are illustrated. For further details on the effects of adsorption at interfaces and molecular aggregation in solution, consult Refs. (27–29).

The limiting structure of BLM is shown in Fig. 6. Ordinarily, when a BLM is formed in aqueous solution containing no dissolved proteins, the polar groups of lipid could be adapted to one of the two likely config-uration at the membrane/water interface. The results of the adsorp-tion studies of fatty acids and fatty alcohols at the hydrocarbon–water

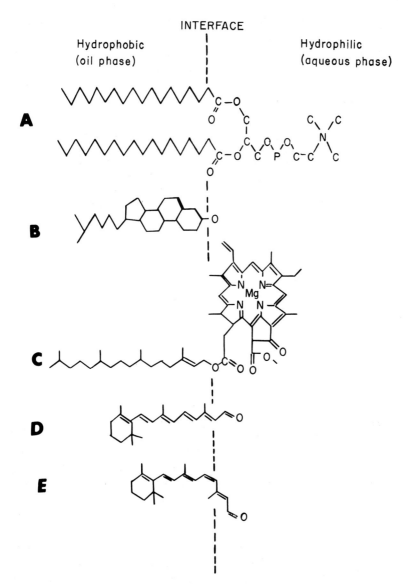

Fig. 5. Orientation of some of the most important biocompounds at an oil–water interface. (A) Phosphatidyl choline (lecithin); (B) cholesterol; (C) chlorophyll *a*; (D) all-*trans* retinal; (E) 11-*cis* retinal. They are shown in an energetically favorable configuration. The dielectric constants of the oil and aqueous phase are approximately 2 and 80, respectively.

interfaces would suggest that the opposite arrangement to that given in Fig. 6 is not favored. This can be shown (very approximately) by the following thermodynamic considerations. The condition for equilibrium is $K = [L_o]/[L_a]$, where L_o and L_a are the concentrations of lipid molecules with the polar groups facing the organic phase and with the polar groups facing the aqueous phase, respectively. Since the ratio of the numbers of

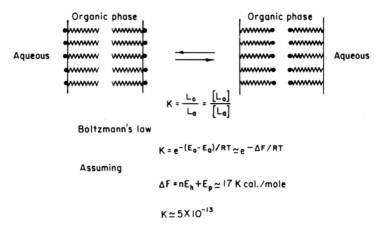

Fig. 6. Estimation of equilibrium constant for the reaction depicted.

molecules of the two orientations is equal to the ratio of their concentrations, we can also write $K = L_o/L_a$. According to Boltzmann's distribution law, L_o and L_a are given, respectively, by $L_o = A \exp(-E_o)/kT$ and $L_a = A' \exp(-E_a)/kT$, where E is the energy for the molecule of a given orientation k is the Boltzmann constant, and T is the absolute temperature. If for simplicity, it is assumed here that $A = A'$, then $K = \exp(-E_o + E_a)/RT = \exp(-\Delta F)/RT$, where ΔF is the standard free energy change. A rough estimate of the magnitude of ΔF for the process may be made by considering the work required to remove the hydrocarbon chains from the organic phase to the aqueous phase. This work may be equated to the energy required for desorption, which has been given by Davies and Rideal (27) to be about 900 cal/mole per methylene group. Similarly, the desorption energy for an $-OH$ group is about 800 cal/mole as estimated by Langmuir (30). If we now assume that the free energy for the process is made up of these two terms and that they are additive [as has been assumed by Rideal for a case of an aliphatic alcohol at the air water interface (27)], this will lead to a value of 17 kcal/mole for ΔF. Therefore, the corresponding K for the process is about 5×10^{-13}. The significance of this crude analysis is that a molecular arrangement contrary to the one depicted in Fig. 2B is highly unlikely.

The arrangement of lipids in BLM illustrated in Fig. 2B is also reminiscent of one of the mesomorphic structures of concentrated soap solutions extensively studied by Luzzati and Husson (31). It has been assumed that BLM adopts the "neat" or smectic mesomorphic form. The polar groups of the lipids face outward in contact with the aqueous solution and the hydrocarbon chains in the BLM are assumed to be in a liquid state. The experimental facts in support of this picture come from the following lines of evidence: (a) The liquid nature of the BLM can be demonstrated by the fact that a BLM may be probed with a fine object (thin wire or hair), which can be moved within the membrane and then withdrawn without rupturing the membrane. (b) The "hydrocarbon-like" interior is deduced from the materials used. (c) Recent water permeability studies have shown that to a first approximation the BLM may be considered a continuous liquid hydrocarbon layer of less than 100 Å thick. The permeability coefficient for water of such a layer would be expected to be about 35 μ/sec as estimated from the solubility and diffusion data (32). (d) The dc resistance of the BLM is about 10^8 Ω-cm^2. This corresponds to a bulk resistivity of about 10^{14} Ω-cm. The resistivity of most wet liquid hydrocarbons is of the same order of magnitude. (e) The dielectric breakdown strength of the BLM is usually in the range of 10^5–10^6 V/cm, which again corresponds closely to that of a long-chain liquid hydrocarbon. Other evidence is provided by spin label experiments using nitroxide free radicals which, when incorporated into a system under study, can be detected by electron spin resonance (33). It has been found that the spin label in a type of lipid vesicle described by Rendi (34) and by Bangham et al. (5, 35) appears to be in a liquid environment (36). In view of these facts it seems reasonable to consider that the interior of the BLM is essentially an ultrathin, continuous layer of liquid hydrocarbon with dissolved water.

D. ELECTRICAL PROPERTIES OF BLM

Experimentally, the electrical properties of BLM are easily measured by placing a calomel electrode on each side of the membrane. The dc method has been most frequently used because of the availability of excellent electrometer amplifiers (26). A simple experimental arrangement is shown in Fig. 7.

For an unmodified BLM formed from either lecithin or oxidized cholesterol in a liquid alkane solvent, the resistance of the membrane is exceedingly large and is greater than 10^8 Ω-cm². A BLM separating two aqueous solutions can be represented by a resistance (R_m) in parallel with a capacitance (C_m). The dc resistance is measured by applying a

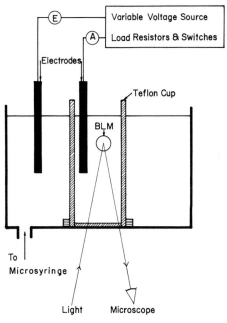

Fig. 7. Arrangement of apparatus used for studying electrical properties of BLM (37).

known voltage through a standard resistor in series with the BLM (see Fig. 8). The result is found by the direct application of Ohm's law. The BLM capacitance (obtained by the well-known dc discharge method (37)) is treated as a parallel plate condenser, i.e.,

$$C_m = \frac{\epsilon A}{4\pi t} \tag{1}$$

where ϵ is the dielectric constant, A is the area, and t is the thickness of the hydrocarbon region of the BLM. Hanai, Haydon, and Taylor (38) carried out a thorough investigation of lecithin BLM and reported a value of

0.38 μF/cm² for the membrane. This value was shown to be independent of applied voltage (up to 50 mV) and frequency between 5×10^{-3} and 5×10^6 cps. It was also shown to be unaffected by the nature and concentration of the surrounding salt solution (NaCl, KCl, and CaCl$_2$). In contrast to the above findings, Rosen and Sutton (*39*) have recently reported that the capacitance was found to be a function of the electrolyte concentration in the bathing solution. Babakov et al. (*40*) have found an

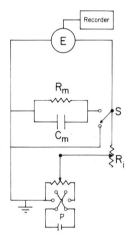

Fig. 8. Circuit diagram for electrical measurements. E = electrometer, R_m = membrane resistance, C_m = membrane capacitance, S = switch, R_i = input resistors, P = voltage divider and switch for applying polarizing potentials (*37*).

increase in BLM capacitance roughly proportional to the square of the potential difference. These results have led Rosen and Sutton to suggest two specific requirements in a model of the BLM: (1) electrical charges must be localized both in the membrane and in the solution rather than smeared out from the interface, and (2) the lipid molecules are not fixed in the plane of the membrane, a suggestion that had been inferred earlier from optical studies (*41*).

The measurement of the BLM capacitance is of special interest and constitutes one of the three methods used at present in assessing the BLM thickness (the other two being optical methods and electron microscopy). By assuming a value for the dielectric constant in Eq. (1), a corresponding "electrical" thickness of the membrane can be calculated. The thickness of the membrane will be considered in later paragraphs on the optical properties of BLM.

In general, current/voltage curves for unmodified BLM are ohmic up to potential differences of about 100 mV. When the impressed voltage exceeds the above value the current of the BLM may change erratically with time, is no longer proportional to the voltage, and results eventually in the rupture of the membrane. The voltage at which the BLM ruptures is called the dielectric breakdown voltage. The breakdown voltage of a BLM usually does not have a definite value but can be measured within ± 50 mV. It is influenced by at least two factors: (1) the concentration of impurities (or modifiers) in the BLM and (2) a change in local conditions such as local heating, interfacial tension change, and thickness fluctuation due to mechanical vibration or other causes. Variations in breakdown voltage have not been systematically investigated. For a BLM formed from egg lecithin and tetradecane dissolved in a chloroform–methanol solution, Miyamoto and Thompson (42) have reported an average value of 198 ± 39 mV. For this particular BLM system the breakdown voltage was found to be independent of the ionic species or concentration of electrolyte used. In the cholesterol–HDTAB (hexadecyltrimethylammonium bromide) BLM system studied by Tien and Diana (37, 43) it was found that the breakdown voltage was dependent both on the electrolyte concentration and the species used. The breakdown voltage of the BLM was 1.3×10^6 V/cm in 10^{-4} M KCl and decreased to 3×10^5 V/cm in 10^{-1} M KCl. It is of interest to note that for a typical insulating oil used in the construction of electrical capacitors and cables the breakdown voltage rarely exceeds 2×10^5 V/cm (44).

The other important electrical properties of BLM are the membrane potential and ionic selectivity. For most unmodified BLM (e.g., formed from either oxidized cholesterol or egg lecithin) in a common electrolyte such as NaCl or KCl, a few millivolts may be detected per tenfold concentration difference. Such BLM are usually poorly ion-selective (cation over anion) and ion-specific (i.e., K^+ over Na^+).

E. THE THICKNESS OF BLM

In the discussion of the structure of biological membranes one of the central problems is the determination of their thickness. This parameter is most conveniently evaluated by electron microscopy and by x-ray diffraction techniques [see Ref. (45) for a recent review]. Similarly, the determination of BLM thickness was among one of the first problems investigated. The thickness of BLM was first estimated to be in the range of 60–90 Å from interference phenomena of the reflected light and from electron photomicrographs (24). The BLM thickness can also be estimated from measurements of electrical capacitance with the aid of Eq. (1). We will be concerned with only the optical methods and electron microscopy of BLM in the following paragraphs.

1. Electron Microscopy of BLM

Electron microscopy is a complicated and involved technique and the problems of both sample preparation and interpretation of results are greatly compounded in the BLM system. In spite of these obstacles, electron micrographs have been obtained by several groups of workers (24, 46, 47). The first of these groups used standard techniques and the resulting photomicrographs showed only single solid lines with an occasional suggestion of two dense lines separated by a lighter region. Henn et al. (46) have published some remarkable electron photomicrographs of BLM formed from a mixture of phospholipid and cholesterol using $La(NO_3)_3$ as a fixing agent, followed by $KMnO_4$ treatment. Microdensitometer tracing gave a thickness ranging from 37.5–116 Å. From their results Henn et al. have suggested that the hydrocarbon solvent may be trapped in the BLM, which could account for the thickness variations. More recently, electron micrographs have been obtained by Blough and Gordon (47). Figure 9 shows a BLM formed from egg lecithin alone. In the central portion of the figure a triple-layered structure is discernible.

Further application of electron microscopy to BLM under different experimental conditions (e.g., in the presence of a protein) is definitely of value since to a very great extent our knowledge concerning the structure of biological membranes is derived from electron microscopy.

2. Optical Properties of BLM

Using optical methods Huang and Thompson have measured quantitatively the thickness of BLM (48). The intensity of light reflected from the BLM is used in conjunction with membrane refractive indices determined from measurements of Brewster's angle. The optical properties of BLM in aqueous solution are fundamentally the same as those of thin transparent solid films (or soap films), hence many of the theoretical considerations on thin films can be applied directly (41). It should be noted that many problems still remain to be solved for these structures. As a first approximation, the simple laws of optics are adequate for our purposes. Here we are chiefly concerned with the evaluation of a lipid membrane thickness in aqueous solution.

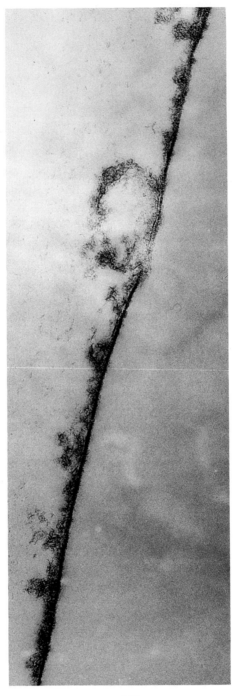

Fig. 9. Electron micrograph of a BLM composed of lecithin. Magnification 110,000 ×. (Courtesy of H. A. Blough and G. Gordon.)

A lipid membrane in aqueous solution may be pictured as a non-absorbing, transparent layer of thickness t submerged in a medium of different refractive index from itself. The "black" appearance of a black lipid membrane may be explained with the aid of Fig. 10.

When a beam of light originates from a source S incident on the membrane (shaded area), the so-called Fizeau fringes, or fringes of equal thickness are observed. (Similar interference phenomena are exhibited by the familiar soap films.) Briefly, the optical path difference between

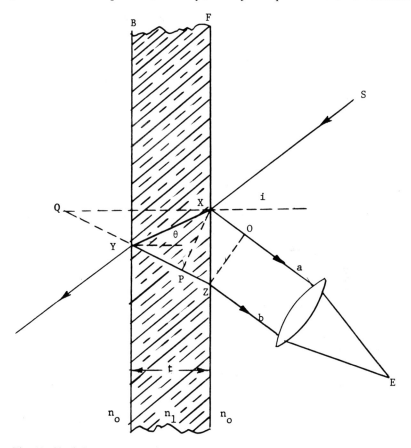

Fig. 10. Shaded area representing a transparent ultrathin film of uniform thickness t and refractive index n_1 immersed in an aqueous solution of refractive index n_0, S is a source of light; i and θ are the angles of incidence and refraction, respectively; a and b are interfering rays, superimposed at E, due to reflections at the two parallel interfaces F and B. OZ is the wave front of the reflected light.

the two reflected rays a and b is $n_1(\overline{QZ}) - n_0(\overline{OX})$, where n_1 and n_0 are the refractive indices of the membrane and aqueous solution, respectively. Since $\overline{QZ} = \overline{PQ} + \overline{PZ}$ and $n_1(\overline{PZ}) = n_0(\overline{OX})$, it can be shown by elementary geometry that the optically equivalent path difference is $n_1(\overline{PQ})$. With the help of Snell's law, which relates the incident and refracted angles with refractive indices, the path difference of the two rays is then equal to

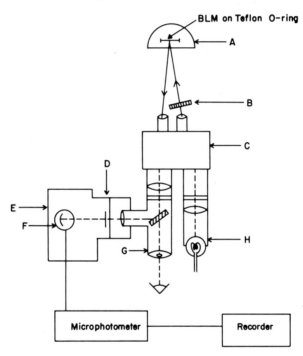

Fig. 11. Apparatus for measuring the reflectivity of a lipid membrane. (A) Cell chamber (see Fig. 12); (B) interference filter; (C) stereomicroscope; (D) shutter; (E) photomultiplier housing; (F) photomultiplier tube (1P21 or EMI 9558Q); (G) viewing tube; (H) light source.

$2n_1t \cos \theta$, where t is the membrane thickness and θ is the angle of refraction in the membrane. The incident beam of light from source S is divided into two parts upon reaching the front water–lipid interface F. The larger of the two parts is the refracted ray, which is again divided into two upon reaching the back lipid–water interface B. The reflected rays a and b, after traveling different paths, are superimposed at E to give the

observed interference phenomena. Further, when the problem of reflection is considered mathematically, it has been shown that there is a
phase change of 0° and 180° for the reflection from the back and front
interfaces, respectively (in the present case $n_1 > n_0$). When the thickness
t falls much below the wavelength of light, destructive interference can
then take place giving rise to the optically "black" appearance.

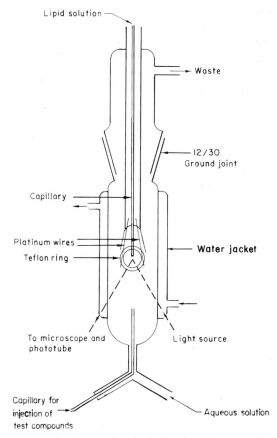

Fig. 12. Details of cell chamber (7).

The evaluation of a BLM thickness by optical methods has been given
in detail (41, 48, 49). The optical methods used are essentially the same
as described for soap film studies (50, 51). An arrangement for reflectance
studies and the cell chamber used for BLM formation are illustrated in
Figs. 11 and 12, respectively.

From a measurement of reflectance R, the thickness of the BLM can be calculated from the well-known Rayleigh equation

$$R = \frac{I}{I_0} = 4r^2 \sin^2 \delta = 4r^2 \sin^2 \left(\frac{2n_1 t \cos \theta}{\lambda} \right) \qquad (2)$$

where $r = (n_0 - n_1)/(n_0 + n_1)$. The reflectance R is defined as the ratio of the reflected energy I to the incident energy I_0.

In Eq. (2) we see that the reflectance of the BLM under consideration depends on the refractive indices of the membrane and the aqueous phase, and on the path difference of the light rays in the membrane (i.e., on the membrane thickness t). The material thickness of the BLM may then be evaluated, provided we can measure the ratio of the light intensities and the refractive index of the BLM.

It should be stated that in using the Rayleigh equation the following specific assumptions are made: (1) the membrane is assumed to be transparent and homogeneous, (2) the reflected light is made up simply of rays a and b, the multiple reflections within the membrane being neglected, and (3) the refractive index of the membrane is that obtained by observing reflectance at the Brewster angle (see the following).

The refractive index of the membrane is determined from measurements of the Brewster angle (52). This method is based on the fact that for light polarized with the electric vector in the plane of incidence, the reflectance of a membrane of refractive index n_1 immersed in a medium of index n_0 will be a minimum at the angle defined by $\tan \phi_p = n_1/n_0$, where ϕ_p is the so-called polarizing angle. This simple, elegant method involves practically no calculation and is only dependent on a knowledge of the refractive index of the medium, which can be easily determined. Further, no absolute intensity measurements are required if one is only interested in the refractive index.

The refractive index of a lecithin BLM has been obtained by several groups of workers using Brewster's law. Huang and Thompson obtained a value of about 1.66 (48), Tien reports a value of 1.60, and recently a much lower value of 1.37 was given by Cherry and Chapman (53). This wide range of results obtained for an apparently similar lecithin BLM suggests the inherent difficulties in applying Brewster angle measurements to an ultrathin structure such as BLM (41). In spite of this uncertainty it is remarkable that the reported thickness values lie between 60–72 Å.

A further comment should be made in connection with the use of Rayleigh's equation [Eq. (2)]. It has been assumed that the BLM under consideration is both homogeneous and isotropic (i.e., possesses identical properties irrespective of direction). Although these assumptions may be used to a first approximation, it is argued that they may not be valid for BLM with "bulky" hydrophilic groups thus far studied (41). A BLM would be more accurately pictured as a triple-layered structure with refractive indices n_p and n_h, where the subscripts p and h refer to the polar and hydrocarbon portions of the membrane, respectively (see Fig. 4). An equation relating reflectance R and thickness of a triple-layered structure has been derived by investigators studying thin dielectric layers (54, 55). The essential steps in the derivation of the equations as applied to the BLM system in aqueous media have been given (41). In general, the thickness is about 10% higher when interpreting BLM reflectance data in terms of the triple-layered optical model (56).

F. THE WATER PERMEABILITY OF BLM

1. Experimental Aspects

One of the phenomena of fundamental importance in biology is material transport. The current hypothesis of the structure of biological membranes makes BLM a suitable model for an investigation of the transport mechanism of water (and other molecules and ions as well) across membranes of biological relevance.

Experimentally, permeability to water in a BLM has been measured by the osmotic method and the tagged water method. In the latter the flow of isotopically tagged water through the BLM is determined in the absence of an osmotic gradient. The permeability coefficient thus measured is termed the diffusion permeability coefficient P_d. In the osmotic method a concentration gradient is created and the net flow of water across the BLM measured (see Fig. 13). The permeability coefficient P_o is calculated from the equation

$$J = \frac{P_o A \Delta \pi}{RT} \tag{3}$$

where J is the water flux, A the area of the BLM, R the gas constant, T the absolute temperature, and $\Delta \pi$ the osmotic gradient (Fig. 14). Huang and Thompson have used both methods in their studies (57) and have found that the ratio of P_o to P_d was much greater than unity, implying the existence of membrane pores. (For further discussion on this important topic, see Ref. 58.) Hanai et al. (59) have also found a difference between P_o and P_d. However, they attributed the difference to inadequate

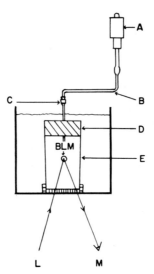

Fig. 13. Apparatus for measuring *trans* bilayer lipid membrane osmosis. (A) Digital pipette; (B) polyethylene or Teflon tubing; (C) syringe needle; (D) stopper; (E) Teflon beaker; (L) light source; (M) binocular microscope (*61*).

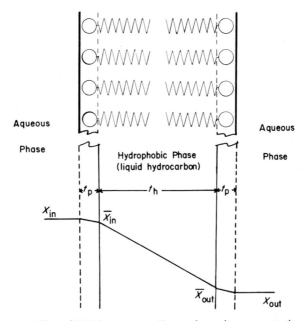

Fig. 14. Permeability of BLM to water. *Top:* schematic representation of an unmodified BLM. *Bottom:* permeation of water under an osmotic pressure gradient. t_p and t_h denote the thickness of polar region and hydrocarbon phase, respectively; X, equilibrium concentrations of water in various phases. Subscripts *in* and *out* refer to the two sides of the BLM.

256

stirring, which gave rise to stagnant layers at the interfaces. Independent studies carried out by Vreeman have also shown that the difference must be due to the existence of unstirred layers at the membrane/solution interfaces (32). Later work of Cass and Finkelstein has demonstrated conclusively the importance of stirring in performing self-diffusion experiments through BLM (60).

It is generally recognized that only a very limited amount of information can be derived from measurements of water permeability coefficient at a single temperature (61). Also, the question of the nature of the interfacial region (i.e., the zone near the solution/membrane interface) upon the water permeability has been raised. With these points in mind, the permeability coefficients of water through the BLM have been measured as a function of temperature and the ionic composition of the bathing medium (61). These data permit an evaluation of the various derived quantities from the absolute reaction rate theory (62). We will discuss these data and examine the effect of ions on water permeation through the BLM and on the nature of the BLM/solution interface.

2. Activation Energy and Theory of Absolute Reaction Rates

a. The Arrhenius Activation Energy, E_a. It is assumed that permeation of water through an unmodified BLM takes place by a molecular process that can be described by an Arrhenius equation of the form

$$\ln P_0 = A - (E_a/RT) \tag{4}$$

where A is a constant, E_a is the activation energy involved in the rate-controlling step in the process, and R and T are the gas constant and absolute temperature, respectively. It would be expected that if the assumption were correct a $\ln P_0$ versus $1/T$ plot would give a straight line. The slope and intercept of the plot yield the activation energy E_a and the constant A (frequency factor). This has been found to be the case for a BLM system studied in the temperature range 22.5–44.0°C (see Fig. 15).

The very high dc resistance implies that the BLM is practically impermeable to charge carriers such as monovalent ions generally present in the aqueous solution used. This impermeability to charge carriers may be shown by a simple calculation. For a 1 cm² BLM with a dc resistance of 10^7 Ω-cm² and an applied potential of 0.1 V, only about 4×10^{-10} equivalent of charge move through the BLM per hour. It therefore seems that an unmodified BLM, behaves essentially like a

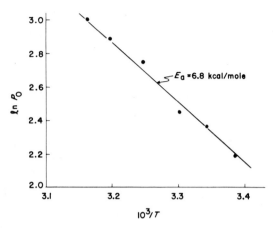

Fig. 15. Osmotic permeability coefficient P_o as a function of temperature. The BLMs used in this study were formed from a solution of oxidized cholesterol in *n*-octane (*61*).

semipermeable membrane. In the discussion that follows this liquid–hydrocarbon-like picture of BLM will be our frame of reference (see Fig. 14).

b. The Osmotic Permeability Coefficient, P_o. The units of P_o (μ/sec) have the dimensions of velocity and we may therefore consider, that $1/P_o$ is a measure of the resistance offered by the BLM to water permeability. The total resistance encountered by a water molecule in transferring from one side of the membrane to the other side may be viewed as consisting of five separate resistances in series. These are two resistances at the two membrane/solution interfaces, one due to the membrane, and two bulk-phase resistances. Since the self-diffusion coefficient of water is several orders of magnitude larger than the other coefficients, we need only to consider the resistances due to the membrane and the membrane/solution interfaces. Therefore, this total resistance may be expressed as

$$1/P_o = R_{sm} + R_m + R_{ms} \tag{5}$$

where the subscripts s and m refer to the solution and membrane, respectively. According to an analysis given by Zwolinski et al. in terms of the Eyring absolute reaction rate theory (*62*), there is a specific rate constant k_i associated with each step (see Fig. 16). Here k_m refers to the rate constant in the BLM, while R_{sm} and R_{ms} are the rate constant for diffusion in the solution/BLM and BLM/solution interfacial region, respectively.

Although the magnitude of P_o provides us with some clues concerning the permeation barrier, it is not possible to assess the relative contribution resulting from each resistance. Hence, nothing more can be said about the rate-controlling step and the mechanism of water permeation through the BLM.

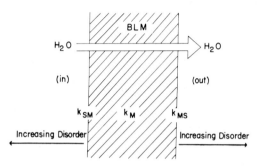

Fig. 16. A model representing the possible rate-limiting steps to water permeation across a BLM. (i) Inner chamber; (O) outer chamber; k_{sm}, rate constant for diffusion in solution/membrane interfacial region; k_m rate constant for diffusion in BLM; k_{ms}, rate constant for diffusion in membrane/solution interfacial region (*61*).

c. The Derived Quantities ΔS^*, ΔH^*, and $\Delta F'$. Application of the theory of absolute reaction rates to diffusion processes through membranes has been made by several investigators. In particular, Zwolinski et al. have extended the theory to the diffusion and permeability of membranes (*62*). We shall follow their treatment in analyzing the water permeability data through BLM.

As indicated in Fig. 16, there are three rate constants involved in the permeation of water across the BLM. From the nature of the interface involved, we may assume that k_{ms} is much larger than the k_{sm} and k_m associated with the solution/membrane interfacial region and the BLM itself, respectively. Zwolsinski et al. have derived an equation relating P_o and other parameters:

$$\frac{1}{P_o} = \frac{2\lambda}{D_{sm}} + \frac{t}{D_m K} \tag{6}$$

where t is the thickness of the membrane, $K = k_{sm}/k_{ms}$, and D_{sm} and D_m are the respective diffusion coefficients for the solution/membrane interface and the membrane. Further, $D_i = k_i \lambda^2$ where i refers to either sm or m. Referring to Fig. 16, three cases may be considered: (1) the membrane limiting, (2) the interface limiting, and (3) the membrane and interface limiting. We shall consider each case separately in the following paragraphs.

(*i*) *Membrane Limiting Case.* The rate-controlling step is diffusion in the BLM. Therefore, k_{ms} is much greater than k_m. In this case the first term on the right-hand side of Eq. (6) is negligible and we have

$$P_0 = \frac{\lambda^2 k'}{t} \qquad (8)$$

where $k' = k_m K$. From the absolute reaction rate theory we can write

$$P_0 = \left(\frac{kT}{h}\right)\left(\frac{\lambda^2}{t}\right)\exp\left(\frac{-\Delta F'}{RT}\right) \qquad (9)$$

or

$$P_0 = \frac{kT}{h}\left(\frac{\lambda^2}{t}\right)\exp\left(\frac{-\Delta H^*}{RT}\right)\exp\left(\frac{\Delta S^*}{R}\right)$$

where $\Delta F'$ is the free energy of activation for permeability, which is interpreted to mean the difference in free energy of the permeating water molecule between its initial position in the aqueous solution bathing the membrane and the top of the highest energy barrier the water molecule must overcome within the BLM. [In the calculations values of $t = 40$ Å or 70 Å and a value of $\lambda = 5$ Å are used. The calculated ΔS^*, ΔH^*, and $\Delta F'$ according to Eq. (9) are given in Table 3.]

TABLE 3

EXPERIMENTAL ACTIVATION ENERGIES AND DERIVED QUANTITIES FROM TRANSITION STATE THEORY OF RATE PROCESSES FOR VARIOUS SYSTEMS (*61*)

System	E_a (kcal/ mole)	ΔS^* (cal/mole/ deg)	ΔH^* (kcal/ mole)	$\Delta F'$ (kcal/ mole)
Water	4.6	2.5	4.0	3.9
n-Hexadecane	3.4	−0.2	2.8	3.4
Cellular membranes	15–22	16–39	14–21	9.7–10.1
BLM[a] Case I[b] ($t = 40$ Å)	6.8	−13.3	6.2	10.4
($t = 70$ Å)	6.8	−12.2	6.2	10.1
Case II[c] ($\lambda = 1$ Å)	6.8	−13.7	6.2	9.6
($\lambda = 3$ Å)	6.8	−15.8	6.2	11.4
Case III[c] ($\lambda = 1$ Å)	6.8	29.1	6.2	−2.4
($\lambda = 3$ Å)	6.8	22.1	6.2	−0.4

[a] BLM thickness. $t = 40$ Å unless otherwise noted.
[b] Case I. The derived quantities calculated using the Eq. (9) where $\lambda = 5$ Å.
[c] Cases I and II. Values calculated according to Eq. (9).

(*ii*) *Interface limiting case.* In this case $k_m \gg k_{ms}$ and Eq. (6) becomes

$$P_o = \frac{\lambda k_{sm}}{2} \tag{10}$$

By plotting $(P_o h/kt)$ versus $1/T$, the various derived quantities were evaluated, and are also summarized in Table 3.

(*iii*) *Membrane and interface limiting.* In this case, since $k_{ms} \simeq k_m$, the permeability coefficient is given by

$$P = \frac{k_{sm}\lambda}{t} \tag{11}$$

The derived quantities are given in Table 3.

On the basis of the $\Delta F'$ calculation, we can rule out Case (iii) since a negative $\Delta F'$ would mean that the activated state actually possesses less energy—a highly unlikely event. (In terms of resistance, $R \leqslant 0$ and therefore $P \to \infty$.)

As mentioned above, the magnitude of the permeability coefficients will not provide information concerning the rate-determining step involved in the process. The temperature dependence of the permeability coefficient by itself also does not seem to be able to decide the controlling step regarding the mechanism of permeation, as can be seen from the calculated values in Table 3, which are remarkably similar. As pointed out by Zwolinski et al., if k_m is only slightly larger than k_{ms}, the ln (Ph/kt) versus $1/T$ plot would exhibit a definite curvature toward the abscissa axis. This case can be ruled out also since the plot is essentially linear. Thus we have only two cases to decide, i.e., whether the rate-controlling step is due to the membrane or the solution/membrane interfacial region. In the case where the interfacial region is the slowest step, any modification of the interfacial region should have a significant effect on the magnitude of P_o. In the following, additional experimental evidence will be given examining the effect of several molecular properties at the biface on BLM to water permeability, which may provide us with a basis to understand the mechanism of water permeation through the BLM.

d. The Entropy of Activation. It can be seen from Eq. (9) that the entropy of activation ΔS^* may be determined if one assumes a reasonable value of λ. In these calculations, the molecular jump λ is assumed to be either 3 or 5 Å (about 1–2 molecular diameters of water). In the third column of Table 3, ΔS^* for various systems is listed. The most striking feature of the tabulated values is that for Cases (i) and (ii) rather large negative entropies of activation are obtained. The values given are very approximate since both λ and BLM thickness t have to be assumed. In Case (ii) the permeation of water is governed by diffusion through the solution/membrane interfacial region; the P_0 is therefore independent of the membrane thickness. According to the theory of absolute rates (62), the negative ΔS^* implies one of two events: (1) the bonding of activated complex with the membrane, i.e., the permeating water molecules are being partially immobilized in the membrane structure or in the interfacial region (iceberg formation); or (2) permeation through the membrane is not the rate-controlling step in the process. Experimental evidence discussed next seems to support the latter case.

3. The Mechanism of Water Permeation

To attempt an interpretation of the above data it is necessary to consider the most likely pathways for water movement through the solution/BLM/solution system, and the nature of liquid/liquid interfaces relevant to this discussion.

a. The Effect of Molecular Properties at the Biface. Up to now we have been mainly concerned with the BLM itself. As mentioned in the preceding paragraphs, the permeability–temperature data alone is incapable of deciding the mechanism of water permeation (with perhaps one exception noted above). Since the interior of the BLM could not be easily modified (apart perhaps from introducing unsaturation in the hydrocarbon chains), the effect most amenable to change will be at the solution/membrane interface. Since BLM can be formed from several surface-active agents in combination with cholesterol, P_0 values can be measured with BLM-forming compounds bearing either positive or negative charges when ionized in solution. In spite of the fact that these BLM possess different bifacial tensions (γ_b) and electrical charges, little change in the values of P_0 is observed. The systems studied, together with P_0 data, are given in Table 4. This information suggests that the effect of bifacial tension of the membrane on P_0 is negligible. Although the chemical composition of the BLM remains unknown, these membranes are believed to be composed predominantly of cholesterol molecules. The more surface-active agents (HDTAB, DAP, or "impurities" associated with oxidized cholesterol) constitute a very small fraction of the BLM structure. The main effect of those agents probably lies in their ability to stabilize the BLM structure (63).

TABLE 4

BIMOLECULAR LIPID MEMBRANES (BLM) GENERATED FROM THREE DIFFERENT LIPID SOLUTIONS AND THEIR PERMEABILITY TO WATER AT 22.5° (61)

BLM formed from	Aqueous phase	P_o (μ/sec)	γ_b (dyne-cm^{-1})
Oxidized Cholesterol in			
n-octane	0.1 N NaCl	8.4 ± 0.5	1.9
Cholesterol in n-dodecane	0.1 N NaCl		
	containing	8.2 ± 0.5	0.2
	0.008% HDTAB^a		
Cholesterol–DAP^a in n-dodecane	0.1 N NaCl	8.8 ± 0 5	1.1
Oxidized cholesterol +			
2.5 × 10^{-5} M valinomycin	0.1 N KCl	1.0^b	—
Oxidized cholesterol +			
2.5 × 10^{-5} M valinomycin	0.1 N NaCl	9.5^b	—

^{a} HDTAB—hexadecyltrimethylammonium bromide. DAP—dodecyl acid phosphate.
^{b} Recent data obtained by Dr. H. P. Ting.

b. *The Effect of Solute at the Biface.* A dramatic effect on P_o has been observed when different solutes were used to generate the osmotic gradient. The data are presented in Table 5. Here, a threefold increase in P_o is obtained when the aqueous solution is changed from NaCl to either NaI or CsCl.

c. *The Nature of Water Flow through BLM With and Without "Pores."* Data from tritiated water and osmotic flux measurements suggests that "pores" may exist in the BLM (57). If pores or channels (presumably for

TABLE 5

THE EFFECT OF SOLUTE USED ON THE P_o FOR WATER FOR BLM GENERATED FROM OXIDIZED CHOLESTEROL IN n-OCTANE AT 22.5° (61)

Aqueous solution	P_o (μ/sec)	dc Resistance (Ω-cm^2)^b
NaCl	8.4	5 × 10^8
NaI	24.7	2 × 10^6
CsCl	25.9	8 × 10^6
NaCl + I$_2$ (saturated)	11.7	3 × 10^3
NaI and I$_2$ (saturated)	9.6	< 10^3
NaCl + dextran (16,800)	24.8^a	—
NaCl + sucrose	17^a	—
Sucrose	53^a	—

^{a} Recent data obtained by Dr. H. P. Ting.
^{b} Measured in 0.01 N solution by Dr. A. L. Diana.

water flow only) do exist in the BLM, the rate of self-diffusion for water should be proportional to the total area of the pores. It is reasoned, therefore, that P_d (self-permeability coefficient) and P_o would be different, with the latter being considerably larger as a result of a hydrostatic pressure or osmotic gradient. As mentioned earlier, however, $P_d \simeq P_o$. In view of the fact that the interior of BLM is liquid hydrocarbon-like, as well as from the evidence resulting from recent studies, the possibility of the existence of pores in the BLM is untenable (60). The fact that $P_o \simeq P_d$ implies that the same pathway must be involved in transporting the water molecules across the BLM system. Previously, a simple solubility–diffusion mechanism has been proposed (32, 59). This is based upon the finding that the magnitude of P_o found experimentally is roughly the same order as that calculated from the solubility and diffusion coefficient data. However, a more plausible mechanism will be given in the following paragraph.

   d. The Energy Barrier Mechanism and the Nature of the Water/Hydrocarbon Interface.   Alternatively, it has been suggested that an energy-barrier mechanism is responsible for water permeation across the BLM system (61). The energy barrier idea was first used by Langmuir and Schaefer and later by others (64, 65) in reference to monolayer penetration by water and simple gas molecules at the air/water interface. Since the existence of "pores" in the BLM appears to be ruled out, it seems that the only other way for the water molecules to get across the BLM system (solution/BLM/solution biface) is by "barging" their way through (61). This means the creation of "holes" along the pathway as the molecules permeate across the system. In terms of the absolute reaction rate theory, only those water molecules striking the interfacial region with sufficient energy will get across the BLM at the biface, implying that the permeating water molecules are at an activated state.

   Information concerning the nature of liquid/liquid interfaces relevant to this discussion is very limited. Franks and Ives (66) suggest that the water/hydrocarbon interface is covered by a layer of icebergs, which are known to surround hydrocarbon molecules in aqueous solution as suggested earlier by Frank and Evans (67). Further work on hydrocarbon/water interfaces has been reported by several other investigators (68). It is generally accepted that the interaction between saturated hydrocarbons and water involves dispersion forces (69). Although no information is available about the BLM systems, from the afore-mentioned studies it seems probable that the presence of polar (or ionic) groups at the biface would enhance the orderliness of the interfacial region—i.e., the bifacial region tends to be more organized than bulk phases. The exact thickness of this interfacial region is not known and could be anywhere from one molecular layer to several hundred. The important point is that this interfacial region is more orderly than the bulk solution. Perhaps a reasonable view to adopt is that there is a gradual change of structure from the BLM/solution interface in which the orientation of the water molecules becomes increasingly disordered as the distance from the interface is increased (see Fig. 16).

Applying these considerations to the water permeation through the BLM system, attention will be focused upon the bifacial region rather than the BLM itself. If the rate limiting step were due to the membrane (i.e. liquid hydrocarbon with or without pores) one would expect the experimental activation energy E_a to be similar or less to that for diffusion. This is not supported by the data collected in Table 3. In fact, E_a for permeation of water in the liquid hydrocarbon (C_{16}) is somewhat lower than for free diffusion in water, whereas P_0 for the BLM system is twice that for n-hexadecane. The higher value for the BLM system, according to the absolute reaction rate theory, means that the permeating water molecules must overcome a higher energy barrier. This could come from the orderliness of the solution/membrane interfacial region.

If the interfacial region were the rate-controlling step, than any disturbance of the region should have a noticeable effect on E_a, which in turn could be reflected in P_0. The data presented in Table 5 are therefore of interest. The permeability coefficients for water are three times larger when NaCl is replaced by either NaI or CsCl as the solute. This marked change of P_0 may be explained in terms of the Gurney order–disorder concept (70). According to this view, liquid water possesses distinctive structural features near its freezing point. Depending upon the size of ionic solute introduced, those that can fit into the structural "vacancies" will not disrupt the water structure. Cs^+ and I^- are classified as structure breakers. Assuming that the results are applicable to the present situation, the introduction of these large ionic solutes (Cs^+ and I^-) tends to break up the ordered interfacial region, thereby lowering the interfacial resistance for water permeation.

It may be argued that Cs^+ or I^- could also modify the BLM structure and produce a lower energy barrier for water permeation. The evidence against this is that although the dc electrical resistance of the BLM is lowered somewhat when these ions are used as the bathing medium, P_0 values are drastically changed. On the other hand it has been found that I_2 greatly reduces the dc resistance of the membrane while P_0 remains practically unchanged. In this connection it is interesting to note that a proteinaceous material of uncertain composition known as EIM (see Sect. IV.D) lowers the dc resistance of the BLM by several orders of magnitude when added to the aqueous phase (24). At the same time it has been found that the permeability coefficient for water was little affected (60). It has been suggested that upon introduction of EIM the structure of the membrane is altered insofar as ionic transport is concerned and it would seem from these results that ionic movement across the BLM is governed by the nature of the membrane, whereas water permeation is determined primarily by the orderliness of the interfacial region. If this explanation is correct, one would predict that any species whose presence in water causes the breaking up of the water structure at the biface will promote water permeation, while electrolyte transport may or may not be affected (61).

The permeability of BLM to other compounds has also been studied. Generally, the values of permeability coefficient obtained for other compounds are appreciably lower than that of water. Specific examples will be considered in later sections dealing with the effects of BLM modifier.

G. BIFACIAL TENSION OF BLM

Although a bimolecular layer (bilayer) lipid membrane separating two aqueous solutions may be considered as two monomolecular layers with the hydrocarbon chains jointed together, the physical properties are quite

TABLE 6

BIFACIAL TENSION OF SOME BIMOLECULAR LIPID MEMBRANES

BLM generated from	γ_1 (dynes/cm)	Ref.
Lecithin in tetradecane CH_3OH and $CHCl_3$	1.0	(71, 72)
Lecithin (egg) in n-dodecane 1% (w/v)	0.9 ± 0.1	(43)
Lecithin	0.73	(3)
Brain lipids	2.2–4.9	(117)
Cholesterol (oxidized) in n-octane 4% (w/v)	1.9 ± 0.5[a]	(43)
Glycerol dioleate in n-dodecane, 1% (v/v)	1.5 ± 0.2	(56)
Dioctadecyl phosphate (DODP) and cholesterol	5.7 ± 0.2[b]	(63)
Hexadecyltrimethylammonium bromide (HDTAB) and cholesterol	0.15 ± 0.5[c]	(63)
Chloroplast extract	3.8–4.5	(166)

[a] Variable depending upon the extent of oxidation.
[b] In distilled water.
[c] HDTAB (0.008%) in 0.1 N NaCl.

different, and they are not simply additive. For instance, the surface tension of a monolayer at the air/water interface is usually about 20 dynes/cm (27). A BLM is unique in that it possesses two identical interfaces, insofar as the dielectric constants of the two bulk phases are concerned. As a first approximation, a BLM has two aqueous solution/

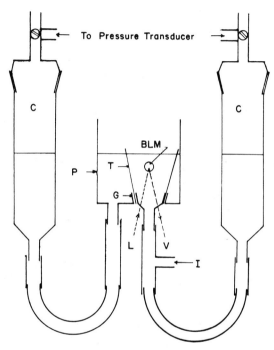

Fig. 17. Schematic diagram of apparatus for measuring bifacial tension of BLM (C) Ballast chamber; (T) Teflon sleeve; (P) glass outer chamber; (G) ground joints; (V) viewing tube; (L) light source; (I) infusion–withdrawal pump (63).

membrane interfaces or a *biface* (the term is introduced to stress the interdependence of the two coexisting interfaces). The bifacial tension of BLM for a variety of BLM systems in general is less than 6 dynes/cm (56) [Table 6]. An experimental arrangement used for this determination is shown in Fig. 17 (63).

1. Elementary Theory

The theory of the bulging method for measuring the bifacial tension assumes that the maximum pressure exists when the BLM (or film) is just hemispherical (*63*). Implicit in the theory is a bifacial tension (γ_b) present in every thin film separating two phases (liquid, solid, or gaseous). Further, the bifacial tension is assumed to be the same at every point in a given film. It can be shown that at equilibrium the work done in changing the radius of BLM bubble R to $R + dR$ is equal to $P \times 4\pi R^2\, dR$ (see Fig. 18). Thus

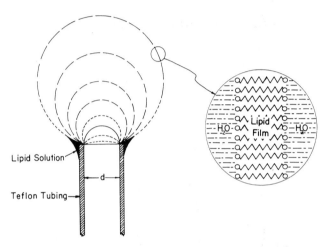

Fig. 18. Drawing showing formation of a spherical BLM separating two aqueous solutions. The pressure difference across the BLM is used to calculate the bifacial tension (*63*).

$$16\pi\gamma_b R\, dR = 4\pi P R^2\, dR \qquad (12)$$

and

$$P = \frac{8\gamma_b}{d} \qquad (13)$$

where $d = 2R$, the diameter of the BLM bubble or the diameter of the opening used in the membrane support, and P is the pressure difference across the BLM. For comparison, the results obtained by different investigators are given in Table 6.

2. The Plateau–Gibbs Border and BLM Stability

There are at least two major problems associated with BLM investigations. First, we do not know precisely what molecular species are present in the membrane. The second problem is the fragility of the structure, which presents considerable experimental difficulties in investigating its mechanical and other properties. Therefore our description of the BLM would necessarily be qualitative in many ways. In our consideration the relationship between the BLM and its Plateau–Gibbs border (i.e., the edge where the BLM is terminated) will be stressed. The Plateau–Gibbs border is believed to be essential for the integrity of BLM structure. It has been frequently observed that the area of BLM does not usually extend over to the entire aperture in the membrane support. That means that the BLM may occupy anywhere from near zero to almost 100% of the total available area at the biface. Figure 19 shows a BLM separating two aqueous phases also in equilibrium with the P–G border (phase III). Thus we have a situation where three interfaces intersect along

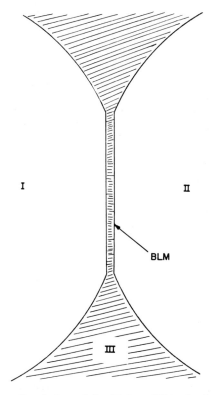

I

II

BLM

III

Fig. 19. A cross-sectional view of the Plateau–Gibbs border supporting a BLM. Numerals I, II, and III denote, respectively, aqueous phases I and II, and the lipid solution at the border.

a line of contact at N (see Fig. 20). According to the theorem of Neumann's triangle (27), an element of the line of length dl through N is subjected to the three forces. At equilibrium the vectorial sum of the three forces must be equal to zero. It can be shown that in the case of a stable BLM the following relationship must be satisfied:

$$\gamma_{\text{BLM}} = 2\gamma_i \cos \theta \qquad (14)$$

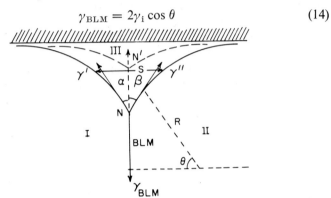

Fig. 20 The relationship between the Plateau–Gibbs border and interfacial tensions. N is the line of intersection of three phases I, II, and III (see Fig. 19). γ' and γ'', interfacial tension of various interfaces; R, radius of the Plateau–Gibbs border. In the case of $\gamma_{\text{BLM}} < 2\gamma(\gamma' = \gamma'')$, the Plateau–Gibbs border will move forward the edge (shaded) and the new intersection point at N'. This situation is described by $\gamma_{\text{BLM}} = 2\gamma' \cos \theta$ (56).

where γ_i is the interfacial tension between the two bulk phases (for a BLM interposed between two solutions of the same composition, $\gamma_i = \gamma' = \gamma''$, see Fig. 20). At the P–G border the interfaces are circular according to the Young–Laplace equation, with radius equal to R. At equilibrium the downward tension (γ_{BLM}) must be balanced by the upward tension γ_v, which may be split into two terms given by

$$\gamma_v = 2\gamma_{\text{ML}} \cos \alpha + (S\gamma_{\text{ML}}/R) \qquad (15)$$

On the right-hand side the first term is the vertical component of the interfacial tension and the second represents the P–G border suction pressure. The horizontal distance S is given by

$$S = 2R(1 - \cos \alpha) \qquad (16)$$

The observable circular area of a BLM is governed by γ_{BLM}, γ_{ML}, and the angle θ. If α is zero, then on substituting S into Eq. (15) we have

$$\gamma_{\text{BLM}} = 2\gamma_{\text{ML}} \qquad (17)$$

In this case no BLM is obtainable. If α is greater than zero, the distance S is

$$2R(\cos \theta - \cos \alpha) \qquad (18)$$

which means that γ_{BLM} must be less than $2\gamma_{ML}$. For this to be true, the P–G border must move toward the rim of the aperture as dictated by Eq. (14). The latter situation is shown in dotted line in Fig. 20 (63). The ideas discussed here may be further developed to provide an alternate means in the evaluation of the tension of the BLM (56).

Summarizing the observations on the γ values of the BLM generated from a number of compounds, the results strongly suggest that a low γ value is a prerequisite for a stable BLM formation. The lower limit is slightly above zero; the upper limit, if it exists, probably does not exceed 9 dynes/cm (56). It can be concluded, therefore, that a positive free energy is necessary for the stability of the BLM, and as long as the integrity of the membrane is maintained, the limiting structure of the BLM shown in Fig. 2B represents the lowest free energy configuration.

3. The Thermodynamics of BLM at the Biface

In this section the classical thermodynamic equations developed by Gibbs for the interface between two bulk phases will be applied to the BLM system (56). We shall assume that each extensive property of the system, such as the free energy F, the entropy S, or the number of moles of each component, is the sum of three parts: (1) the contribution of the aqueous phase I, (2) the similar contribution of aqueous phase II, and (3) the BLM, which has a small but definite uniform thickness. For the BLM, the general variation of the Helmholtz free energy is given by

$$dF_{BLM} = -S'_{BLM}\, dT - P\, dV_{BLM} + \gamma\, dA + \sum_i \mu\, dn \qquad (19)$$

where μ and n are the chemical potential and the number of moles in each phase, respectively. By rearranging these equations we can obtain the Gibbs–Duhem equation for the BLM:

$$S'_{BLM}\, dT - V_{BLM}\, dP + A\, d\gamma + \sum_i n\, d\mu = 0 \qquad (20)$$

If Eq. (20) is divided throughout by A, it becomes

$$S_{BLM}\, dT - t\, dP + d\gamma + \sum_i \Gamma\, d\mu = 0 \qquad (21)$$

where S_{BLM} is the entropy per unit area of BLM and Γ is the so-called interfacial excess. It is evident that Eq. (21) is a form of the Gibbs adsorption equation and will be considered further in connection with the BLM structure. We can also write the change of γ with temperature as

$$-\frac{d\gamma}{dT} = S_{\mathrm{BLM}} - t\frac{dP}{dT} + \sum_i \Gamma\frac{d\mu}{dT} \qquad (22)$$

The meaning of ΔF may be seen from the following thermodynamic argument (for simplicity the change of energy to be considered refers to that of a unit area of the BLM). The formation of a BLM may be imagined to take place in three steps as follows:

$$\text{Liquid hydrocarbon (O)} + \mathrm{H_2O(W)} \rightarrow \mathrm{W/O/W} + \Delta F_1 \qquad (23)$$

$$\mathrm{W/O/W} + \text{lipid} \rightarrow \text{monolayer} + \text{Plateau–Gibbs border} + \Delta F_2 \qquad (24)$$

$$\text{Monolayer} + \text{P–G border} \left< \begin{array}{ll} \text{no BLM } (> 9 \text{ dynes/cm)} & (25a) \\ \mathrm{BLM} + \Delta F_3 & (25b) \end{array} \right.$$

where W/O/W represents the water/oil/water biface and ΔF is the interfacial free energy change involved in each step (75). The process represented by Eq. (23) is seen to be similar to that of spreading an oil on a water surface. The free energy change for the reaction is given by

$$\Delta F_1 = \gamma_{\mathrm{O/S}} + \gamma_{\mathrm{O/W}} - \gamma_{\mathrm{W/S}} \qquad (26)$$

where $\gamma_{\mathrm{O/S}}$, $\gamma_{\mathrm{O/W}}$, and $\gamma_{\mathrm{W/S}}$ are, respectively, the interfacial tension at the oil/solid interface, interfacial tension at the oil/water interface, and the interfacial tension at the water/solid interface. It can be shown that $\Delta F_1 \simeq 0$. ΔF_1 is therefore immaterial to the energetics of BLM formation. The energy changes accompanying the reactions given by Eqs. (24) and (25) are determined by

$$\Delta F_2 = \gamma_{\mathrm{ML}} - \gamma_{\mathrm{O/W}} \qquad (27)$$

and

$$\Delta F_3 = \gamma_{\mathrm{BLM}} - \gamma_{\mathrm{ML}} \qquad (28)$$

where γ_{ML} denotes the interfacial tension of the monolayer at the biface. The net result of these two steps is the summation of Eqs. (24) and (25b); hence the free energy change for the overall process is the sum of the two ΔF values; thus

$$\Delta F_2 + \Delta F_3 = \Delta F_i = \gamma_{\mathrm{BLM}} - \gamma_{\mathrm{O/W}} \qquad (29)$$

The overall reaction is

$$\mathrm{W/O/W} \text{ biface} + \text{lipid} \rightarrow \mathrm{BLM} + \Delta F_i \qquad (30)$$

Therefore, the ΔF_i represents the work done in the formation of a BLM and bringing the surface-active lipids from the bulk phase(s). Obviously the major part of the work done is the migration of surface-active lipid molecules to the W/O/W biface. On the other hand, the work required in the formation of a BLM from the existing monolayers at the W/O/W biface is a relatively small quantity by comparison. This large driving force (ΔF_2) toward the formation of monolayers at the W/O/W biface with attendant reduction of $\gamma_{O/W}$ is obviously a prerequisite for the formation of BLM. However, the low γ alone is not a sufficient condition for the formation of a stable BLM as has been shown experimentally (63).

4. Application of Gibbs' Adsorption Equation to a BLM System

As mentioned in the section on the P–G border and BLM stability (p. 268), a major problem is that there is no precise knowledge concerning the chemical composition of the membrane. At present, the use of direct chemical analysis for BLM composition is not feasible owing to the very small areas and thickness of the membrane. One approach to this problem would be by resorting to the use of Gibbs' adsorption isotherm [Eq. (21)] and identifying γ with γ_{BLM}. In order to calculate the excess concentration of lipid molecules in the BLM, the usual extra-thermodynamic assumption would have to be made that the only species in the membrane would be the surface-active lipid molecules. Since the Gibbs equation relates γ_{BLM} and the concentration of the dissolved material, a knowledge of the interfacial concentration of adsorbed molecules would permit an evaluation of the area occupied by BLM-forming molecules. From this information certain deductions concerning the structural aspects of the membrane might be made. We will now present results of an investigation of BLM generated from the cholesterol–dodecane–HDTAB (hexadecyltrimethylammonium bromide) system as a specific example. In order to establish whether there is any physical parameter governing black film formation and its stability, γ_i values were measured as a function of surfactant concentration for the dodecane–HDTAB–H_2O and cholesterol–dodecane–HDTAB–H_2O systems. As can be seen in Fig. 21, γ_i was found to decrease linearly in both cases as a function of log [HDTAB]. However, stable black films were formed only when cholesterol was present. Although not consistently observed, the stable black film was most easily produced when the HDTAB concentration was $9 \times 10^{-6}\ M$ or higher. Above $4.5 \times 10^{-4}\ M$, however, no films were obtainable. It is interesting to note that the extrapolated values are of the same order as the critical micelle concentration (63).

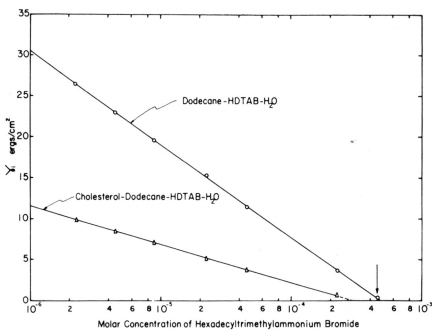

Fig. 21. Interfacial free energies of cholesterol BLM as a function of HDTAB concentration (see text and Ref. *63*).

With the afore-mentioned assumptions in mind, Eq. (21) may be written as

$$d\gamma_{\mathrm{BLM}} = RT\Gamma_{\mathrm{C}} \, d\ln\,[C] + RT\Gamma_{\mathrm{H}} \, d\ln\,[\mathrm{HDTAB}] \qquad (31)$$

under constant temperature and pressure conditions, where Γ_{C} and Γ_{H} refer to the interfacial excess concentrations of cholesterol and HDTAB, respectively. (For simplicity we have used the concentrations instead of activities. Also for dissociable surfactant HDTAB a factor of 2 should be taken into consideration.) The respective interfacial excess may be obtained from the slope of γ/concentration plots in the usual manner. It should be stated that ideally one would like to obtain the γ/concentration plot for one component while holding the concentration of the other constant. However, this has not been possible owing to difficulties in producing BLM from either cholesterol or HDTAB alone. Thus the data even for this simple BLM system are not complete.

The following is presented to show that structural information may be obtained in spite of the limited data. It should be mentioned that the values for the area occupied per molecule are expressed in terms of minimum area because of the lack of activity coefficient data and drastic simplifications used in the application of Gibbs' equation. In the case of HDTAB in the presence of cholesterol, the area is calculated to be about 200 Å² from the slope of the γ_{BLM} versus [HDTAB] plot at a constant concentration of cholesterol. If it is assumed that the area occupied by the respective species remains unchanged in the membrane, this would mean that the molar ratio of the cholesterol to that of HDTAB is about 3:1 in the BLM (HDTAB area = 80 Å²). To account for this particular

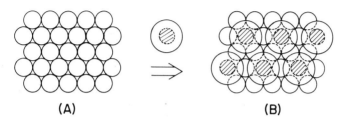

(A) **(B)**

Fig. 22. BLM produced from cholesterol–HDTAB system. (A) A front view of a sheet of cholesterol molecules at the water/oil/water biface. (B) Adsorption of HDTAB stabilized the system leading to a BLM formation. ○, Cholesterol molecules; ●, HDTAB molecules (56).

ratio, a most likely arrangement would be that of a hexagonal packing with six cholesterol molecules around each HDTAB. It is suggested that the slenderness of the HDTAB hydrocarbon tail could enable it to fit into the space between the cholesterol molecules, with the head group of HDTAB sticking out at the membrane/solution interface. A probable structure of BLM produced from cholesterol-HDTAB as deduced from the above data is illustrated diagrammatically in Fig. 22. It seems likely that the insertion of HDTAB into the cholesterol "bilayer leaflet" may be the reason for the BLM stability (56).

5. Electrocapillary Curves of BLM

An attempt has been made recently in the author's laboratory to examine the effect of applied voltage on the bifacial tension of BLM, a situation analogous to that of measuring the classical electrocapillary curve (28). Some preliminary results are shown in Fig. 23. It is of interest to note that in contrast to the shape of classical electrocapillary curves the bifacial tension is found first to decrease. After reaching a minimum it

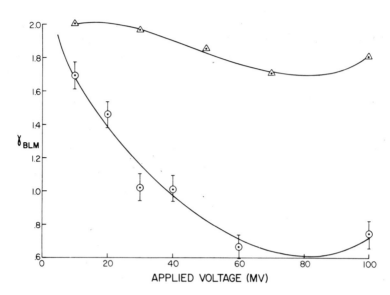

Fig. 23. Electrocapillary curves of BLM. △, Lecithin BLM in $10^{-3}\,M$ NaCl; ◯, ox. chol. BLM in $10^{-1}\,M$ NaCl.

begins to increase at higher voltages. The significance of this phenomenon is uncertain and will be further investigated.

In concluding this discussion it may be useful to point out that application of classical methods to the BLM system (e.g., the Gibbs equation and electrocapillarity) is of definite value. These methods may be helpful in establishing the detailed structure of the BLM and its associated electrical double layers at the biface.

H. THEORETICAL CONSIDERATIONS

In the past several years attempts have been made by a number of investigators to analyze the bimolecular leaflet model from theoretical points of view. Among the many interesting question raised are: (1) the molecular basis of BLM stability, (2) the nature of forces operating between the constituent molecules and their surroundings, (3) the calculation of dielectric constant of BLM and its relationship to thickness measurements, and (4) the origin of BLM in nature. In view of the mathematical complexities involved in these theoretical analyses, it is only possible to give in most cases a brief summary of the results obtained in these interesting studies.

1. Stability of BLM and Intralayer Interactions

Using the model shown in Fig. 2B, a BLM is considered to be a layer of oriented lipid molecules sandwiched between two aqueous electrolyte solutions.

In Sect. II.E it has been shown that a BLM consists of two lipid molecules oriented perpendicularly to the biface with their hydrophobic portion toward each other. Qualitatively, the stability of the BLM structure may be understood in terms of two opposing forces: van der Waals forces and steric repulsive forces. The latter forces come into play when the two hydrocarbon chains interact at a thickness of less than 100 Å (49). Vaidhyanathan and Goel (76) have discussed at length some aspects of the electrostatic energy of charged ions acting across a BLM. In their analysis they assume that two hydrocarbon layers of semi-infinite thickness, one having electrolyte solution on the left and the other having electrolyte solution on the right, are brought together with the hydrocarbon phase facing each other, thereby forming a BLM. It is postulated that the relevant electrostatic energy is the difference between the electrostatic energy of the hydrocarbon layer with electrolyte solutions on both sides and the sum of electrostatic energies of the two semi-infinite hydrocarbon layers described above. The nature of lipid in the BLM is not considered except to note that it has a very low dielectric constant value. With these assumptions in mind, Vaidhyanathan and Goel have computed the electrostatic energy as a function of the thickness of the BLM and ionic strength of surrounding solutions, which are taken to be identical in composition across the membrane.

In principle, it first appeared that these calculations could easily be accomplished. However, it turned out to be exceedingly difficult due to the changes in ion distribution that occur as the two semi-infinite layers approach each other. To avoid this difficulty it had to be further assumed that the perturbation in the distribution of charged species near the interfaces was negligible. If this crucial assumption is granted Vaidhyanathan and Goel state that the electrostatic energy relevant for the stability of the BLM is only due to the interaction of ions across the membrane. The calculated energy (identical with the free energy) for the assumed model is said to be of the same order of magnitude as that observed experimentally. From the electrostatic free energy versus thickness plot, a minimum is shown to exist corresponding to the measured BLM thickness (see Sect. II.E).

In contrast to the work cited above, Ohki and Fukuda (77) have presented a different picture regarding the stability of BLM. Using the second-order perturbation theory of Hirschfelder et al. (78), Ohki and Fukuda have considered the interaction between oriented lipid molecules in the BLM, which is pictured as two layers of dipoles. In their calculation they have taken into consideration various interactions such as dipole–dipole, dipole–induced dipole, and induced dipole–induced dipole. The main conclusion from these authors' work is that if the dimension of the BLM is small there is a net repulsive force between the permanent dipole layers. This interaction may be greater than the sum of the dipole–induced dipole and induced–dipole interactions. In general, the stability is owing to the attractive van der Waals–London dispersion forces.

Very recently, important work has been carried out by Parsegian and Ninham (79). These workers have calculated van der Waals forces for BLM and lipid–water systems using the powerful Lifshitz theory (80) which formerly was considered not applicable for calculating short range forces in the absence of complete spectral data. The main findings of Parsegian's work relevant to the BLM are as follows: (a) there is a large contribution to van der Waals forces through the infrared vibration spectrum of water, (b) solute that increase the index of refraction of water decrease the van der Waals forces across the BLM, and (c) the presence of a BLM between two aqueous solutions cannot be described by a simple index of refraction correction factor. It is suggested that the approach using the Lifshitz theory will be more successful than the method of pairwise summation in predicting variation of dispersion forces with addition of materials to mixture in condensed phases (e.g., a BLM separating two solutions).

2. The Dielectric Constant of BLM

As outlined in Sect. II.E, the thickness of BLM can be measured by three different methods. The electron microscopy of BLM is a complicated and involved procedure and the results obtained subject to artifacts. The two other methods (electrical capacitance and optical reflectance) are generally preferred but important parameters (dielectric constant and refractive index of the BLM) have to be assumed in order to use the equations. It should be mentioned also that for a membrane of about 2 molecules thick the meaning of dielectric constant (and refractive index) is ambiguous, since these quantities refer to macroscopic systems. Notwithstanding this difficulty, Ohki (81) has attempted to calculate these quantities for the BLM.

Insofar as is known, amphipathic molecules are necessary in the formation of stable BLM (see Fig. 5 and Ref. 41). Therefore, if one considers a BLM to be composed of oriented lipid molecules, the dielectric constant and refractive index of the BLM must be anisotropic. Starting from this assumption, Ohki has calculated the dielectric constant of oriented layers according to Kirkwood's theory (82). It is known from the early work (80) that the dielectric constant and refractive index are related to each other by the equations

$$n_0 = (\epsilon_\perp)^{1/2} \tag{32}$$

$$n_e^{-2} = \frac{\sin^2 \theta}{\epsilon_\parallel} + \frac{\cos^2 \theta}{\epsilon_\perp} \tag{33}$$

where n_0 = the refractive index for an ordinary wave,

n_e = the refractive index for an extraordinary wave,

θ = the angle which the wave vector intersects the optical axis,

$\epsilon_\parallel$ = the dielectric constant parallel to the optical axis,

$\epsilon_\perp$ = the dielectric constant perpendicular to the optical axis.

Ohki has investigated the polarization of a nonpolar dielectric in a homogeneous field. The calculation of the local field is made both for an oriented hydrocarbon layer in bulk and for an ultrathin layer of infinite extent. The dielectric constant is obtained as a function of the distance between the hydrocarbon chains and the thickness of the molecular layer. For an interchain distance of 4.47 Å and thickness 60 Å, the results are $\epsilon_\parallel^b = 2.49$ and $\epsilon_\perp^b = 1.74$, and $\epsilon_\parallel^f = 2.61$ and $\epsilon_\perp^f = 4.03$, respectively (the superscript b refers to the bulk and f to the film). For a slightly larger distance (5.00 Å), the values are $\epsilon_\parallel^f = 2.77$ and $\epsilon_\perp^f = 2.28$. By a judicious combination of experimental data Ohki has recently calculated the values for a lecithin BLM (81). These calculated values and those reported by other investigators are given in Table 7. It may be pointed out that the calculated values apply only to the hydrocarbon portion of the BLM, as has been explicitly assumed in Ohki's model. This is similar to the model used in the capacitance method. However, the results are seen to be in better agreement with the optically determined values. In the latter case the polar groups of the lipids are included in the assessment of BLM thickness (41).

TABLE 7
THEORETICAL AND EXPERIMENTAL CONSTANTS FOR A LECITHIN BLM

	Calculated (81)	Capacitance (38)	From optical measurements Ref. (48)	From optical measurements Ref. (41)
Thickness, t (Å)	54[a]	48[a]	72[b]	69[b]
Dielectric constant	2.65	2.07 (estimated)	2.76	2.66
Refractive index, $n_b = (\epsilon)^{1/2}$	1.63	1.44	1.66	1.60

[a] Hydrocarbon portion only (38).
[b] Hydrocarbon portion plus polar groups, which have been estimated to be about 5–13 Å per polar group (41).

3. Aggregation of Lipids and the Spontaneous Formation of BLM in Nature

As mentioned in the introduction the BLM came into being as a result of prior developments in surface and colloid science. The monolayer technique of Langmuir and the Gorter–Grendel bimolecular leaflet concept have been crucial in the formulation of the current theories of biological membrane structure. In this connection the behavior of lipids in aqueous solution should also be mentioned since it is pertinent to an understanding of BLM formation and stability. As early as 1913, McBain (83) suggested that amphiphatic compounds (e.g., soaps and lipids) aggregate in solution in the form of micelles. Two types of micelles were suggested by McBain: spherical and lamellar. In either one of these micellar forms the limiting structure (see Fig. 2B) is frequently idealized as composed of a bimolecular layer of amphipathic molecules (or ions). Thus the formation of a BLM in aqueous solution provides a direct demonstration of the existence of the lamellar micells postulated by McBain. The energetics of micelle formation have been considered by numerous investigators (84–86). The theory suggested by Debye (87) is sufficient for the present qualitative discussion, which is extended to a consideration of the spontaneous formation of BLM in nature (88).

According to Debye's theory, to form a lamellar micelle such as the one illustrated in Fig. 2B, work has to be done when the charged head groups are brought together. On the other hand, energy is released by bringing hydrocarbon tails in contact. At equilibrium the repulsive long-range Coulomb forces due to the charged heads and the short-range attractive van der Waals forces among the hydrocarbon tails must be balanced. Debye suggested that if the curve for the total energy W is

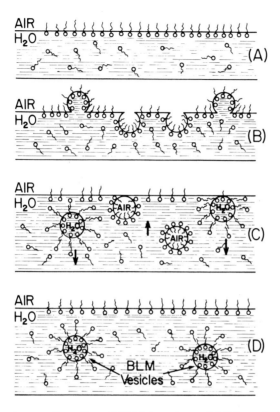

Fig. 24. Spontaneous formation of BLM in nature (A) Schematic illustration of amphipathic molecules (e.g., lipids) arranged in a monolayer at an air/water interface. (B) Turbulence at the interface. (C) Formation of two types of spherical micelles. Those with air trapped inside are not stable and tend to move toward interface. (D) Stabilization of spherical micelles having water trapped inside in the aqueous phase. The BLM visicles are stabilized as a result of interplay between short-range van der Waals forces and long-range forces (*88*).

plotted as a function of the number of molecules N, a minimum for a certain value $N = N_0$ will exist, at which point the energy W_0 of the micelle is negative. This implies that the micelle is more stable than N_0 separate molecules.

With the above ideas in mind it is suggested that in nature the formation of a BLM can take place according to the scheme illustrated in Fig. 24. Figure 24 depicts a situation where insufficient lipid molecules (e.g., lecithin) are present in the aqueous phase. According to the Gibbs adsorption equation [Eq. (21)], a greater concentration of lipid molecules will be located at the air/H_2O interface. Suppose that a turbulence is created at the interface—for example, a wind blowing as has been proposed by Goldacre (89)—two possible configurations may result as shown in Fig. 24B. There are two types of spheres: (1) those with aqueous solutions entrapped inside and (2) those with an air-filled interior, both of which could then submerge into the bulk phase as pictured in Fig. 24C. Those spheres with air inside are unstable and will float upward toward the air/H_2O interface. Initially, those spheres with water trapped inside also lack stability from an energetic point of view. However, the presence of lipid molecules in the bulk phase could easily stabilize the spheres by forming micelles in accordance to Debye's theory described above. The formation of BLM vesicles similar to the ones reported by Bangham et al. (5) is thus achieved (Fig. 24D).

Conceivably, the BLM vesicles—a type of primitive cell—could be generated in this fashion in nature by a process of selective adsorption. The membrane could be modified to exhibit both specificity and catalytic characteristics. For instance, the adsorption of chlorophyll molecules (and other pigments) could transform the BLM vesicles into primitive quantum converters. Could these modified BLM vesicles be the first cells evolved? This and other intriguing questions can now be legitimately asked. In fact, the formation of BLM vesicles of the type described above has been considered by Goldacre (89) and Calvin (90) in speculating about the origin of life.

It should be stated that the formation of BLM vesicles in the absence of protein and other macromolecules does not rule out the possibility of generating nonlipid BLMs. In view of the critical role generally assigned to proteins in the structure and function of biological membranes, it is entirely possible that BLM may be formed from protein and other bio-polymers alone under appropriate conditions. For readers interested in this approach the experiments and views of Green and Perdue (91), Fox (92), and Oparin (93) are of great interest.

III. The Effects of Modifiers on the Intrinsic Properties of BLM

From the data presented it can be seen that the physical properties of an unmodified BLM such as those formed from either phosphatidyl choline (lecithin) or oxidized cholesterol in an alkane solvent are strikingly similar to those expected of a layer of liquid hydrocarbon of equivalent thickness. These properties may be considered as the *intrinsic* properties of unmodified BLM. Before considering the effects of modifying agents (modifiers) on the intrinsic properties it is useful to tabulate the known properties of unmodified BLM and those extrapolated values for an ultrathin layer of liquid hydrocarbon. Table 8 gives a comparison of the

TABLE 8

COMPARISON OF PROPERTIES OF UNMODIFIED BLM AND A LAYER OF LIQUID HYDRO-
CARBON OF EQUIVALENT THICKNESS

Property	Unmodified BLM (experimental)	Liquid hydrocarbon (extrapolated)
Thickness (Å)	40–130	100
Resistance (Ω-cm^2)	10^8–10^9	10^8
Capacitance (μF-cm^{-2})	0.3–1.3	$\sim$1.0
Breakdown voltage (V/cm)	10^5–10^6	10^5–10^6
Dielectric constant	2–5	2–5
Refractive index	1.4–1.6	1.4–1.6
Water permeability (μ/sec)	8–24	$\sim$35
Interfacial tension (dynes/cm)	0.2–6	$\sim$50
Potential difference per 10-fold concentration of KCl (mV)	$\sim$0	$\sim$0
Electrical excitability	None	None
Photoelectric effects	None	None

properties. It is evident from the data given that an unmodified BLM appears to be a poor model for the biological membrane. For instance it is well known that biological membranes are ion-selective. In the case of nerve membrane, electrical "excitability" is one of the most unique features (see Sect. IV.D). The inadequacies of an unmodified BLM as a model for the biological membrane were realized from the results obtained in the initial experiments.

In an attempt to modify the intrinsic properties of the BLM, several hundred compounds were evaluated. Among these the following groups of materials have been tried: common proteins, enzymes, surfactants, fermentation products, vitamins, tissue extracts (e.g., retina), and a variety of organic and inorganic compounds (24). This broad and preliminary testing of materials has led to the discovery of a modifier of uncertain composition (still not known to date) termed "excitability inducing material" (EIM) that not only dramatically reduced the BLM resistance but induced electrical "excitability" (24). Beginning with EIM, a number of modifying agents (or modifiers) has been discovered which, when present in the BLM, impart new properties that are of biological interest. The BLM modifiers can be broadly divided into 5 categories: (1) those altering the passive electrical properties, (2) those changing the mechanical properties, (3) those conferring ion selectivity, (4) those inducing excitability, and (5) those generating photoelectric effects. The effects of these five categories of BLM modifiers will be considered under separate headings after a discussion of two standard lipid solutions from which most BLM have been generated. The intrinsic properties listed in Table 8 refer mainly to these unmodified BLMs.

A. The Composition of Standard BLM-Forming Solutions

Phosphatidyl choline (lecithin) in a saturated hydrocarbon solvent (e.g., n-decane or in a more complex solvent mixture) developed by Huang et al. (71) and Hanai et al. (38) constitutes the first of two standard BLM-forming solutions commonly used. This solution is simply prepared by dissolving purified egg lecithin (1–2%) in the solvent. The BLM formed from phospholipid solution in general are not very durable (lifetime about 30 min or less) owing to the sensitivity of the lipid oxidation (71).

The second standard BLM-forming solution is that of oxidized cholesterol in n-octane (94). This lipid solution, although of unspecified composition (see Fig. 25), is prepared as follows:

A 4% solution of cholesterol in n-octane is heated at the reflux temperature for 6.5 h in a three-neck flask. The flask is fitted with a thermometer, a condenser, and a gas dispersing tube (fritted glass, medium porosity) through which pure oxygen is introduced. The rate of oxygen flow is about 90 ml/min. (A shorter duration works equally well if the solvent has a higher boiling point or a faster rate of oxygen flow is used.) The resulting solution is colorless with white precipitate upon cooling. The supernatant is used directly to form BLM.

The oxidized cholesterol solution, prepared according to (26), gives exceedingly stable BLM over a wide temperature range (10–50°) and in various electrolyte concentrations (e.g., from distilled water to saturated NaCl or KCl). This lipid solution has been employed in a number of studies (39, 61, 94–100). It may be useful to consider here the likely composition of the oxidized cholesterol solution.

The ease of oxidation of cholesterol by molecular oxygen under suitable conditions has been discussed in detail by Bergstrom (*101*). The mechanism of oxidation of cholesterol in colloidal state (in aqueous solution) postulated by Bergstrom may be extended to the present case. Briefly, it is believed that the 7-hydroperoxide (B) acts as an intermediate as shown in Fig. 25.

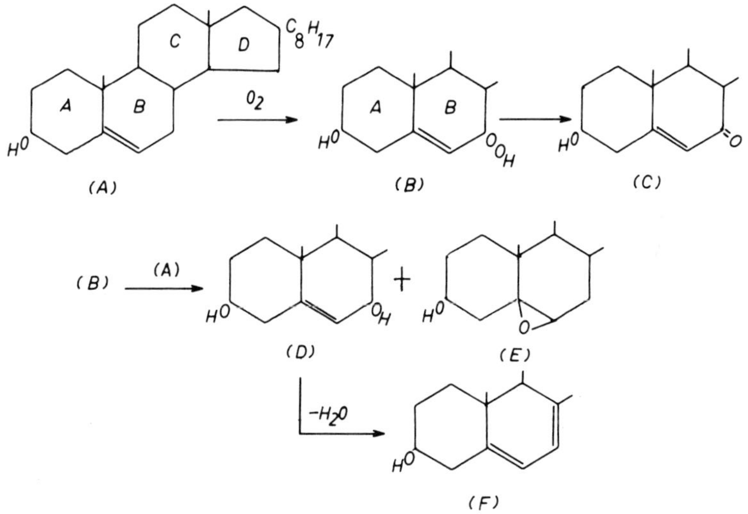

Fig. 25. Probable mechanism of oxidation of cholesterol (A) by molecular oxygen in an aliphatic hydrocarbon solvent. (B) 7-Hydroperoxide intermediate; (C) 7-keto-cholesterol; (D) 7-hydroxylcholesterol; (E) 8-epoxycholesterol; (F) 7-dehydrocholesterol.

It should be stressed here that the actual composition of the black lipid membrane under consideration is not known precisely. It seems probable, however, that small quantity of the solvent used could be present in the BLM, although recent results of Henn and Thompson (*102*) suggest that the molar ratio of decane to phospholipid can be as high as 10:1. As discussed in Ref. (*49*) the solvent molecules, if present in the BLM, can serve as "space fillters," thereby enhancing the stability of the membrane through van der Waals interactions. Although the optimum concentration (and composition) of a lipid solution for stable black film formation is accurately known, it seems most unlikely that the lipid solution used represents the actual composition in the BLM as already discussed in Sect. II.G.4.

B. Modifiers That Alter Basic Electrical Properties

The general arrangement of the cell for the measurements of electrical properties may be represented as follows:

$$\text{Reference} \mid \text{Aqueous} \mid \text{BLM on Teflon} \mid \text{Aqueous} \mid \text{Reference} \atop \text{electrode} \quad \text{solution} \quad \text{support} \qquad \text{solution} \quad \text{electrode} \quad (34)$$

Both ac and dc methods can be used for such measurements, although the dc method has been more frequently employed (26). The basic electrical properties of the BLM generally include the resistance, capacitance, and dielectric breakdown voltage. For an unmodified oxidized cholesterol BLM in 0.1 M KCl or NaCl, the following typical values are usually obtained: resistance is greater than 10^8 Ω-cm^2, capacitance is about 0.6 μF/cm^2, and dielectric breakdown voltage is around 300 mV. The current/voltage curve at low applied voltages obeys Ohm's law. Frequently, a low resistance is noted just prior to membrane breakdown. As has been mentioned above (see Table 8), these properties are remarkably similar to those of a liquid hydrocarbon layer of equivalent thickness. The intrinsic electrical properties of a BLM can be modified by a variety of materials. These include simple inorganic and organic ions, surfactants, polypeptides, and proteins. In addition, physical parameters such as temperature and applied electrical field can also alter the electrical properties of BLM.

1. Inorganic Ions

For BLM formed from brain lipids and phospholipids such as egg lecithin, common electrolytes such as NaCl and KCl have minimal effect on the basic electrical properties (24). Hanai et al. have reported that in lecithin BLM the electrical properties are not altered when the temperature is varied from 20–37° or when the concentration of any electrolyte (NaCl, KCl, HCl, and CaCl$_2$) is varied from 10^{-3} to 10^{-1} N (38). The first simple electrolyte that produced a dramatic effect on BLM was KI, which was reported by Lauger et al. (103). They noted that when the KCl solution on one side of the BLM was replaced by isotonic KI the resistance dropped by some three orders of magnitude (i.e., from about 5×10^8 to 5×10^5 Ω-cm^2). In a later report (104) the resistance of membrane was observed to decrease by seven orders of magnitude when I$_2$ in combination with I$^-$ was added to one side. Two mechanisms have been proposed to account for the observed phenomenon. In the first proposal the formation of a molecular complex in the BLM is assumed, which is partially dissociated into fixed lecithin I^+ and mobile I^- ions. The latter is capable of moving across the BLM under the influence of an external field. In the second proposal a mechanism involving the reaction

$$\tfrac{1}{2}I_2 + e \rightarrow I^- \tag{35}$$

has been suggested where the BLM is assumed to be an electronic conductor. The I^- ion moving toward the opposite direction of the electric field is discharged on one side of the BLM. The "free" electron that migrates across the membrane is reduced on the other side. Although both suggestions are reasonable, the possibility of the electronic mechanism was greatly decreased because no photoelectric effect was detected even with intense illumination (104). This is due to the fact that most organic semiconductors are also good photoconductors—for example, the anthracene–I_2 complex studied by Kallmann and Pope (105). Rosenberg and Jendrasiak have also investigated iodine modified BLM (96) and have suggested that a charge–transfer complex might be formed in the membrane, which could participate in the conduction process. The suggestion of Rosenberg and Jendrasiak is based upon the spectroscopic evidence that the formation of donor–acceptor complexes between iodine and lecithin or cholesterol was involved (106). Contrary to the above suggestions, Finkelstein and Cass favored an ionic conduction mechanism involving the polyiodides such as I_3^- and I_5^- (107).

Recent experiments using oxidized cholesterol BLM provide additional interesting observations. In the presence of I^-–I_2, the oxidized cholesterol BLM behaved like an ideal iodide electrode (108). The response to I^- can be easily brought about by incorporating I_2 in the lipid solution prior to BLM formation. This phenomenon has been found to be independent of BLM-forming solution since identical results can be obtained with BLM formed from chloroplast lipids and synthetic surfactants. These results seem to support the suggestions proposed by earlier workers that a BLM in the presence of I^-–I_2 complex is effectively permeable to iodide ions. The current/voltage curves for an asymmetrical system using KI–I_2 and $Na_2S_2O_3$ are shown in Fig. 26. The I/V curve shown in Fig. 26 was obtained with the aid of an automatic recording polarograph (Polarecord Model E, Metrohm Ltd.). The presence of $Na_2S_2O_3$ on one side of the BLM generated a potential difference of about 45 mV. It is evident that the I/V curve is no longer linear but still entirely symmetrical. The presence of I_2 in the BLM could facilitate the selective iodide transport, which in turn might manifest itself in the development of a concentration potential in accordance with the Nernst equation. On the other hand, an alternative explanation based upon redox–electrode reaction is also plausible (see Sect. III.D).

The principal current carriers in an unmodified BLM have been discussed by Miyamoto and Thompson (*42*). They state that the principal conduction pathway is through the BLM itself. For alkali cations, only

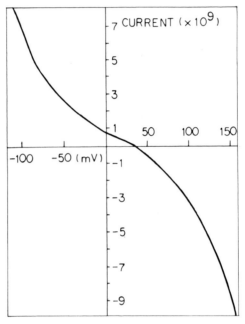

Fig. 26. Current/voltage curve of an oxidized cholesterol BLM in 0.1 M KI + 4×10^{-4} M iodine. In addition, the inner chamber contained 5×10^{-3} M $Na_2S_2O_3$.

Li^+ altered the BLM resistance. On the other hand, the chlorides of di- and trivalent ions such as Cd^{2+}, Cu^{2+}, and Mn^{2+} produced an increase in resistance in some preliminary experiments. In contrast, Fe^{3+} or Fe^{2+} produced a decrease in the resistance of the same BLM. Miyamoto and Thompson have noted the nonlinearity of the I/V curves and suggested that the existence of ionic association may occur within the BLM (*42*). The degree of association of ions is known to decrease with increasing field strength across the membrane. According to Bjerrum's theory of ion association (*109*),

$$M^- + C^+ \rightarrow M^- — C^+ \tag{36}$$

The dissociation constant for the reaction is given by

$$K^{-1} = \frac{4N\pi}{1000} \left(\frac{e^2}{DkT}\right) Q(b) \tag{37}$$

where N, D, k, T, and e have their usual significance. In Eq. (36) M^-—C^+ stands for an undissociated complex. The quantity $Q(b)$ is a definite integral that is a function of (b) defined by the following equation

$$b = \frac{e^2}{DkTa} \tag{38}$$

where a is the distance between the ion pair. In principle one should be able to calculate the dissociation constant of the ion pair in the hydrophobic region of the BLM. The Bjerrum theory of ion association has been quantitatively confirmed for aqueous solutions and nonaqueous systems by the work of Kraus and Fuoss (110). Whether the theory can be extended to BLM systems remains to be seen. The evidence at hand suggests that when modifiers or ionizable species are present in the BLM system the current/voltage curves are no longer linear (Fig. 26). It seems probable that the dissociation of ion-pairs under the condition of intense field (1–5×10^5 V/cm) may be responsible for the nonlinear I/V behavior. In fact, the production of additional charge carriers as a function of the electric field (the Wien effect) has been considered recently in conjunction with nerve excitation (111). The possibility that a similar mechanism is also operative in the BLM system under the influence of modifiers and electric field seems quite likely.

2. Organic Compounds

The electrical resistance of cholesterol BLM (stabilized by hexadecyl-trimethylammonium bromide) is affected by the concentration and the anionic species used (37). For most electrolytes studied, the dc membrane resistance decreases by several orders of magnitude when the concentration is about 0.1 M. The order of specific effect of anions in lowering membrane resistance has been found to be $I^- > Br^- > SO_4^{2-} > Cl^- > F^-$. The observed effect is discussed qualitatively in terms of the interaction between the quaternary ammonium groups and anionic species at the membrane–water interface. The results of a large number of measurements are presented in Fig. 27.

The dc resistance of lecithin and phosphatidylserine BLM as a function of pH, electrolyte composition and electrolyte concentration has been

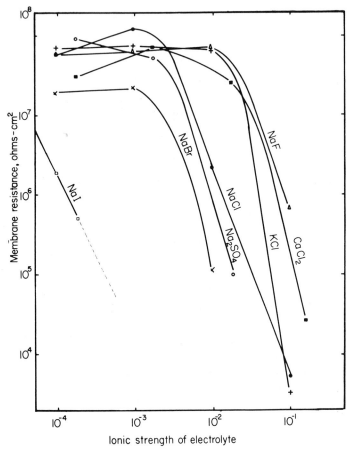

Fig. 27. Resistance of black cholesterol membrane in aqueous medium containing 0.008% HDTAB and various salts (pH 6.85–6.95) (*37*).

investigated by Ohki and Goldup (*112*). The resistance of these BLM depends greatly upon pH, the particular ions present, and their concentration in solution. For the lecithin BLM, the membrane resistance has a maximum at about pH 4, whereas a value between pH 3–4 is reported for phosphatidylserine BLM. Furthermore, addition of divalent ions (e.g., Ca^{2+} and Mg^{2+}) to the bathing solution alters greatly the behavior of the BLM resistance with respect to pH. Noguchi and Koga have also investigated a variety of phospholipid BLM (*113*). The lipids used include lecithin, cephalin, and phosphatidylserine. Their results differ significantly from those reported by Ohki and Goldup. The maximum for all these phospholipid BLM lies somewhere between pH 7–8. In addition, two other peaks (for conductance) are also present at pH values 4–5 and pH 9. The reason for these discrepancies is not known. Perhaps the difficulty lies in the formation of reproducible BLM (26).

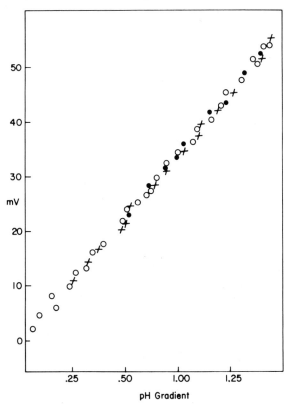

Fig. 28. Membrane potentials of cholesterol BLM in aqueous solution containing 0.002% of N-lauryl, myristyl-β-aminopropionic acid separating two aqueous solutions at different pH values.

When a pH gradient exists across lecithin BLM, membrane potentials of about 60 mV/pH have been observed by Noguchi and Koga (113). This would appear to be the first published report that an unmodified BLM can respond to H$^+$ without the presence of a modifier (see Sect. III.D). The BLMs formed from cholesterol-N-lauryl, myristyl-β-aminopropionic acid (Deriphat 170C, General Mills Chemical Division,

Kankakee, Illinois) are known to exhibit H^+ selectivity (J. Beggs, unpublished results, 1967). However, in this particular BLM system the slope is only about 45 mV/pH in the range examined (Fig. 28).

The effects of electric fields across a BLM can also alter its basic properties (39, 40, 100). The first two groups of workers showed that with the application of 100 mV across a phospholipid BLM the total capacitance is increased by 3–12% and is proportional to the square of the applied voltage. Using oxidized cholesterol BLM, White (100) has found that the BLM capacitance is related to voltage by a simple relation:

$$C_m = C_0 + \beta V_a^2 \tag{39}$$

where C_m = the BLM capacitance in $\mu F/cm^2$

C_0 = the capacitance at the zero voltage

β = a constant for a given temperature

V_a = the applied voltage.

White has also found that both C_0 and β are linear in temperature. As temperature decreases, β decreases and C_0 increases. In addition to the effects of applied voltage, Rosen and Sutton (39) have also noted that the percentage change of BLM capacitance is dependent on the electrolyte concentration. A minimum was detected at concentrations of 0.1 M of uni-univalent electrolytes (e.g., KCl, NaCl, and KI). For the same lecithin BLM in a 2–2 electrolyte ($MgSO_4$) the minimum was at 0.025 M. These minima are attributed to a change of ionic distribution near the membrane/solution owing to the presence of phosphate and trimethylammonium groups. Evidence in support of this explanation is that little capacitance change as a function of electrolyte concentration was found in the case of oxidized cholesterol BLM in which the ionizable groups are presumably absent (see Fig. 25).

It should be mentioned that both the lipid composition and the solvent used have a pronounced effect on the electrical properties of BLM. For example, lecithin BLM made from n-decane solvent has a resistance some two orders of magnitude higher than those formed from chloroform–methanol–tetradecane mixture (71), a point noted by Bangham (5). In this connection it is of interest to compare the value of sorbitan tristearate BLM reported by Bradley and Dighe (114). This BLM was found to be ohmic up to 300 mV and the average resistance was 5×10^6 Ω-cm^2. The lipid solvent was chloroform. Thus it seems probable that the use of a polar lipid solvent such as chloroform or alcohol may be an important factor responsible for the reduction of BLM resistance. The effect of lipid composition on the BLM resistance has been dramatically demonstrated by van Zutphen and van Deenen (*115*). By incorporating 2 mole% of lysolecithin (prepared by hydrolyzing lecithin with snake-venom phospholipase A) in the lipid solution, a very significant decrease in the BLM resistance was detected (from about 10^7 to 10^5 Ω-cm^2). It was found that it was not possible to form stable BLM when the lipid solution contained about 20% of lysolecithin (see Sect. III.C). On the other hand, Leslie and Chapman (*116*) have found that by incorporating either β-carotene or all-*trans* retinene into lecithin BLM the electrical conductivity is not unduly different from that of the unmodified BLM. These workers have also experimented with vitamin K and ubiquinone but found that no stable BLM could be obtained.

The effect of surfactants on the electrical properties of BLM has also been investigated by Hashimoto (*117*). Some of his findings are consistent with those reported earlier (*37, 63*). In addition, Hashimoto finds that the resistance is nearly independent of temperature (7–45°C) and pH and suggests that the structural change of the membrane may be caused by both electrostatic and hydrophobic interactions of ionic surfactants with the membrane (see also Ref. 37) and that the electrostatic interaction is necessary for the change in the BLM structure.

In addition to surfactants, other chemically well defined compounds have been used. Liberman and Babakov (*118*) and Liberman and Topaly (*119*) have investigated the effects of tetrachlorotrifluorobenzimidazole (TTFB) and tetraphenylboron (TϕB) on phospholipid BLM. They suggest that TTFB and TϕB may serve as charge carriers for H$^+$, and the H$^+$ permeability is voltage-dependent. Further, they compare the impedance diagrams of modified BLM with those obtained from the muscle fiber with intracellular recording. The similarity of the impedance diagrams is offered as evidence in that the physical mechanism responsible is the same in both cases. More recently, LeBlanc has also examined the TϕB modified BLM system (*120*). He interprets the complex current/voltage curves of these membranes in terms of the superposition of large, decaying current transients on top of a saturating, steady-state current/voltage characteristic. At low concentration ($< 3 \times 10^{-4}$ M) of the TϕB on either side of the BLM, the permeability to TϕB is so large that transport is limited mainly by diffusion in the bathing solution. At higher concentrations the TϕB flux is found to be independent of concentration. It thus appears that there is an upper limit to the number of TϕB that can enter the membrane. Further, LeBlanc states that space charge is a likely cause of the observed behavior (*120*).

Finally, mention must be made of the interesting observation of Ochs and Burton (*121*). They have found that a standard BLM formed in the usual manner can respond to mechanical vibration that has caused a capacitance change. A relationship is noted between the BLM capacitance change and the waveform of the vibration.

C. Modifiers That Change the Mechanical Properties

Under this heading consideration will be given to effects of modifiers on the stability, thickness, permeability, and bifacial tension of the BLM. The mechanical stability of the lecithin BLM is usually increased by the addition of cholesterol and its derivatives. In addition, the presence of cholesterol increases the capacitance of lecithin BLM to about 0.6 $\mu F/cm^2$ (*122*). This change may be due either to the thickness reduction or to the dielectric constant increase [see Eq. (1)]. Finkelstin and Cass (*107*) have found that permeability to water is markedly dependent on the molar ratio of cholesterol to phospholipid in the BLM-forming solution. For example, P_0 is about 42 μ/sec for a brain phospholipid BLM without added cholesterol, whereas P_0 is only about 8 μ/sec when the molar ratio of phospholipid/cholesterol is 1:8. A similar effect has been noted with ergosterol. This decrease in permeability to water is attributed to an increase in the viscosity of the hydrocarbon region of the BLM, resulting in a decrease of the diffusion coefficient of water within this phase. It is interesting to note that the presence of cholesterol causes a decrease of the apparent area occupied by lecithin molecules in monolayer systems (*123*). The condensing effect of cholesterol is therefore likely to reduce the available "free" volume for permeating water molecules.

The presence of modifiers such as proteins can have profound effect on the BLM stability and bifacial tension. Early workers (*24*) have observed that BLM formed from lipids alone can have appreciable interfacial tension and in this respect resemble drum-taut air soap films, which shatter violently when broken. A bulge formed by hydrostatic pressure upon the BLM will flatten out again upon release of the pressure. In contrast, BLM formed from total lipids containing proteolipids can show very little interfacial tension. Once bulged, this type of BLM does not always return to its planar configuration after the pressure is equilized. Instead the membrane can become floppy, moving back and forth with the convection currents in the bathing solution (*24*). Such BLM rupture less abruptly when broken and sometimes give the appearance of tearing rather than shattering. The oxidized cholesterol BLM exhibit similar characteristics. Evidently the presence of protein is not an absolute requirement for these properties (*63*). It should be mentioned that the presence of proteins in the aqueous solution usually makes the formation of stable BLM exceedingly difficult if not impossible. This, however, may not necessarily imply that protein modified BLM are inherently less stable. It seems more likely that proteins and their related compounds are also highly surface-active materials that can modify the BLM support to such an extent that adhesion is difficult for the lipid solution.

The instability of a BLM can be caused by at least two other factors. Van Zutphen et al. (*124*) found that in BLMs formed from lecithin and cholesterol, antibiotic Filipin caused a marked decrease in stability. No effects of this compound were observed with cholesterol-free BLM. This finding implies that drugs may selectively interact with a particular membrane constituent. An asymmetrical distribution of ions can also impair BLM stability. Papahadjopoulos and Ohki (*125*) have made BLM of phosphatidylserine (PS). The PS-BLM can be formed even at pH 9.5 with Ca^{2+} (100 mM) present on both sides. However, if Ca^{2+} is added only to one side of the BLM, it ruptures at a concentration of Ca^{2+} as low as 1 mM. Papahadjopoulos and Ohki explain that the instability of their BLM is due to the difference in interfacial energy between the two opposing sides of the membrane. Under an asymmetrical situation such as described above, it is suggested that molecules or clusters of molecules will flip from one side to the other. In doing so they could increase the permeability of the membrane which under extreme conditions cause rupture.

The total protein derived from red blood cells has been tried on lecithin BLM by Maddy et al. (*126*). The principal effects are a reduction of the bifacial tension of the membrane and an increase in optical reflectance. The latter effect has also been observed by Smekal (*127*). Using the optical set-up illustrated in Fig. 11, Smekal has carried out a series of BLM–protein interaction experiments. In these experiments an oxidized cholesterol BLM is first formed on the Teflon loop and its reflectivity measured. A solution of protein is then injected into the cell and the reflectivity of the membrane is again measured. The following proteins were tested in this manner: alcohol dehydrogenase, insulin, γ-globulin, chymotrypsin, trypsin, and ribonuclease. In the case of alcohol dehydrogenase the thickness of the membrane increased from 50 to about 100 Å The other proteins increased the membrane thickness to about 80 Å. In most of these BLM–protein interaction experiments, the pH of the bathing solution had a marked effect on membrane reflectivity.

D. Modifiers That Confer Ion Selectivity

As described in Sect. III.B, the KI–I$_2$ modified BLM has a resistance of about 3 orders of magnitude lower than that of standard BLM formed from either lecithin or oxidized cholesterol. Lauger et al. (*103*) have further shown that in addition to resistance reduction, the iodide–iodine modified BLM are highly selective to I$^-$; i.e., a 59 mV potential difference appears across the membrane per 10-fold concentration gradient of iodide, the presence of iodine being required however. The work of Lauger et al. has been confirmed by Finkelstein and Cass (*107*), Jendrasiak (*98*), and Rosenberg and Bhowmik (*106*). Finkelstein and Cass interpret their results in terms of a carrier mechanism in which the postulated carriers are thought to be polyiodide complexes.

In all previous work, iodine and/or iodide were introduced into the aqueous solution. The BLM in these cases are presumably being modified by the additives through a sorption process. Recently, iodine has been incorporated directly into the BLM-forming solutions (oxidized cholesterol or chloroplast extract). The BLM modified in this fashion have been found to respond ideally to iodide. The presence of other ions such as Cl$^-$, SO$_4^{2-}$, or F$^-$ did not interfere with the "electrode" response to iodide ions (*108*). The behavior of I$_2$-containing BLM is reminiscent of a metallic electrode reversible to its ion (e.g., Ag electrode and Ag$^+$). The development of the membrane potential in this case does not seem to require a concentration gradient across the BLM. It seems therefore that I$^-$ or its complexes need not be transported across the membrane in order to give rise to a membrane potential.

The intrinsic properties of lecithin and oxidized cholesterol BLM show little ion selectivity (apart from H$^+$ and I$^-$ noted above). This is consistent with the expected properties of the ultrathin hydrocarbon layer of the BLM. These inert properties of BLM, however, can be modified by a number of neutral macrocyclic molecules such as those listed in Table 9. The cyclic peptide valinomycin, for instance, has been found to confer a selective order among the alkali metal cations as follows: Rb$^+$ > K$^+$ > Cs$^+$ > Na$^+$ ≃ Li$^+$ (*128, 129*). In order to calculate the selectivity coefficient K, the BLM is considered to be similar to an ion-exchange system (e.g., cation-exchange resins) in which there exists an equilibrium of the type

$$\text{BLM—C}_1^+ + \text{C}_2^+ \rightarrow \text{BLM—C}_2^+ + \text{C}_1^+ \qquad (40)$$

where C$_1^+$, and C$_2^+$ are exchanging cations. The selectivity coefficient K is obtained from measurements of membrane potential and calculated with

TABLE 9
MODIFIERS THAT AFFECT ION SELECTIVITY IN BLM AND NATURAL MEMBRANES

Compound	Observed effect	Ref.
Valinomycin	Increased K^+ uptake	(136)
		(131)
		(35)
		(128)
		(129)
Amphotercin B	Anion selective	(107)
Nystatin	Resistance reduction, anion selective	(107)
I^-–I_2 complex	Resistance reduction, membrane potential	(103)
		(108)
ATPase	$Na^+ > K^+$	(99)

the aid of an empirical equation (130):

$$E = E' + \frac{RT}{zF} [(C_1^+)^{1/n} + K^{1/n}(C_2^+)^{1/n}]^n \tag{41}$$

where E' and n are constants with $n \geqslant 1$. The effect of valinomycin was first noted by Bangham et al. (35) and by Chappell and Crofts (131) and Henderson et al. (132) in microvesicles. The observations of Mueller and Rudin (128), and Lev and Buzhinsky (133) has been extended by other investigators. Andreoli et al. (134) have shown that the action of valinomycin is independent of the lipid composition. It should be mentioned that the mechanism by which valinomycin produces its effect on BLM and certain biological membranes is not clearly understood. At present, at least two different mechanisms have been proposed. Mueller and Rudin (128) favor the creation of specific pores in the BLM whereas Lardy et al. (135) and Pressman (136) suggest that valinomycin may act as molecular carriers. Recently, Eisenman et al. have suggested a number of possible mechanisms to account for the observed effects (137).

In addition to I^--I_2, Amphotericin B and Nystatin make BLM anion-selective. Finkelstein and Cass (107) have reported that bi-ionic potentials up to 55–60 mV can develop across nystatin-treated BLM, separating equal molar concentrations of NaCl and sodium isethionate. The observed selectivity among anions appears to be dependent on ionic size, however.

The BLMs generated from chloroplast extract exhibit photoelectric effects (see Sect. III.F and IV.B). The photoelectric phenomena in BLM provide strong evidence that electronic charge carriers are involved in the membrane (*138*). This type of BLM should therefore be capable of functioning as a redox electrode similar to the platinum electrode. To test this idea a chloroplast BLM (Chl-BLM) separating two buffered solutions is formed. To one side of the BLM a known quantity of Fe^{2+} is added, producing a few millivolts across the membrane. The Fe^{2+} is then titrated with $KMnO_4$. Figure 29 shows a typical titration curve of Fe^{2+} with $KMnO_4$. It is evident that a titration curve of classical form is

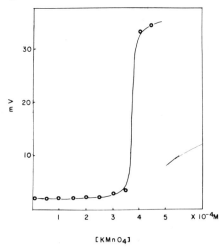

Fig. 29. Potentiometric titration of Fe^{2+} with $KMnO_4$ using chloroplast BLM as the indicator electrode. The rapid rise occurs at the equivalent point (obtained by S. P. Verma).

obtained with a rapid rise at the equivalent concentration. This evidence, together with the iodine-containing BLM, suggests that a BLM (after suitable modification) can behave like a redox electrode. In other words, a modified BLM can be considered either as a reversible electrode (i.e., a nonmetal and its corresponding anions) or as an inert electrode analogous to a platinum electrode that merely serves as a conductor for facilitating electron transfer.

E. Modifiers That Induce Electrical Excitability

Electrical excitability may be described as non-ohmic and unusual electrical behavior displayed by specially treated BLM. The best known but still incompletely understood modifier called EIM (excitability inducing material) can induce electrical excitability (*24, 139*)—i.e., an EIM-modified black lipid membrane whose resistance has been lowered by 3–4 orders of magnitude will respond to constant current pulses. The membrane resistance is voltage-dependent, and at a definite threshold potential increases regeneratively by a factor of 5–10. These unusual properties are reminiscent of the action potentials observed in frog nerves and alga *Valonia*. A further description of EIM-induced phenomena will be given in Sect. IV.D.

It should be mentioned here that in addition to EIM a number of chemically defined compounds can also induce electrical excitability in BLM. The most outstanding example is alamethicin, a cyclopeptide antibiotic. This compound, in the hands of Mueller and Rudin (*97*), develops a cationic conductivity in BLM made of mixed brain lipids, egg lecithin, or oxidized cholesterol. The most unusual finding is that when alamethicin is brought together with a minute quantity of protamine and in the presence of an ionic gradient it develops characteristics of negative resistance. Further, Mueller and Rudin have observed delayed rectification, bistable changes of membrane potential, and single or rhythmic firing. In BLM formed from oxidized cholesterol and dodecyl acid phosphate (*43*) the action potentials resemble those occurring in frog skin and in EIM-modified membranes described above (see also Sect. IV.D).

F. Modifiers That Generate Photoelectric Effects

Photoelectric effects may be broadly divided into three categories: (1) the classical electron emission of solids by light, (2) the production of voltage across a barrier by light, and (3) the marked increase in electrical conductivity in a material by light. Underlying all these phenomena is the generation of *electronic* charge carriers upon illumination. In the following paragraph a brief comparison is made for a photoactive BLM composed of chloroplast pigments and a standard silicon solar cell. Section IV.B gives a detailed account of photoelectric phenomena in BLM.

A silicon barrier cell of standard construction is used in place of the BLM. The light response of the solar cell as a function of time is shown in Fig. 30. In the same figure the response of a Chl-BLM to light is also shown. It is seen that the silicon solar cell behaves like the so-called "hard"-type cell and the Chl-BLM belongs to the "soft" type, since in the latter it takes several seconds for the photocurrent to reach the maximum value. The softness of the Chl-BLM may be due to a number of factors; the large resistance and capacitance of the BLM may be responsible for the increase in response time. Nevertheless, the similarity of the silicon solar cell and the Chl-BLM is striking. The observation of photo-emf (open-circuit voltage induced by light) together with the action spectrum of the Chl-BLM (see Sect. IV.B) provides strong evidence for the separation of *electronic* charges in the membrane. It is also conceivable

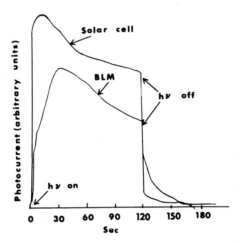

Fig. 30. Comparison of chloroplast BLM with a standard silicon solar cell with regard to light response under prolonged illumination (*174*).

that electron donors and acceptors, when present in the aqueous phase, can serve as efficient sources of charges for the membrane. In fact, a photo-emf of greater than 45 mV has been observed in a Chl-BLM in the presence of hydroquinone and Fe^{3+} on opposite sides of the bathing solution. Tien, unpublished observation, 1968).

The results obtained for the photoactive BLM under asymmetrical conditions can be explained by a simple model based upon the Langmuir adsorption theory. The basic assumption is that the photogenerated charges (e.g., electrons) are attracted by the electron traps situated at the solution/BLM biface, the density of the traps being greater on the side containing the added electron acceptor. The equation for the open-circuit photo-emf is given by

$$E_{op} = \frac{2.303kT}{e} \log (n_{out}/n_{in}) \qquad (42)$$

where k is the Boltzmann constant, e is the electronic charge, n_{out} and n_{in} are, respectively, the number of filled electron traps situated on the opposite sides of the biface (subscripts out and in denote the outer and inner side). It can be readily shown by a kinetic argument similar to the one given by Langmuir that n_{out} may be expressed in terms of the rate constant and light intensity, i.e.,

$$n_{out} = \frac{IN_{out}}{I + k_{out}} \tag{43}$$

where I is the light intensity, N_{out} is the number of available traps, and k_{out} is the ratio of rate constants. Similarly, one can write

$$n_{in} = \frac{IN_{in}}{I + k_{in}} \tag{44}$$

Substituting Eqs. (43) and (44) into Eq. (42) we have

$$E_{op} = \frac{2.303\,kT}{e} \log \left[\frac{N_{out}(I + k_{in})}{N_{in}(I + k_{out})} \right] \tag{45}$$

At low light intensities it can be shown that Eq. (45) reduces to

$$E_{op} = \psi \log kI \tag{46}$$

when $N_{out} > N_{in}$ and $k_{out} > k_{in}$. In Eq. (46) ψ denotes $2.303kT/e$ and $k = 1/k_{in}$. In the limiting case of high light intensities, Eq. (45) becomes

$$E_{op} = \psi \log (N_{out}/N_{in}) \tag{47}$$

i.e., for a given photoactive BLM system under saturating light intensity, a maximum photo-emf will be obtained, in which E_{op} depends solely upon the ratio of traps across the biface.

In the case discussed above, the resulting electrons and positive holes across the BLM may initiate redox (oxidation–reduction) reactions. It seems probable that these reactions will be demonstrated in suitably modified BLM to mimic certain aspects of biological processes such as the photophosphorylation in the plant chloroplasts. The incorporation of other modifiers that generate photoelectric effects, such as retinals and other carotenoid pigments, will be considered in Sect. IV.E.

Before concluding this section, mention should be made of the results obtained recently with the use of a fluorescent probe for studying BLM-protein interactions (127). Recently, Azzi et al. (140) and Rubalcava et al. (141) have used ANS (8-anilino-1-naphthalene-sulfonate) in studies of biological membranes. It has been known that the fluorescence properties of ANS and its derivatives are dependent upon the polarity and steric structure of the environment (*142*). In aqueous solution, ANS is practically nonfluorescent but fluoresces with appreciable quantum yields when dissolved in organic solvents and when associated with proteins (*143*). These facts make ANS an ideal probe for investigating BLM–protein

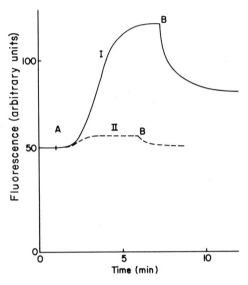

Fig. 31. Fluorescence of BLM treated with 8-anilino-1-napthalene sulfonate (A) and bovine serum albumin (B). Curve I, ANS added before BSA. Curve II, BSA added first (*127*).

interactions. In our experiments BLM were formed from (1) oxidized cholesterol or (2) a mixture of oxidized cholesterol and egg lecithin on a Teflon loop (see Fig. 12). A low pressure mercury lamp with a selected interference provided excitation at 365 mμ. The protein chosen for this investigation was bovine serum albumin (BSA).

The results obtained by Smekal are shown in Fig. 31. The intensity of the fluorescence response depends on which of the two compounds (BSA and ANS) is introduced into the aqueous phase first. In Curve I of Fig. 31, ANS is adsorbed into the BLM first, followed by the addition of BSA. The fluorescence response is about three times larger than in the second case (Curve II) where BSA is adsorbed into the BLM before ANS is added. Washing the BLM with 0.1 M KCl (point B) decreases the fluorescence to half of its original intensity in the first case (ANS followed by BSA), while in the second case (Curve II) the fluorescence returns almost to the base line. The maximum of pH-dependent fluorescence falls between pH 5.6 and 7 with a sharper rise of the curve on the acidic side; however, the change is not significant. When BLM was previously saturated with ANS, addition of BSA did not show significant change in fluorescence. Also, BLM formed from oxidized cholesterol alone does not give any fluorescent emission following addition of ANS and BSA (or vice versa). The following conclusions can be drawn from Smekal's results. First, the presence of phospholipids is necessary for ANS adsorption. Second, the lack of enhanced fluorescence when ANS is added after BSA suggests that this protein must interact with the phospholipid portion of the BLM in such a way that the hydrophobic regions are not available for ANS penetration.

The preceding results indicate that it is feasible to study the binding between lipid and protein by the use of a fluorescent probe. This method may be of great value in studies of lipid–protein interactions, which hitherto have been largely neglected owing to a lack of suitable *in vitro* techniques.

IV. Black (Bilayer) Lipid Membranes as Models of Biological Membranes

As briefly described in the introduction, our knowledge concerning biological membranes comes from electron microscopy and biochemical and physicochemical studies. Biochemical analysis reveals the composition of the membrane, whereas the physicochemical approach provides us with possible structural arrangements in terms of known biocompounds. Electron microscopy gives us the visual images of biological membranes. At present, the resolving power and experimental techniques of electron microscopy are such that skillful interpretations are necessary. Nevertheless, the technique indicates that the most important biological membranes thus far investigated give an appearance of a triple-layered image. This consists of two denser regions sandwiched in a lighter region. The overall thickness of this limiting structure is about 50–100 Å (*144*). From a functional point of view, biological membranes may be divided into five basic types as depicted in Fig. 1. We are here concerned with the use of suitably modified BLM systems as experimental models for the five basic types of biological membranes (Table 10).

TABLE 10
BLM as Experimental Models for Biological Membranes

Basic component	Modifier	Model for
A. Phospholipids + cholesterol	Proteins	Plasma membrane (*32, 99, 115, 128, 149, 159*)
B. Phospholipids and sulfolipids	Chlorophylls and other pigments, proteins	Thylakoid membrane (*138*)
C. Oxidized cholesterol + phospholipids and surfactant	ATPase, proteins	Mitochondrial membrane (*99, 171, 172*)
D. Brain lipids	EIM,[a] alamethicin	Nerve membrane (*18, 97, 139, 193*)
E. Phospholipids + oxidized cholesterol	Carotenoid pigments	Visual receptor membrane (*202, 206*)

[a] EIM, excitability inducing material (*24, 190*).

A. BLM as a Model for the Plasma Membrane

Among the five basic types of biological membranes listed (Table 10 and Fig. 1), the plasma membrane has been, and still is, the most frequently studied living membrane. It seems probable that all the other biological membranes may have evolved from the plasma membrane (*9*), although this has not been conclusively demonstrated.

1. Brain and Phospholipids

The most familiar example of a plasma membrane is that of erythrocytes or red blood cells. Insofar as is known, the function of the plasma membrane, in addition to serving as a phase boundary between the cytoplasm and its environment, is to create and maintain the interior content of the cell by the active transport of ions and nutrients, and to eliminate the waste products. The plasma membrane is also known to be antigenic. Despite the detailed characterizations of chemical composition and *in vivo* behavior, the physical mechanisms by which ions and other materials interact with the plasma membrane are not well understood. At present, even the structure of the red cell membrane has not been definitely established (*145*).

Because of the obvious importance of the plasma membrane in living systems, a number of workers have used BLM as a model for the plasma membrane. The most interesting work using BLM as a surface membrane has been reported by Del Castillo et al. (*146*). The BLM are used as surface sites for antigen–antibody and enzyme–substrate reactions. The interaction is detected by monitoring the BLM impedance change. In the experiments involving antigen–antibody reactions, crystallized human and bovine serum albumins were used as antigens; immune serums were used as antibodies. To study the effect of enzymatic reactions, trypsin, chymotrypsin, glutamic acid dehydrogenase, cholinesterase, and other enzymes were employed in the work. Del Castillo and co-workers found that addition of albumin or serum alone had no observable effect on the BLM. However, addition of one compound after the other has been introduced caused a sudden reduction in membrane impedance, then a slow return to control level. A similar response was obtained when small amounts of substrate were added to a solution in which the membrane had been previously exposed to the appropriate enzyme. The decrease in impedance lasted only about 0.6 sec or less. The response of an enzyme-treated BLM to its substrate can be demonstrated repeatedly. Confirmatory experiments have been carried out by Howard and Burton (*147*). The application of BLM in this fashion for quick detection of antigen–antibody reactions is of obvious diagnostic interest.

Del Castillo has also used the BLM system for elucidating the synaptic mechanism at the presynaptic membrane. Experiments using BLM for such purposes will be summarized in a later section.

TABLE 11

PERMEABILITY OF BLMs AND NATURAL MEMBRANES TO VARIOUS COMPOUNDS

Compound	Permeability coefficient (μ/sec)	Ref.
Urea	0.042	(*32*)
	0.001–0.03	(*153*)
	0.006	(*150*)
	1.9 (erythrocyte)	(*32*)
Glycerol	0.046	(*32*)
	0.015 (erythrocyte)	(*32*)
Erythritol	0.0075	(*32*)
	0.05 (erythrocyte	(*32*)
Sorbitol	3×10^{-4})	(*150*)
Glucose	10^{-4}	(*150*)
Thiourea	0.012–0.046	(*150*)
	0.63 (modified with Amphotercin B)	(*153*)
Indole-3-acetic acid	0.006–0.2	(*152*)
Indole-3-ethanol	2.3–3.1	(*148*)
Tryptamine	0.02–0.7	(*148*)

In addition to water the plasma membrane is known to be permeable to other compounds such as alcohol, glucose, and urea (*32*). It has been found that in general the diffusion permeability coefficient P_d is much smaller than the osmotic permeability coefficient P_o (see Sect. II.F). In order to understand this difference and to elucidate the complex phenomena occurring in the biological membrane, Vreeman has carried out a large number of experiments using BLM as a model (*32*). The results of the work of Vreeman and others (*148–150*), together with those obtained with the plasma membrane of various cells, are given in Table 11. Vreeman has found that unmodified lecithin BLM are permeable to urea, glycerol, and erythritol, but poorly permeable to Na^+ and mannitol. As mentioned earlier, the apparent difference between P_o and P_d is due to unstirred layers at the biface. The impermeability to sodium ion as found by Vreeman is consistent with the electrical properties of BLM.

2. Red Blood Cell Extracts

Wood et al. have formed BLM from human red blood cells (*150*). They have found that the electrical properties and water permeability of their membranes are similar to values reported earlier (see Table 2). In addition, they measured the urea, sorbitol, and glucose permeability. Their value of the glucose permeability coefficient is some 10^6-fold less than that of the human red cell. Wood et al. suggest that sugar transport involves membrane constituents other than lipids in the plasma membrane since the BLM they studied were devoid of constituents such as membrane protein. Some of this work has been repeated by Jung and Snell using spherical BLM (*151*).

A similar conclusion—that BLM formed from red cell lipids are radically different from the membrane of intact red cells—has also been reached by Andreoli et al. (*149*). These workers have determined the transport numbers from the steady-state membrane potential difference. The BLM were formed from sheep erythrocytes in isopropanol–chloroform solution. The transport numbers are, respectively, 0.80–0.85 for Na^+ and K^+ and 0.20–0.15 for Cl^-, implying that BLM of this type are quite cation selective. However, in terms of relative permeability to these ions, the intact red cell or plasma membrane is about 10^6 times more permeable to Cl^- than to Na^+ or K^+. Further, the natural membrane is more specific for K^+ than Na^+. Andreoli et al. suggest that cation selectivity of red cell BLM is due to the negatively charged phospholipids (e.g., phosphatidyl ethanolamine) and that the protein components of natural membranes may be more important factors governing the ionic permeability properties.

Urea and thiourea permeabilities of two types of BLM have been reported by Lippe (152). The results of Lippe's work, given in Table 11, are in agreement with those obtained by Vreeman (32). The effects of antibiotic amphotericin B on thiourea permeability of BLM have also been reported by Lippe (153). One of the interesting findings is that amphotericin B at low concentration significantly alters the permeability of cholesterol–HDTAB BLM, while the lecithin BLM are unaffected. It is suggested that the presence of phospholipids and their possible interaction with cholesterol do not seem to be necessary for amphotericin B action.

3. Escherichia coli Lipids

Correlation of membrane function with lipid composition has been difficult because of the presence of other lipid materials external to the plasma membrane (154). In the case of bacterial cells, the lipids appear to be located almost exclusively in the plasma membrane. In addition, the problem of isolation has been made easier because mitochondrial, microsomal, and nuclear membranes are absent in bacteria. In bacterial cells such as E. coli the lipid composition is devoid of the common phosphatidyl choline (lecithin) and cholesterol. In spite of these differences in lipid composition, E. coli also develops a high K^+ and low Na^+ interior when nutrient solution is low in K^+ and high in Na^+. In view of the simplicity of E. coli lipids, BLM formed from extract of E. coli have been investigated (155).

In these experiments, E. coli strain K-12 were grown at 37°C in a culture medium. The cells were harvested in the logarithmic phase of growth and treated first with 5% cold trichloroacetic acid in 80% ethanol. The treated cells were dissolved in 20 ml/g chloroform:methanol (2:1) and the mixture was blended in a Waring blender for 5 min. After centrifugation the supernatant was flash evaporated and the residue was redissolved in chloroform and filtered. The supernatant was again flash evaporated and the final residue was dissolved in a hydrocarbon solvent. Dodecane was used in preliminary experiments; however, it was found that black membranes formed faster and more smoothly in octane, and this solvent was subsequently used on all extractions. The dc electrical resistance of E. coli BLM was 10^7–10^8 Ω-cm^2. After correcting for urea, usually 1 mm^2, the corrected resistance value was 10^5–10^6 Ω-cm^2. The resistance increased slightly in more dilute salt solutions, roughly a twofold increase for every tenfold dilution in the range of 10^{-3}–10^{-1} M salt solution. The change in resistance as a function of temperature has been measured; from a plot of log R versus $1/T$ using the Arrhenius equation, the activation energy was calculated to be 18.6 kcal/mole. The data are shown in Fig. 32.

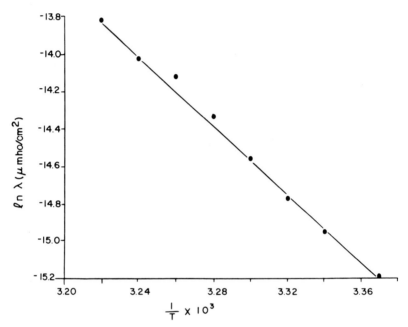

Fig. 32. Variation of log specific conductivity with inverse absolute temperature for a BLM made of *E. coli* lipids (*155*). $\lambda = \lambda_0 \exp(-E_a/RT)$; $E_a = 18.6$ kcal/mole.

Two compounds that have a dramatic effect on most black membrane systems were tested on *E. coli* BLM. Both iodine and EIM caused drops in resistance down to the level of 5×10^3 Ω-cm². The effect with iodine was immediate, while the addition of EIM took 1–2 min to show its effect. In an effort to determine qualitatively the net charge on the BLM surfaces, various anionic and cationic surface active materials were added to one side of the membrane and the transient transmembrane potential was measured. The cationic materials tested were hexadecyltrimethylammonium bromide (HDTAB) and protamine sulfate; the anionic agents were sodium dodecyl sulfate and casein. Of these compounds, both of the positively charged surface-active materials produced large transmembrane potentials when added to the aqueous solution on one side of the membrane. Addition of HDTAB to one side of the BLM produced a transmembrane potential that was positive on that side. The potential developed a few seconds after the material was added, the short delay being due to the time required for the stirring motion to bring the material to the membrane interface. The potential developed across the membrane was transient and began decaying in an exponential manner after about 10 sec.

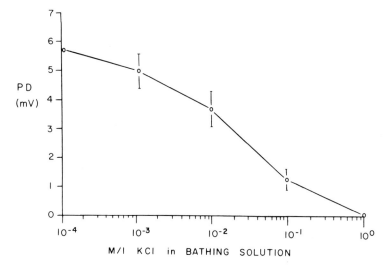

Fig. 33. The effect of protamine sulfate (0.68 mg inner chamber) on *E. coli* BLM potential separating KCl solutions at various concentrations (*155*).

The total time course of the potential was in the range of 1–3 min. The transmembrane potential increased with increasing concentrations of HDTAB until the detergent caused the membrane to rupture. Similar results were produced with protamine sulfate except that the potential developed was much smaller (see Fig. 33).

The effect of these cationic surface-active materials was much more pronounced when the salt concentration in the aqueous solution was low. The negatively charged materials had little effect, although casein was not tested at high concentrations because of its very limited solubility in neutral salt solution. It was concluded from these experiments that the net interfacial charge on the *E. coli* BLM was negative (*155*).

The effect of pH gradients on the transmembrane potential was also measured and the results of one set of experiments are given in Fig. 34. The absolute pH values ran from 3.8 to 7.7 and gradients were formed with varying pH values on both sides. The transmembrane potential that developed was relatively stable during the experiments. The slope of the line is somewhat less than the 59 mV/tenfold concentration difference,

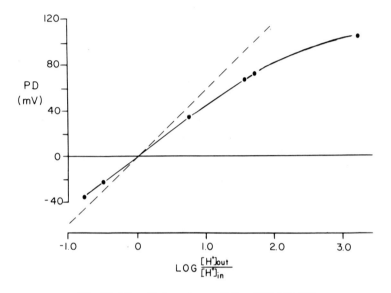

Fig. 34. The pH dependence of *E. coli* BLM (*155*).

shown by the dashed line. In the presence of NaCl or KCl gradients, or for BLM separating NaCl from KCl solutions, no stable transmembrane potentials were detected. It appears that the data can be interpreted in two ways. The obvious explanation is that the potential is due to selective permeability of the membrane to H^+. In this case the hydrogen ion concentration is higher on the outside of the membrane than on the inside. Hydrogen tends to diffuse across the membrane and this causes the inside electrode to show a positive potential. The slope of the line should be about 59 mV/pH and fall off at higher gradients. This is as actually observed (Fig. 34).

However, another interpretation of this data is equally tenable. This alternate interpretation is based on well-known principles of surface chemistry and is similar to the explanation of the effect of surface active materials. Initially, the same solution is present on both sides of the membrane. Then the pH on one side is changed by adding either HCl or KOH. If HCl is added to the outer chambers the ratio H_{out}^+/H_{in}^+ would be greater than 1. This hydrogen ion added to the outer solution reduces the negative charge on the membrane surface by combining with the carboxyl or phosphate groups presumably present at the biface. Consequently, the

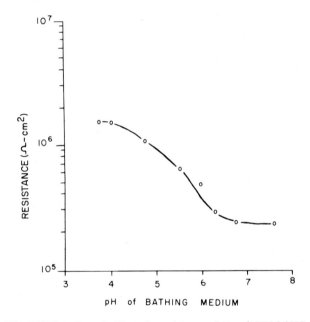

Fig. 35. The effect of pH on the resistance of *E. coli* BLM (*155*).

potential on the outer surface is reduced. The net potential difference is positive on the inside and hence a transmembrane potential would be developed. It has been known for some time that this phenomenon corresponds in exactly the same manner as a diffusion potential does to a hydrogen ion gradient. The prerequisite is that alterations of the charge on one surface cause the potential development instead of diffusion of one specific ion.

In order to determine which alternative is operating, the resistance of the BLM is measured as a function of pH. If the potential were due to diffusion of hydrogen ion, one would expect that by increasing the concentration of hydrogen ions the resistance of the membrane would decrease. This is in direct contradiction to actual observation (Fig. 35). When the concentration of hydrogen ion was lowered, the resistance of the membrane increased.

These data are reminiscent of experiments performed with the hydrogen electrode. For many years it was thought that glass electrodes used in pH measurements worked because the glass was premeable only to H^+. When standard 0.1 N NaCl was present on the inside of the electrode and some other [H^+] was present on the outside, diffusion of H^+ caused the potential that was seen. However, it was shown by Hubbard and Rynders (156) that if the 0.1 N HCl on the inside were replaced by mercury or any metal that could make electrical contact, the same results could be obtained. In this case there was no diffusion, and the potential measured had to be due to a difference in surface potentials. Ling and Kushnir (157) showed that when the pH electrode was coated with a thin layer of collodion, a material that has been shown to bind potassium, the electrode became K^+ sensitive. This clearly was not a diffusion potential because the glass was the same and only the surface had been changed.

The results obtained by Diana and Tien (155) provide another example of a system in which an ion gradient and a transmembrane potential are related by a Nernst-type equation—i.e., about 59 mV/tenfold change of concentration—and yet the evidence tends to indicate that the cause of the potential is a difference in surface charge, or adsorption potential, across the membrane, and not selective permeability to H^+.

Relevant to the above discussion is the work reported by Colacicco, who has studied the electrical potentials at an oil/water interface (158). He found that the sign of the potential in his system is determined by the sign of the ionizable group at the interface and the magnitude of the potential decreases with increasing electrolyte concentration. Some of Diana's findings are consistent with the observation made by Colacicco and can therefore be similarly interpreted.

In an attempt to reconstitute a bacterial plasma membrane, Redwood et al. (159) have carried out some interesting experiments. They have found that by treating phospholipid BLM with a solubilized nonparticulate ATPase from Streptococcus faecalis spheroplast membranes, a two- to fourfold decrease in BLM resistance is observed. This action on BLM appears to be dependent on the presence of Mg^{2+} and upon the concentrations of both Na^+ and K^+ in the bathing solution. Redwood et al. suggest that the decreased resistance in the BLM results from its interaction with the ATPase. Jain et al. (99) earlier reported a similar study (see Sect. IV.C).

B. BLM as a Model for the Thylakoid Membrane

Life may be represented by an infinity sign as illustrated in Fig. 36. The coupled circles represent a steady state transducing system of life. The two basic processes of life are photosynthesis and respiration, which are carried out, respectively, in the chloroplasts and mitochondria. In photosynthesis the sun's energy in the form of photons promotes electrons in the ground state (valence band) to a higher energy state (conduction

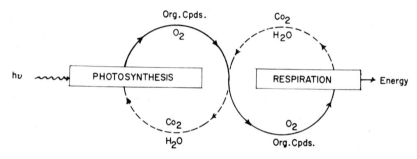

Fig. 36. An infinity sign illustrating the two basic living processes: photosynthesis and respiration.

band). Before falling back to the ground state, the free energy of the electron is converted into chemical-bond energy. The process of photosynthesis in terms of electron flow has been suggested by Szent-Györgyi (160). The organic compounds and oxygen produced by photosynthesis are utilized by other living systems (i.e., animals) that are incapable of making direct use of photon energy of the sun. The conversion of food stuff into chemical-bond energy by animals, termed respiration, takes place in a membranous structure called mitochondrion (see the following section). In a mitochondrion, electrons are again promoted to higher states, the driving force being provided by combustion processes of organic compound (i.e., food stuff) and oxygen. The chemical energy is stored in the pyrophosphate bonds of adenosine triphosphate (ATP), which is the universal currency of life. The use of BLM as a model for the mitochondrial (or cristae) membrane will be described in Sect. IV.C. The present section is concerned with BLM as a model for the thylakoid membrane of the chloroplast.

In simplest terms, photosynthesis can be considered as a process by which the green plant reduces CO_2 to carbohydrates and oxidizes water to oxygen with the aid of sunlight. The process itself can be subdivided into two distinct types of reactions: (1) a photophysical process and (2) a series of dark reactions. The dark reactions, which have been elucidated in detail by Bassham and Calvin (161), appear to be mainly a biochemical problem. The photophysical process or the primary step of photosynthesis is not completely understood and constitutes an area of much current research. Before describing the use of a modified BLM as a model for the thylakoid membrane of the chloroplast, a brief account of the postulated structure of thylakoid membrane is in order.

Evidence derived from electron microscopy, x-ray diffraction, and birefringence as well as circular dichroism studies strongly indicates that the photosynthetic apparatus is composed of highly organized lamellar structure. For instance, a simple model of a chloroplast thylakoid by Menke (162) consists of two protein layers separated by a monomolecular layer of lipids. Other models have been proposed by Weier and Benson (163) and Muhlethaler (17). The salient feature of all of these models is their oriented lipid core onto which other important cellular constituents such as proteins may interact through either ionic or van der Waals attraction or both. Within the chloroplast the usual picture is that of a granum composed of an ordered array of lamellar membranes. Each lamellar membrane is believed to separate two phases forming the so-called inner and outer spaces. The thickness of this thylakoid membrane is estimated to be about 70–120 Å (17, 164). This high degree of orderliness and lamellar organization of thylakoid membranes has led to the suggestion by Bassham and Calvin (161) that a crystalline lattice containing chlorophyll molecules and other compounds may be involved in the photosynthetic apparatus. They further suggest that such an organized structure of light-sensitive pigments may possess photoconductive properties resembling those of organic semiconductors. Arnold and Sherwood have in fact observed semiconductive properties in dried chloroplast and chromatophore preparations (165).

A series of experiments has recently been undertaken to investigate (1) whether a BLM could be constituted from photoactive pigments and (2) the possibility of using such a BLM system as an energy-transducer in the study of certain aspects of photophysical and photochemical processes (138). The results of these experiments are summarized in the following paragraphs.

1. The Basic Properties of Chloroplast BLM

We shall see that BLM formed from photoactive compounds such as chlorophyll and retinal exhibit two interesting phenomena: (1) photovoltaic effect and (2) photoconductivity. As a model for the thylakoid membrane of the chloroplast, BLM of this type are generated from various sources such as extracts of spinach leaves, *Chlorella*, and chlorophylls obtained from commercial sources.

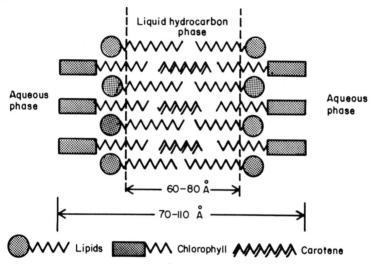

Fig. 37. Schematic model of possible molecular organization for a BLM of chloroplast pigments. The membrane shown here is formed on a Teflon support separating two aqueous solutions. When formed in this manner, the transverse electrical and transport properties of the membrane can be measured and controlled chemical investigation can be carried out.

In the absence of exciting light the dc resistance of the Chl-BLM (Chl denotes either chlorophyll or chloroplast) is $2-10 \times 10^5$ Ω-cm^2 and the dielectric breakdown voltage is generally much less than 200 mV. Current/voltage curves of Chl-BLM appear to be linear up to 150 mV. The thickness of the membrane was measured in the apparatus of Fig. 11. Preliminary experiments give a thickness of 105 Å, which could differ from the true value by as much as 50% since assumed values of refractive index were used in the calculation (*166*). Taking the experimentally determined thickness at its face value, the BLM generated from chloroplast pigments are pictured as similar to those of liquid crystals in two dimensions (Fig.

37). It is suggested that in the case of chlorophyll BLM the hydrophobic portion (phytyl group) of the molecule extends inward while the hydrophilic head (porphyrin) is situated at the aqueous solution/membrane interface. The high dc resistance of the membrane is attributed to the liquid hydrocarbon-like interior. Several orientations of the porphyrin head group (shown in rectangles) at the biface are possible. Since the interior of the BLM is believed to be liquid-like, the porphyrin group may be in a "dynamic" state, i.e., the porphyrin plate may sometimes lie parallel to the biface, somtimes perpendicular, and at other times somewhere in between. However, a more or less perpendicular orientation is favored based upon the bifacial tension data (see the following). With this picture in mind, the observed thickness of the membrane could be accounted for by the depth of anchoring of the phytyl chain and by the amount of other lipids such as carotene located in the interior of the membrane.

The bifacial tensions of the BLM formed from chloroplast pigments in n-octane and in n-octane–butanol mixture are, respectively, 4.5 ± 0.1 and 3.8 ± 0.2 (dynes/cm). The results are significantly higher than those of other BLM produced from natural lipids. It appears also that the pigment solvent used has a noticeable effect on the results. Whether or not the solvent molecules are incorporated into the BLM structure is not certain. The lower value in the case of the octane–butanol mixture strongly suggests, however, that the bifacial region is modified as a result of the presence of additional interfacial-active species. Extending the equations and reasonings used in the monolayer studies at the air/water interface to the bilayer lipid membranes at the water–oil–water biface, the bifacial pressure to which the constituent pigments (and other interfacial-active species) are subjected is given by

$$\pi_i = \gamma_0 - \gamma_i \tag{48}$$

where π_i is the bifacial pressure and γ_0 and γ_i are the interfacial tension between liquid hydrocarbon solvent and water and the bifacial tension of the BLM, respectively. Hence, the bifacial pressure is of the order of 45–50 dynes/cm. At the observed bifacial pressure one would expect therefore that the pigment molecules (as typified by cholrophylls) in the membrane would occupy their limiting areas, i.e., the molecules in the BLM are closely packed. On the basis of the bifacial tension data it seems likely that the porphyrin plates of the molecules are oriented more or less perpendicularly to the biface, as illustrated in Fig. 37.

The permeability to water of Chl BLM has been measured osmotically and is calculated using the equation

$$P_0 = 0.925 RTP' \tag{49}$$

where $P' = J/\Delta\pi$ and R and T have the usual significance. J is the net volume flow of water in time dt and $\Delta\pi$ is the osmotic pressure difference across the BLM. The conversion factor 0.925 is used to express P_0 in μ/sec units (61). The P_0 value for Chl-BLM is about 50 μ/sec, about six times higher than for an oxidized cholesterol BLM. At present it is difficult to make any meaningful interpretation concerning the relatively high permeability to water by the Chl-BLM. As far as is known no data exist for the thylakoid membranes (comparable to Chl-BLM) although a number of investigations have been made on the permeability and osmotic properties of the chloroplast membrane (167, 168). The permeability coefficient of Chl-BLM can be altered by several factors such as illumination and chemical agents. In a later section we will consider the topic of light-induced water permeation. Chl-BLM are highly specific to H^+ below pH 8, and, in the pH region 4–6 a potential difference of theoretical value has been observed (Fig. 38). This implies that the surface of Chl-BLM is negatively charged (presumably due to ionizable sulfolipids and phospholipids). The dark conductivity of Chl-BLM has been measured as a function of temperature and, as expected, the conductivity increased with increasing temperature. In the temperature range studied (16–30°) the membrane conductivity followed the familiar Arrhenius equation

$$\lambda = \lambda_0 \exp\left(-\frac{E}{RT}\right) \tag{50}$$

where λ_0 is a constant. From a plot of measured λ versus the reciprocal of absolute temperature, the quantity E is evaluated from the slope in the usual manner; a typical plot is given in Fig. 39. For this particular Chl-BLM in 0.1 M NaCl, E is 16 kcal/mole (166).

It should be noted that almost all rate processes observed experimentally can be fitted into a form of Arrhenius equation, and in the absence of additional information it is difficult to give a unique interpretation of E. Therefore E may be interpreted to mean either the activation energy required for the charge transport or the creation of charge carriers or both. For instance, the dark conductivity of BLM in general and Chl-BLM in particular (Fig. 38) could be due to the presence of H^+. It is generally known that unmodified BLM are freely permeable to water, but

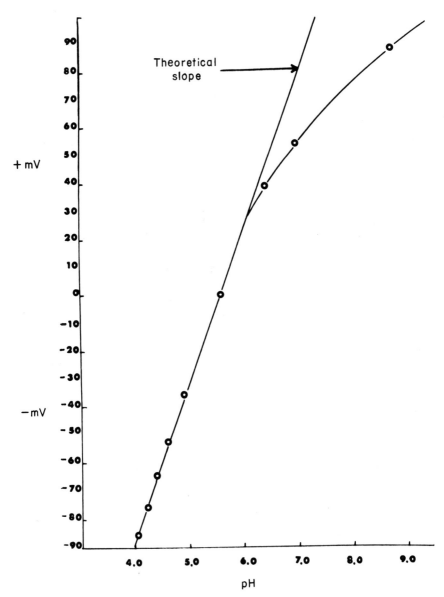

Fig. 38. Chloroplast BLM responses to H⁺. The response follows the Nernst equation up to pH 6, i.e., the membrane behaved as a hydrogen electrode.

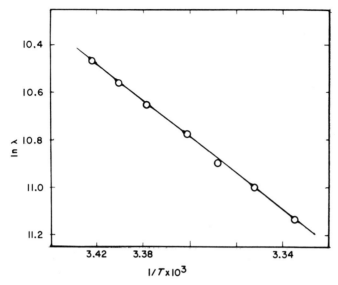

Fig. 39. Temperature dependence of membrane resistance. The slope yields $E =$ 16 kcal/mole.

they are poorly permeable to ions as evidenced from dc resistance measurements (usually in the range $10^5–10^9$ Ω-cm^2). A simple calculation shows that the amount of water present in the BLM can provide enough charge carriers in the form of H^+ and OH^- ions to account for the dark current of the BLM. The experimental activation energy value of 0.7 eV for the chlorophyll BLM lends some support for such a protonic conduction mechanism (*169*).

2. The Effect of Uncouplers on Chl-BLM

The separation of charges by light in chloroplasts is believed to be among the first steps involved in the primary process (i.e., quantum conversion) of photosynthesis. Chemical compounds such as 2,4 dinitrophenol (DNP) can uncouple electron transfer from ATP formation in the electron transport chain. The similar effect of DNP on oxidative phosphorylation in the mitochondria is well known (*170*). DNP and a number of other uncouplers have been tested on lecithin and brain lipid BLM (*171, 172*, see Sect. IV.C). The effect of DNP, *p*-trifluoromethoxy-carbonylcyanide-phenylhydrazone (FCCP), and dicoumarol on the electrical conductivity of Chl-BLM has been examined recently (*108*).

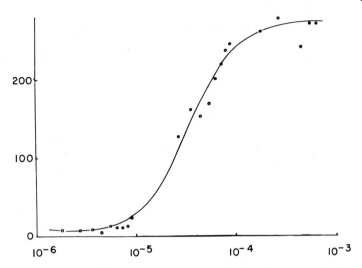

Fig. 40. The effect of FCCP concentration on a BLM made of spinach chloroplast extract. Ordinate, current measured at 60 mV applied voltage (obtained by N. Kobamoto).

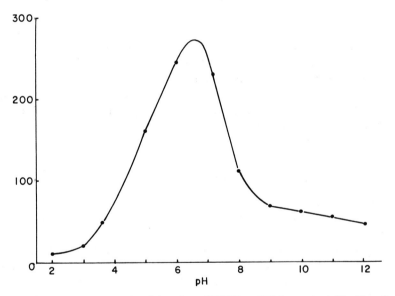

Fig. 41. The pH dependence of the effect of FCCP on BLM current at 60 mV applied voltage.

The concentration dependence of the effect of FCCP is shown in Fig. 40. Below 10^{-5} M little change in the conductivity was observed. Beyond this concentration the conductivity increased rapidly as a function of increasing FCCP concentration until 10^{-4} M. Similar curves have been obtained with DNP and dicoumarol, although the effect took place at different concentrations. The order of effectiveness when compared at the same concentration is FCCP $>$ dicoumarol $>$ DNP. The pH-dependence of the effect of FCCP is shown in Fig. 41. The peak corresponds roughly to the pK value of the uncoupler, which suggests that H$^+$ is the charge transported across the membrane. The uncoupler, which has appreciable lipid solubility, serves as the carrier of hydrogen ion. It is interesting to note that the pH dependence effect of the uncoupler does not appear to depend on the composition of the lipid solution used for BLM formation since the lecithin BLM reported by Liberman and colleagues (*172*) display similar characteristics.

3. Light-Induced Water Flow across Chl-BLM

As previously mentioned (p. 317), the Chl-BLM exhibit significant higher water permeability than, for example, the BLM formed from oxidized cholesterol. In addition, an increase in water flow when the membrane is illuminated has been observed (*173*). The arrangement for the light-induced water flux measurement is shown in Fig. 42. In

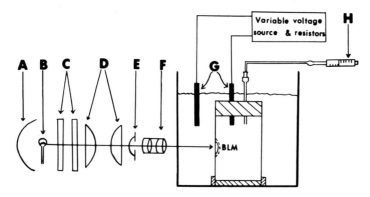

Fig. 42. Schematic diagram of experimental setup for measurements of trans bilayer lipid membrane osmosis, electroosmosis, and light-induced water flow. (A) Reflector; (B) 650-W "Sun" lamp; (C) water jacket containing saturated CuSO$_4$ solution; (D) condenser lenses; (E) heat-absorbing filter; (F) projection lenses; (G) Ag/AgCl electrodes; (H) micrometer (*173*).

order to prevent the thermal effect, the light is passed through a layer of CuSO₄ solution before reaching the BLM. A typical plot of volume change versus time, showing light-induced water flow, can be described as follows. The initial slope is 0.04 μl/min due to the osmotic gradient. Upon illumination, the slope is changed to 0.07 μl/min; when the light is turned off, the initial rate of flow is again observed. If an electrical potential difference

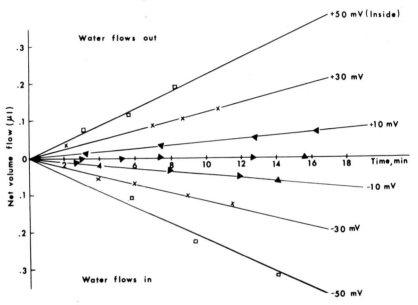

Fig. 43. Electroosmosis across a chloroplast BLM. Note that the water transport rate increases as a function of applied voltage. The direction of water flow depends upon the polarity of the electrodes (*173*).

is applied across such a BLM, electroosmosis is observed. The volume flow of water as a function of applied voltage for a typical run is illustrated in Fig. 43, it is evident that the flow rate increases with increasing applied voltage. The direction of the water flow depends upon the polarity of the field. The net flux of water is in the compartment where the potential is negative.

This interesting phenomenon of light-induced water flow can be decreased or abolished completely if an inhibitor or uncoupler of photophosphorylation is present. It is found, however, that high concentrations of uncouplers such as 2,4-dinitrophenol (DNP) and FCCP tend to rupture the BLM. When 10^{-5} M of DNP or 10^{-7} M of FCCP is added to the aqueous phase, both the photovoltaic effect and light-induced water flow are abolished. Another compound, DCMU (3,3,4-dichlorophenyl-1,1-dimethylurea) at 10^{-7} M, is only partially effective. The results are given in Fig. 44.

The findings described above can be explained in terms of semiconductor physics and classic electrokinetics. When a beam of light excites the BLM,

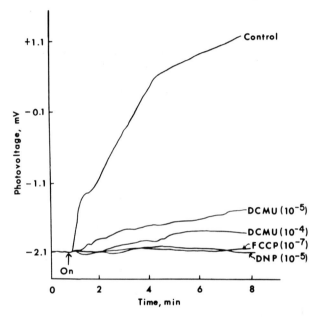

Fig. 44. The effect of inhibitor (DCMU-3,4-dichlorophenyl-1,1-dimethylurea) and uncouplers (FCCP-p-trifluoromethoxycarbonylcyanide phenylhydrazone; DNP-2,4-dinitrophenol) on the photovoltaic effect (173).

electrons and holes are produced. If it is assumed that the electrons and holes thus produced have different lifetimes and mobilities, a separation of charges in the BLM would result, leading eventually to a potential difference across the membrane. Since the chloroplast BLM is shown to be negatively charged, it is therefore poorly permeable to anions. The passage of current through it will produce local concentration changes at the biface. The concentration gradient thus produced could result in an osmotic water flow, in addition to the movement of hydration water associated with current-carrying cations. The light-induced voltage across the BLM is believed to be the primary driving force responsible for the water flow. Externally applied voltage should therefore accomplish the same purpose, as has been experimentally demonstrated (see Fig. 43).

The development of membrane potential across Chl-BLM may take place in two steps: (1) electronic charge carriers are first generated by light and (2) the field across the BLM causes the charges of different sign to move toward the opposite side of the biface, thereby giving one side

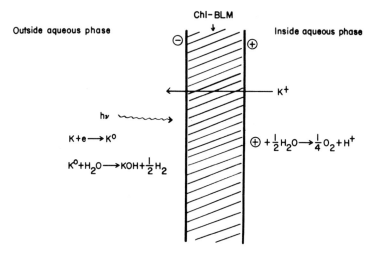

Fig. 45. A suggested scheme for light-induced pH change across Chl-BLM. The discharge of K^+ causes the pH on that side to be higher.

of the biface an oxidizing character and the other a reducing character (see Fig. 45). Instead of producing an unknown entity [H] depicted as in Fig. 45, potassium ion may be reduced. If so, the pH on the side where K^+ is neutralized would be higher. Since Chl-BLM is highly sensitive to H^+, the resulting pH difference could then give rise to the observed membrane potential. Figure 45 gives a pictorial description of the suggested redox reactions across the BLM.

4. The Photoelectric Effects in Chl-BLM

The Chl-BLM is of special interest in that upon illumination with light, two basic photoelectric effects can be elicited. A simple experimental arrangement used to study these new phenomena in BLM is

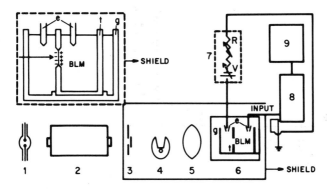

Fig. 46. Apparatus for obtaining photoelectric spectrum of BLM. (1) dc xenon compact arc lamp; (2) grating monochromator; (3) diaphragm and shutter; (4) rotating disc; (5) collimating lens; (6) BLM cell assembly (see insert at top left—*e*, electrodes; *t*, Teflon chamber; *g*, glass outer chamber); (7) variable voltage source; (8) electrometers; (9) recorder (*174*).

shown in Fig. 46. The photoelectric effects have been found to be dependent both on light intensity and wavelength (*138*). Figure 47 shows the measured photocurrent as a function of light intensity. In the particular Chl-BLM studied, the photocurrent is saturated at high light intensity. The variation of photocurrent with applied voltage for a Chl-BLM is shown in Fig. 48. At zero applied field the photocurrent is still observable, even when the small electrode potential is balanced out. The photocurrent varies linearly with the applied voltage. The latter increases the photocurrent when its direction is the same as the photocurrent (i.e., same polarity as the BLM), it decreases the photocurrent when it opposes the BLM, and it is possible to find an applied voltage that can balance completely the emf of the BLM. When this happens, the number (also mobility and lifetime) of the electrons moving in one direction is exactly equal to

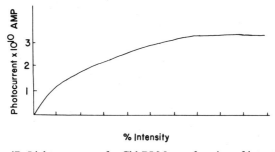

Fig. 47. Light response of a Chl-BLM as a function of intensity.

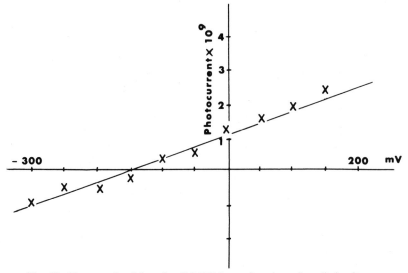

Fig. 48. Photoconductivity of a Chl-BLM as a function of applied voltage.

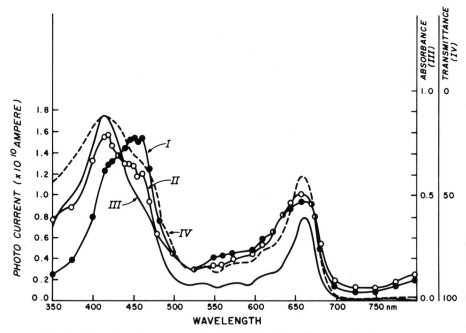

Fig. 49. Photoelectric spectrum of a Chl-BLM. Curve I, uncorrected curve; curve II, curve I corrected with curve I of Fig. 2; curve III, absorption curve of a 100 dilution of the stock solution in a 0.1-cm path length cell (first left-hand side scale); curve IV, $(1 - T)$ curve. The transmittance T is calculated from curve III. T, 10^{-A} where A is the absorbance *(174)*.

the number of holes moving in the opposite direction and the photoelectric effect is not observable. However, when the applied voltage exceeds this critical value, the photocurrent is reversed. For a Chl-BLM in contact with electron acceptors at its dark side, this critical voltage can exceed 100 mV (*174*).

Using the apparatus shown in Fig. 46, the action spectrum of a Chl-BLM together with its absorption spectrum is shown in Fig. 49. Within the experimental error of detection instruments, the two spectra are

$$Chl + h\nu \longrightarrow Chl^* \qquad (1)$$

$$Chl^* + Chl \xrightarrow[\text{Ionization}]{\text{Exciton migration}} Chl^\ominus + Chl^\oplus \quad (2)$$

$$Chl^\ominus + H_3O^+ \longrightarrow Chl + [H] + H_2O \quad (3)$$

$$Chl^\oplus + \tfrac{1}{2} H_2O \longrightarrow \tfrac{1}{4} O_2 + Chl + H^+ \quad (4)$$

Fig. 50. A possible mechanism for the photolysis of water in Chl-BLM.

practically identical. This finding is consistent with the observation that the spectral excitation curve of most organic photoconductors (semiconductors) is generally found to be very similar to that of the absorption spectrum (*174*).

The observation of the photovoltaic effect (open-circuit voltage) coupled with the photo-action spectra of the Chl-BLM provides ample evidence that a separation of charges (electrons and holes) by light is taking place in Chl-BLM. However, the mechanism of charge generation as well as the mechanism of charge transport in the membrane is still obscure, although a number of possible mechanisms have been mentioned (*169*). A possible scheme suggested earlier is depicted in Fig. 50. The explanation of the equations in Fig. 50 is as follows: In Eq. (1) the photo-active pigment as typified by chlorophyll (Chl) is excited by light. The excited molecule is ionized by an unknown mechanism (possibly via exciton migration), leading to the production of a free electron and a positive hole. Assuming the illuminated side is negative with respect to the "dark" side, it is suggested that the separated electron is localized at the aqueous solution/membrane interface (possibly in the porphyrin ring). Equations (3) and (4) of Fig. 50 represent in essence the photolysis of water using the highly organized lipid membrane structure as a barrier to separate the products of oxidation and reduction (*138*).

C. BLM as a Model for the Mitochondrial (Cristae) Membrane

The reverse of photosynthesis is known as respiration, a process which occurs in all living organisms. The overall reaction for respiration can be written as

$$[CH_2O] + O_2 \rightarrow CO_2 + H_2O + \Delta F \qquad (51)$$

where ΔF represents a quantity of useful energy for metabolism. The majority of biochemists believe that the process is accomplished in a series of steps, involving the so-called respiratory or electron-transport chain, in which a flow of electrons from CH_2O through a system of spatially organized cytochromes and dehydrogenases to O_2 takes place. Since living organisms do not operate like heat engines, the ΔF liberated in Eq. (51) has to be stored in a form that is readily available when required. This is usually "trapped" in the so-called energy-rich chemical bonds in a compound such as ATP. A number of questions arise: (1) exactly how is energy converted and trapped in a chemical bond and (2) what is the nature of this energy-transducer. There are no definite answers at the present. Nevertheless, a brief description of the known biochemical and cytological aspects of respiration serves a useful introduction.

The basic operational unit of respiration resides in the mitochondrion, frequently called the power plant of the cell. The fundamental structure of mitochondria as revealed under the electron microscope appears to consist of two or double membranes (16). The outer membrane encloses the mitochondrion whereas the inner membrane folds voluminously back and forth across the interior giving rise to compartments known as cristae. Hence the inner membrane is often called the cristae membrane. The sequences of events taking place in the mitochondrion have been described in considerable detail by biochemists since 1925 (175). The overall process consists of three separate but closely related steps: (1) the production of ATP during the oxidation of the coenzymes, (2) the rearrangement and oxidation of carbohydrates to form reduced coenzymes, (3) the oxidation of these coenzymes by molecular oxygen.

Both the outer and cristae membranes are believed by some investigators to be a lipid bilayer of the Gorter–Grendel type with adsorbed protein layers and serve as structural framework for the mitochondrion. The overall thickness of each membrane is again less than 100 Å. It is generally recognized that the cristae membranes, as in the case with thylakoid membranes of chloroplasts, are the site of energy transduction. It is believed that in the process of electron flow, formation of a compound between the reductant (electron donor) and the oxidant (electron acceptor) takes place with the free energy of the electron stored in the chemical bond (e.g., ATP).

The most intriguing and still unanswered question posed earlier is: "What is the mechanism of electron flow in the electron transport chain as postulated by biochemists?" In order to explain some of the processes taking place in the mitochondrion, a hypothesis known as "the chemiosmotic theory" has been proposed (170, 176). Mitchell and Moyle (177) have shown that ATP hydrolysis is accompanied by an acidification of the bathing solution. This process has two components: one is reversible and sensitive to chemical agents such as 2,4-dinitrophenol (DNP); the other is irreversible and insensitive to DNP. Hydrogen ions are pumped out of the mitochondria during ATP hydrolysis and re-equilibrate slowly. According to Mitchell's theory, a pH gradient must exist across the membrane (cristae or thylakoid membrane) in that the separation of charge due to redox reactions can bring about a change in the so-called "hydrodehydration" reaction with an ATPase, which activates the ATP, ADP, P_i, and water reaction. Further, the chemiosmotic theory is claimed to explain a number of phenomena associated with an electron-transport system, membrane potential, charge separation, photo- and oxidative phosphorylation, and ion exchange (170).

The chemiosmotic hypothesis is very popular among biochemists but by no means universally accepted. The attractiveness of this hypothesis stems from its simplicity, and it suggests further experiments. Concerning the antagonists' view of the chemiosmotic hypothesis, it should be mentioned that the work done by Dilley and Vernon (178), Cockrell et al. (179), Chance et al. (180), Thore et al. (181), and McCarty (182) are of interest. Briefly, these authors' experiments show that proton gradients are not necessary for phosphorylation. They do not, however, exclude the possibility that the "proton-motive force" has been reduced to one of its hypothetical components as postulated by Mitchell.

From a biophysical point of view, the use of whole mitochondria (or chloroplasts) to test a "molecular" theory does not seem satisfying. The translocation of ions such as protons across a complex system such as a chloroplast is in itself ill-defined. The interactions between various fluxes (ion and water movements, electron and hole transport, etc.) are far too complex to be amenable to a simple analysis. It must be concluded that the chemiosmotic hypothesis as it stands today remains as a hypothesis. Whether the postulated mechanism occurs in the cristae membrane is uncertain. Mitchell suggests that his proposal is a useful working hypothesis. At the present time direct tests with the mitochondrial membranes are difficult. Experimental evaluation of the hypothesis using a simpler model system is therefore in order.

Bielawski et al. (171) were among the first to employ a BLM system to test one aspect of Mitchell's hypothesis by studying the effect of 2,4-dinitrophenol (DNP) on the BLM conductivity. The BLM were formed from a solution of purified lecithin, cholesterol, and n-decane. They have found that the uncoupling agent (DNP) at a concentration of 10^{-3} M decreases electrical resistance of BLM to less than 0.4% of the original values (i.e., from 1.5×10^7 to 6.1×10^4 Ω-cm^2).

The effects of uncoupling agents of oxidative phosphorylation on BLM have been published in a series of papers by Liberman, Mokhova, Skulachev, Topaly, Tsofina, and Jasaitis (119, 172, 183). In their experiments, BLM have been formed from brain lipids, phospholipids, and mitochondrial lipids. The major findings of their work are summarized below:

(a) In general, the uncoupling agents (uncouplers) raise the conductivity of BLM by several orders of magnitude, and the effect depends strongly on the pH of bathing solution, being most effective at the pK value of the uncoupler.

(b) The effectiveness of the uncouplers studied follows a series: tetrachlorotrifluoromethylbenzimidazole (TFB) > p-trifluoromethyloxycarbonylcyanide phenylhydrazone (FCCP) > DNP > salicylic acid > acetoacetic ester. The order of effectiveness is similar to that found in the mitochondria.

(c) The current/voltage curves of the BLM in the solution containing uncoupler are nonlinear and depend on pH and the buffer capacity of the solution.

(d) In the presence of TFB the BLM behaves as a hydrogen electrode. The modified BLM is said to be more permeable to H$^+$ (see Fig. 38).

(e) The uncoupler is considered a carrier of H$^+$. The operation is described as follows: on one side of the biface a proton reacts with the uncoupler in the BLM; this uncoupler-H$^+$ complex diffuses across the membrane and to the other side of the biface discharging the H$^+$; the uncoupler returns to the other side.

Liberman et al. (172) suggest that their findings agree with the scheme of oxidative phosphorylation as envisioned by Mitchell and conclude that BLM are a good model for the mitochondrial membrane.

The pH-dependence of the effect of TFB and other uncouplers on the electrical conductivity of the BLM described above has been studied in detail by Sotnikov and Melnik (*184*). They were interested in establishing the relationship between the pK value of the modifier (e.g., DNP) and the conductivity of the BLM. Two series of compound were examined. In the first series of substituted nitrophenols, the general formula is

$$NO_2 \overset{R}{\diagdown} NO_2$$

where R = —NH$_2$, —NH$_3^+$, —COOH, and —NO$_2$. Sotnikov and Melnik have found that at about 10^{-3} M the effectiveness of these compounds in lowering the conductivity follows their pK values. The conductivity for picric acid modified BLM was about 10^4 mho-cm^{-2}. The other series of compounds studied were the imidazole and triazole derivatives. No experimental data are given in their paper but they state that a similar effect was observed.

A most interesting finding using BLM as a model for the mitochondrial membrane (or plasma membrane) has been reported recently by Jain et al. (*99*). These investigators formed BLM from a mixture of oxidized cholesterol and a synthetic surfactant (*63*). In the experiments of Jain et al. the BLM is formed in the usual manner. ATP is added to one side of the BLM, and this is followed by the addition of a preparation of membrane-bound ATPase (Na–K dependent fraction). After these additions, a decrease in the membrane resistance is observed. The current flowing across the BLM is monitored by a method described by Ussing and Zerahn (*185*). The experimental arrangement of Jain et al. for measuring zero voltage current across the membrane is shown in Fig. 51. They observed a positive current flow across the BLM, interpreted as being due to the transport of Na$^+$ which is analogous to that found in the natural membrane. The current flow in their reconstituted system is said to be dependent on the presence of Na$^+$ in the bathing solution, and the magnitude of current on ionic strength. Further the presence of ouabain (a cardiac glucoside), a compound known to block Na$^+$ transport in red cells, inhibits the current flow across the ATPase-modified BLM. It seems probable that by introducing ATPase into the bathing solution "molecular sodium pumps" could have been inserted into the BLM, which may account for the observation. If so, this implies that ATPase (or BLM-modified ATPase) is actually the "sodium pump" described by numerous investigators in the literature (*186*). In any case, Jain et al. have demonstrated a unique application of BLM for membrane biochemistry. The reconstituted system described above may be used equally well as a model for the plasma membrane (Sect. IV.A).

It seems clear that from the foregoing paragraphs (and Sect. IV.B) the next logical experiment is to attempt ATP synthesis using ATPase-modified BLM systems. The driving force for *in vitro* ATP synthesis may be provided either by a membrane potential (e.g., pH gradient as suggested by Mitchell) or by light. In the former case the reaction is equivalent to

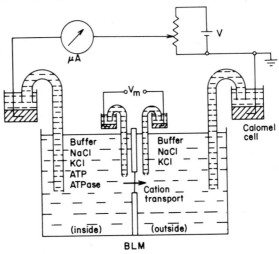

Fig. 51. Cell assembly for measuring zero voltage membrane current across BLM (*99, 185*).

oxidative phosphorylation. The latter reaction is that of photophosphorylation. The type of experiments envisioned can be summarized by the following equations:

Oxidative phosphorylation

$$\text{ADP} + \text{P}_i \xrightarrow[\text{Electrochemical gradient}]{\text{ATPase} + \text{BLM}} \text{ATP} + \text{H}_2\text{O} \tag{52}$$

Photophosphorylation

$$\text{ADP} + \text{P}_i \xrightarrow[\text{light}]{\text{ATPase} + \text{BLM}} \text{ATP} + \text{H}_2\text{O} \tag{53}$$

where ADP is adenosine diphosphate and P_i is inorganic phosphate.

The most difficult part of these experiments lies perhaps in effecting ATPase and BLM interaction. The source and purification of ATPase present another obstacle. Nonetheless, the ATPase (F_1) isolated by Racker (*187*) should be a useful starting material since it has been shown to contain high ATPase activity. It remains to be seen whether the suggested experiments can be successfully performed.

D. BLM as a Model for the Nerve Membrane

Beginning with the observation of Galvani who reported in 1791 that a frog's leg contracted when touched with metallic wires, the phenomenon of nerve excitation has been and still is the focal point of research in electrophysiology. This sustained interest stems from the belief that information processing and transmission in the nervous system are intimately connected with nerve excitation. At present, our understanding of this fascinating subject is far from complete. With this limitation in mind we will begin the subject by sketching the overall process of nerve excitation.

Like other biological process, the phenomenon of nerve excitation has to be considered in conjunction with a membranous structure. A typical nerve fiber or axon is very much like a long tube filled with a salt solution of the same total concentration as that in the outer surrounding solution, although the chemical compositions of the two solutions are very different from each other. The structure of the axon separating the inside from the outside is thought to be similar to the membranes described earlier (see Fig. 1). This very thin oily membrane acts mainly as an insulating layer but with a very small portion of the surface area selectively permeable to ions and other materials. With a pair of microelectrodes of the Ling–Gerard type placed across the membrane (188), a potential in the order of 50–100 mV can usually be measured. The inside solution is negative with respect to the outside. This measured potential is known as "resting" membrane potential. When a region of the membrane is stimulated in some way (such as by a short current pulse) the ionic permeability of the membrane is altered. During the excitation, according to the Hodgkin–Huxley theory (189), the membrane becomes more permeable to Na^+ ions than to other ions.

At the present time, because of the difficulty of experimentation with neural nerve membranes, our understanding of membrane excitability is still incomplete, especially from the viewpoint of membrane structure and function. Several groups of investigators have utilized the BLM systems to understand the phenomenon of action potentials and the nature of synaptic receptors.

As discussed in Sect. III.B, the high resistance of BLM can be lowered several orders of magnitude by a number of compounds. This is also found to be the case with EIM when approximately 0.025 mg/ml (depending upon the EIM purity) is added to the BLM made from brain lipids.

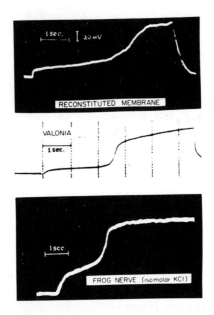

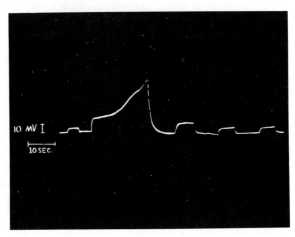

Fig. 52. Potential recordings of responses to rectangular current pulses applied through a high resistance (*139*).

In addition, the EIM induces a voltage-dependent conductance similar to that of excitable cells. Mueller et al. (*18*) initially described this electrical transient behavior as follows:

". . . (the addition of EIM) lowers the resistance and regularly induces the following gating reaction to dc stimulation. At a sharp and generally unidirectional threshold voltage, which can be varied by Ca^{2+} between 15 and 50 mV, the resistance shifts fivefold to a new steady value after a latency which changes inversely with applied potential. Recovery is prompt, the phenomenon repeatable and indistinguishable in detail from the behavior of the excitable alga, *Valonia*, in the 'variable resistance' state and is similar to the resistance increase of the skin 'action potential' and to the frog nerve 'action potential' in isotonic KCl . . ."

Some of the experimental results obtained by the earlier investigators are reproduced in Fig. 52.

For the purpose of observing electrical transient phenomena in BLM, the membrane may be formed from brain lipid extract, oxidized cholesterol, or oxidized cholesterol plus dodecyl acid phosphate (DAP) mixture. This is usually done in the following manner. After formation of the BLM separating two identical solutions (e.g., 0.1 *M* NaCl buffered with histidine at pH 6.8), a small amount of EIM solution is added to the inner chamber. The effect of EIM on the resistance of the BLM is measured approximately every minute. Depending upon the effectiveness of stirring it takes anywhere from a few seconds to several minutes for the BLM to interact with the membrane, leading to the reduction of resistance to about $10^4–10^6$ Ω-cm². Voltages of increasing strength are then applied across the BLM until a nonlinear voltage increase appears during the pulse. This phenomenon usually requires a short latency period and the resistance remains at the final level until the potential is turned off. The latency period is inversely related to the applied voltage, and return to the control level follows a sigmoid time course. If a second potential is applied before the resistance returns to control, a second increase in resistance is measured, but with a decreased latency period. Figure 53 illustrates a typical run in which the events outlined above are shown (*190*).

Mueller and Rudin have proposed a model to explain the kinetics of resistance changes in the excitable membranes (*191*). They assumed that the membrane contains a finite number of parallel channels, each of which may be in a high or low resistance state, depending on the free energy difference between the two states and the applied potential. They further assumed that the transition is a first- or second-order reaction and that the applied potential supplies kinetic energy to a divalent charge in the direction of the field. The "action potentials" are thought to be primarily the result of the voltage controlled permeability changes, or gating of the EIM channels, which in the absence of a resting potential can generate two negative resistance regions in the steady state I–V curve and the corresponding time–voltage transitions under constant currents.

Bean and colleagues (*192, 193*) have since carried out an intensive study on the EIM-modified bilayer system. They confirmed that EIM is a

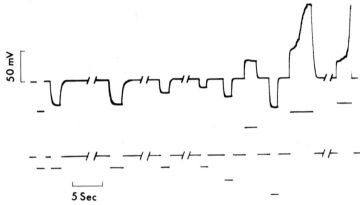

Fig. 53. Potential recordings of responses to applied rectangular current of an EIM modified BLM composed of brain lipids and other additives (*190*).

proteinaceous compound destroyed by a short digestion with trypsin or chymotrypsin (*192*). The increase of BLM conductivity is considered at length. The effect of EIM could manifest itself in several ways: (1) EIM could create pores in the membrane and (2) could act as a carrier. The latter idea is not favored in view of the size of EIM (molecular weight $\sim 10^5$). They have shown that the pores (or channels) generated by EIM can be "blocked" by the action of a proteolytic enzyme. In another series of experiments, Bean and co-workers have examined the influence of lipid composition on the electrical transients of the BLM (*192*). Saturated hydrocarbon solvents such as decane appeared to inhibit EIM action when phospholipids were used as the major lipid component. In contrast, EIM reaction was not noticeably affected in the oxidized cholesterol BLM system.

Bean et al. (*192*) also carried out investigations on the effect of ions on BLMs excitability. Phospholipid BLM are usually unstable outside of the pH range 5–10. Within this range, both brain lipid and sphingomyelin BLM show a shift of threshold voltage with changing pH. On the other hand, divalent cations such as Mg^{2+} and Ca^{2+} have little effect upon the BLM excitability if they comprise the major cation in the bathing solution. The Zn^{2+} ions have an entirely different effect, however, as they tend to abolish completely the EIM-induced properties. A similar effect is observed with Al^{3+}.

Recently, Mueller and Rudin (194) have found conditions in which action potentials can be developed in BLM produced from a 2.5%

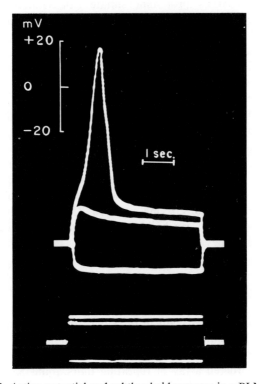

Fig. 54. Action potential and subthreshold response in a BLM (194).

purified sphingomyelin solution in 5 mM histidine chloride buffered to pH 6.8 (Fig. 54). To one compartment, 10–50 mmoles/liter K_2HPO_4 or K_2SO_4, 10^{-4} g/cm³ EIM, and 10^{-4} g/cm³ protamine sulfate are added. Under these conditions an action potential may develop in response to an applied pulse of current. Sometimes it is possible to initiate rhythmic firing (spontaneous development of action potentials of approximately half a second duration). It is noted that when the salt gradient is produced with monovalent anions, a bistable resistance results, similar to that described above. It is further observed that EIM and protamine sulfate may be present together on the same side of the membrane or on opposite sides from each other. Rudin and Mueller interpret their data in terms of the classical theory of Hodgkin and Huxley (189) that EIM develops channels in the membrane that selectively conduct cations and may be in a high or low resistance state. The resistance of a portion of these channels to anions is altered by protamine sulfate. In connection with these interesting observations, mention must be made of the findings obtained by Monnier (195) and Shashoua (196). These investigators used membranes prepared according to their own methods which, although 2–3 orders of magnitude greater in thickness, also exhibit rhythmic firing under applied voltages.

Continuing their interesting observations on the enzyme-modified BLM (Sect. IV.B), Del Castillo and colleagues (197) recently utilized BLM as models for transmitter–receptor systems. They found that the transverse impedance of such membranes, presumably coated with protein molecules, decreases when the specific antigens or substrates are introduced into the system. These responses have been found completely reversible and can be repeated as long as the membrane shows appreciable impedance. Further, the responses of enzyme-modified BLM have also been found to be highly specific and can be blocked by agents known to inhibit the activity of the enzymes (198).

E. Carotenoid BLM as a Model for the Visual Receptor Membrane

The retinals in an outer segment of retinal rod are probably associated with opsin and surrounded by lipids (199). Although the detailed information is still lacking, the structure of the outer segment of rods appears to be a highly ordered array of repeating units about 55 Å thick (200). Similar to the four types of biological membranes already described, the lamellar membrane of retinal rods is also interpreted in terms of a Gorter–Grendel bilayer model (201). With regard to visual excitation, Wald (199) has suggested that light absorbed by a rhodopsin molecule in the outer segment membrane isomerizes the retinal chromophore, producing metarhodopsin ($\lambda_{max} = 498$). This leads to a series of reactions in which a substantial protein conformation change may be involved in turn leading to the formation of metarhodopsin ($\lambda = 380$). It is postulated that visual pigments are an integral part of a highly organized lamellar structure in the rod outer segment membrane that could result in significant changes in volume and electrical polarization. These changes in turn generate the nerve impulse by a mechanism similar to that proposed for the nerve membrane. However, as in the case of photosynthesis, the primary step of light conversion in the visual process is not clearly understood. The study of a well defined model system is therefore of interest.

Recently, a series of experiments has been initiated with the aim of constituting BLM containing carotenoid pigments and rhodopsin. Some preliminary findings are summarized in the following paragraphs (202).

The lipid solutions used for BLM formation consist of various carotenoids (e.g., all-*trans* retinal, 9-*cis* retinal, all-*trans* retinol, and β-carotene), phospholipids, and oxidized cholesterol dissolved in liquid alkanes. The dark dc resistance of the membrane is ohmic, ranging from 10^6 to 10^7 Ω-cm^2, which is about 2–3 orders of magnitude lower than that of carotenoid-free BLM. Upon illumination with white light (DFG tungsten lamp, Sylvania) a maximum photo-emf of about 6 mV is observed in a BLM containing a mixture of all-*trans*–retinal and β-carotene. All carotenoid BLM are found to be photoconductive. In addition, the photoresponses of these carotenoid BLM are complex as the experimental conditions are altered. Depending upon external factors and the carotenoid pigments used (e.g., the presence of modifiers, and applied voltage),

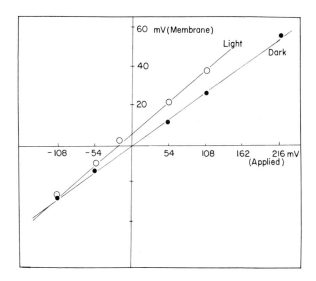

Fig. 55. dc dark and photoconductivity in a carotenoid BLM (202).

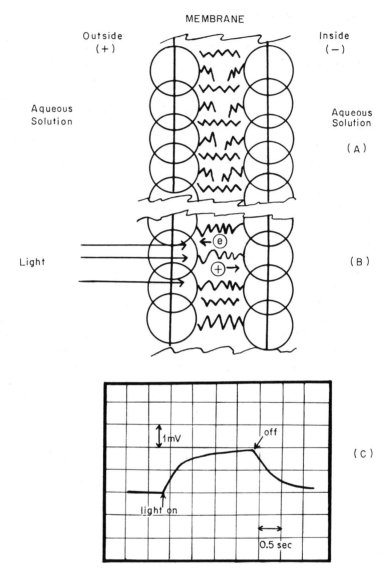

MEMBRANE

Outside
(+)

Inside
(−)

Aqueous
Solution

Aqueous
Solution

(A)

Light

(B)

1 mV

off

(C)

light on

0.5 sec

Fig. 56. A photovoltaic hypothesis of visual excitation. (A) Rod outer segment membrane separating two aqueous solutions. The membrane is polarized with inside negative. Circles represent protein moiety (opsin) and zigzag lines represent phospholipids (e.g. phosphatidylethanolamine) and chromophores (e.g. retinal). The hydrocarbon chain of chromophores are shown in *cis* configuration. Light is assumed to produce two effects: isomerization of chromophore from *cis* to *trans* configuration and generation of charge carriers. These events are shown in (B). It is suggested that charge carriers migrating in the direction of respective electrodes cause membrane depolarization, thus initiating the nerve excitation. (C) Photoelectric response in an experimental black lipid membrane (BLM) containing all-*trans* retinal (*206*).

the voltage/time curves can vary from a simple monophasic response to a typical biphasic waveform not unlike those found in vertebrate retina (203–204).

Photoconductivity in these carotenoid BLM has also been investigated. Figure 55 presents the results obtained for an all-*trans* retinal BLM, whereas Fig. 56c is an oscilloscope trace showing the membrane response to white light. Also shown in the figure is a suggested scheme for visual excitation.

It is observed that the sign of the photo-emf of the BLM containing all-*trans* retinal and β-carotene is independent of the polarity of the applied potential, whereas the magnitude is dependent. For instance, the ratio of the photo-emf obtained by applying 54 mV positive at the illuminated side of the membrane to the same voltage of negative sign is 4. The ratio of i^+/i^- becomes larger with increasing field strength. This observation suggests that the majority charge carriers are holes. At very high field strength (2.5×10^5 V/cm) the sign of the photo-emf can be reversed (see Fig. 55).

The relevance of carotenoid BLM as an experimental model for the visual receptor membrane may be viewed from the standpoint of visual organelles (see Fig. 1). As has been mentioned, the structure of the outer *sac* membrane is believed to be similar to that of the unit membrane (201). Thus, the carotenoid BLM is one of the ideal systems that can be used to investigate both the initial energy transduction mechanism from photons to electrons and holes and also the triggering mechanism for ionic permeability across the membrane. For example, Hagins et al. (205) have reported that the initial membrane depolarization of the sacs of the rod outer segment is farily localized, while the outer plasma membrane depolarization occurs with a large latency and delocalization. Further experiments with the carotenoid BLM are in progress with the objective of understanding the mechanisms of excitation and energy conversion. It is hoped that the results may be useful in analyzing the complicated electroretinogram (203).

V. Summary and Conclusions

In this chapter an attempt has been made to present the unified view that all biological membranes are composed of a hydrophobic bilayer lipid core. Hence, experimental black lipid membranes (BLM) with suitable modification are useful model systems for the five basic types of biological membranes. These are the plasma membrane of individual cells, the thylakoid membrane of chloroplast, the cristae membrane of mitochondrion, the nerve membrane of axon, and the outer segment sac membrane of retinal rod. The main points of this chapter together with certain conclusions are given below:

A. Black or bilayer lipid membranes (BLM) of bimolecular thickness can be formed in aqueous media from a variety of materials. These materials include brain lipids, proteolipids, purified phospholipids, oxidized cholesterol, synthetic surfactants, chloroplast extracts, and carotenoid pigments.

B. The intrinsic properties of unmodified BLM (e.g., formed from either lecithin or oxidized cholesterol) are similar to those expected of an ultrathin layer of liquid hydrocarbon (see Table 8).

C. Upon the introduction of additives (modifiers) to the BLM system the intrinsic or passive properties of the BLM can be drastically modified. The BLM modifiers may be divided into five groups: (1) those altering the passive electrical properties, (2) those changing the mechanical properties, (3) those conferring ion selectivity, (4) those inducing electrical excitability, and (5) those generating photoelectric effects.

D. The modified BLM systems at present constitute realistic experimental models for the various types of biological membranes. Biologically relevant phenomena such as excitability, ion selectivity, antibody–antigen reaction, active ion transport, and photoelectric effects have been demonstrated. The light-induced phenomena in photoactive BLM provide strong evidence for the existence of electronic charge carriers in the membrane.

E. The formation of BLM separating two aqueous solutions permits for the first time direct characterization of the postulated bimolecular leaflet model of the biological membranes in physical chemical terms. It remains to be seen if a variety of outstanding problems in membrane biophysics and biochemistry can be elucidated with the use of experimental BLMs. A partial list of these outstanding problems includes light conversion in photosynthesis and vision, energy migration, active transport, ion selectivity and specificity, generation and conduction of nervous impulse, redox reactions and electron-transfer, oxidative and photophosphorylation, DNA replication, protein synthesis, immunological and pharmacological reactions, lipid–protein interaction, and membrane structure.

Having summarized the work thus far carried out on BLM and having listed some of the important research topics in membrane biophysics and biochemistry, it should be mentioned that our understanding and knowledge of the structure of membranes depends to a large extent on prior developments in surface and colloid chemistry. As one of the leading investigators, T. Teorell (*207*), concluded at a recent symposium; ". . . In particular, we need more knowledge about the physical and chemical properties of the oriented multilayers, which, in a sense are two-dimensional liquid crystals. We must know more about their stability conditions, about phase transitions . . . I would predict that surface chemistry . . . may again become popular."

REFERENCES

1. Burton, R. M., *Proceedings of Symposium on Lipid Monolayer and Bilayer Models and Cellular Membranes*, American Oil Chemists' Society, May 1967. Also published in *J. Amer. Oil Chem. Soc.* **45**, 107, 201, 297 (1968).

2. Castleden, J. A., *J. Pharm. Sci.* **58**, 149 (1969).

3. Henn, F. A., and Thompson, T. E., *Ann. Rev. Biochem.* **38**, 241 (1969).

4. Rothfield, L., and Finkelstein, A., *Ann. Rev. Biochem.* **37**, 463 (1968).

5. Bangham, D., in *Progress in Biophysics and Molecular Biology* (J. A. V. Butler and D. Noble, eds.), Vol. 18, Pergamon, New York, 1968, pp. 29–95.

6. Kajiyama, M., *Biophysics* (*Japan*) **8**, 1 (1968).

7. Tien, H. T., and Diana, A. L., *Chem. Phys. Lipids* **2**, 55 (1968).

8. Dowben, R. M., (ed.), *Biological Membranes*, Little, Brown, Boston, Mass., 1969.

9. Robertson, J. D., *Protoplasma* **63**, 218 (1967).

10. Davson, H., and Danielli, J. F., *Permeability of Natural Membranes*, Cambridge Univ. Press, 1943.

11. Overton, E., *Vierteljahresschr. Naturforsch. Ges. Zurich* **44**, 88 (1899).

12. Gaines, G. L., *Insoluble Monolayers at Liquid-gas Interfaces*, Wiley-Interscience, New York, 1965.

13. Gorter, E., and Grendel, F., *J. Exp. Med.* **41**, 439 (1925).

14. Branton, D., and Park, R. B., *Selected Papers on Biological Membrane Structure*, Little, Brown, Boston, Mass., 1968.

15. Lucy, J. A., in *Biological Membranes* (D. Chapman, ed.), Academic Press, New York, 1968, pp. 233–288.

16. Sjostrand, F. S., *Rev. Mod. Phys.* **31**, 301 (1959).

17. Muhlethaler, K., in *Biochemistry of Chloroplasts* (T. W. Goodwin, ed.), Academic Press, New York, 1966, pp. 49–64.

18. Mueller, P., Rudin, D. O., Tien, H. T., and Wescott, W. C., *Circulation* **26**, No. 5 (Pt. 2), 1167 (1962).

19. Rudin, D. O., Tien, H. T., and Wescott, W. C., Unpublished studies (1957–1959); see *J. Amer. Oil Chem. Soc.* **45**, 201 (1968).

20. Boys, C. V., *Soap Bubbles—Their Colours and the Forces Which Mould Them*, with a Preface by S. Z. Lewin, Dover Publications, Inc., New York, 1959.

21. Lawrence, A. S. C., in *Surface Phenomena in Chemistry and Biology* (J. F. Danielli, K. G. A. Pankhurst, and A. C. Riddiford, eds.), Pergamon Press, London, 1958, pp. 9–17.

22. Stoeckenius, W., *J. Biophys. Biochem. Cytol.* **5**, 491 (1959).

23. Mueller, P., Rudin, D. O., Tien, H. T., and Wescott, W. C., *J. Phys. Chem.* **67,** 534 (1963).
24. Mueller, P., Rudin, D. O., Tien, H. T., and Wescott, W. C., in *Recent Progress in Surface Science*, Vol. 1, Academic Press, New York, 1964, pp. 379–393.
25. Pagano, R., and Thompson, T. E., *Biochem. Biophys. Acta* **144,** 666 (1967).
26. Tien, H. T., and Howard, R. E., in *Techniques of Surface Chemistry and Physics* (R. J. Good, R. R. Stromberg, and R. L. Patrick, eds.), Marcel Dekker, New York, in press.
27. Davies, J. T., and Rideal, E. K., *Interfacial Phenomena*, Academic Press, New York, 1961.
28. Adamson, A. W., *Physical Chemistry of Surfaces*, Wiley-Interscience, New York, 1967.
29. Goddard, E. D., *Molecular Association in Biological and Related Systems*, Advances Chemistry Series, No. 84, American Chemical Society, Washington, D.C., 1968.
30. Langmuir, I., *J. Amer. Chem. Soc.* **39,** 1848 (1917).
31. Luzzati, V., and Husson, F., *J. Cell. Biol.* **12,** 207 (1962).
32. Vreeman, H. J., *Kon. Ned. Akad. Wetensch.* **69,** 542 (1966).
33. Griffith, O. H., and Waggoner, A. S., *Account Chem. Res.* **2,** 17 (1969).
34. Rendi, R., *Biochim. Biophys. Acta* **84,** 694 (1964).
35. Bangham, A. D., Standish, M. M., and Watkins, J. C., *J. Mol. Biol.* **13,** 238 (1965).
36. Libertini, L. J., Waggoner, A. S., Jost, P. C., and Griffith, O. H., *Proc. Nat. Acad. Sci. U.S.* in press (1969).
37. Tien, H. T., and Diana, A. L., *J. Colloid Interface Sci.* **24,** 287 (1967).
38. Hanai, T., Haydon, D. A., and Taylor, J., *Proc. Roy. Soc.* **281A,** 377 (1964).
39. Rosen, D., and Sutton, A. M., *Biochim. Biophys. Acta* **163,** 226 (1968).
40. Babakov, A. V., Ermishkin, L. N., and Liberman, E. A., *Nature* **210,** 933 (1966).
41. Tien, H. T., *J. Theor. Biol.* **16,** 97 (1967).
42. Miyamoto, V. K., and Thompson, T. E., *J. Colloid Interface Sci.* **25,** 16 (1967).
43. Tien, H. T., and Diana, A. L., *Nature* **215,** 1199 (1967).
44. Kok, J. A., *Electrical Breakdown of Insulating Liquids*, Wiley-Interscience, New York, 1961, p. 11.
45. Schmitt, F. O., and Samson, F. E., *Neurosci. Res. Progr. Bull.* **7,** 277 (1969).
46. Henn, F. A., Decker, G. L., Greenawalt, J. W., and Thompson, T. E., *J. Mol. Biol.* **24,** 51 (1967).
47. Blough, H. A., and Gordon, G., to be published (Blough, private communication, 1969).
48. Huang, C., and Thompson, T. E., *J. Mol. Biol.* **13,** 183 (1965); **16,** 576 (1966).
49. Tien, H. T., and Dawidowicz, E. A., *J. Colloid Interface Sci.* **22,** 438 (1966).
50. Mysels, K. J., Shinoda, K., and Frankel, S., *Soap Films*, Pergamon Press, New York, 1959.
51. Corkill, J. M., Goodman, J. F., Haisman, D. R., and Harrold, S. P., *Trans. Faraday Soc.* **57,** 821 (1961).
52. Heaven, O. S., *Optical Properties of Thin Solid Films*, Butterworths, London, 1955.
53. Cherry, R. J., and Chapman, D., *J. Mol. Biol.* **40,** 19 (1969).
54. Vasicek, A., *Optics of Thin Films*, North-Holland Publishing Co., Amsterdam, 1960.
55. Duyvis, E. M., "The Equilibrium Thickness of Free Liquid Films," Doctoral Dissertation, Utrecht (1962).

56. Tien, H. T., *J. Gen. Physiol.* **52**, 125s (1968).
57. Huang, C., and Thompson, T. E., *J. Mol. Biol.* **15**, 539 (1966).
58. Rich, G. T., Shaafi, R. I., Romualder, A., and Solomon, A. K., *J. Gen. Physiol.* **52**, 941 (1968).
59. Hanai, T., Haydon, D. A., and Taylor, J., *J. Theor. Biol.* **11**, 370 (1966).
60. Cass, A., and Finkelstein, A., *J. Gen. Physiol.* **50**, 1765 (1967).
61. Tien, H. T., and Ting, H. P., *J. Colloid Interface Sci.* **27**, 702 (1968).
62. Zwolinski, B. J., Eyring, H., and Reese, C. E., *J. Phys. Chem.* **53**, 1426 (1949).
63. Tien, H. T., *J. Phys. Chem.* **71**, 3395 (1967).
64. Langmuir, I., and Schaefer, V. J., *J. Franklin Inst.* **235**, 119 (1943).
65. Blank, M, and LaMer, V. K., *Retardation of Evaporation by Monolayers*, Academic Press, New York, 1962, pp. 59–66.
66. Franks, F., and Ives, D. J. G., *J. Chem. Soc. (London)* 741 (1960).
67. Frank, H. S., and Evans, M. W., *J. Chem. Phys.* **13**, 507 (1945).
68. Drost-Hansen, W., *Ind. Eng. Chem.* **57**, 28 (1965), and reference therein.
69. Fowkes, F. M., *J. Phys. Chem.* **67**, 2538 (1963).
70. Gurney, R. W., *Ionic Processes in Solution*, McGraw-Hill, New York, 1953, Chapters 3 and 16.
71. Huang, C., Wheeldon, L., and Thompson, T. E., *J. Mol. Biol.* **8**, 148 (1964).
72. Thompson, T. E., in *Cellular Membranes in Development*, Academic Press, New York, 1964, pp. 83–96.
73. Pagano, R., and Thompson, T. E., *J. Mol. Biol.* **38**, 41 (1968).
74. Haydon, D. A., and Taylor, J. L., *Nature* **217**, 739 (1968).
75. Tien, H. T., *J. Phys. Chem.* **72**, 2723 (1968).
76. Vaidhyanathan, V. S., and Goel, N. S., *J. Theor. Biol.* **21**, 331 (1968).
77. Ohki, S., and Fukada, N., *J. Theor. Biol.* **15**, 362 (1967).
78. Hirschfelder, J. O., Curtis, C. F., and Bird, R. B., *Molecular Theory of Gases and Liquids*, Wiley, New York, 1954.
79. Parsegian, V. A., and Ninham, B. W., to be published (Parsegian, private communication, 1969).
80. Landau, L. D., and Lifshitz, E. M., *Electro-dynamics of Continuous Media*, Addison-Wesley, Reading, Mass., 1960.
81. Ohki, S., *J. Theor. Biol.* **19**, 97 (1968).
82. Kirkwood, J. G., *J. Chem. Phys.* **4**, 592 (1936).
83. McBain, J. W., *Trans. Faraday Soc.* **9**, 99 (1913).
84. Reich, I., *J. Phys. Chem.* **60**, 257 (1956).
85. Phillips, J. N., *Trans. Faraday Soc.* **51**, 561 (1955).
86. Booij, H. L., and Bungenberg de Jong, H. G., *Protoplasmatologia* **I**, 2 (1956); see also H. L. Booij in *Intracellular Transport* (K. B. Warren, ed.), Academic Press, New York, 1966, pp. 301–317.
87. Debye, P., *J. Chem. Phys.* **53**, 7 (1949).
88. Tien, H. T., *J. Amer. Oil Chem. Soc.* **46**, A113 (1969).
89. Goldacre, R. J., *Surface Phenomena in Chemistry and Biology*, Pergamon Press, London, 1958, pp. 278–298.
90. Calvin, M., *AIBS Bull.* **12**, 29 (1962).
91. Green, D. E., and Perdue, J., *Proc. Nat. Acad. Sci. U.S.* **55**, 1295 (1966).
92. Fox, S. W., *Science* **132**, 200 (1960).

93. Oparin, A. I., *Life: Its Nature and Development*, Academic Press, New York, 1964.
94. Tien, H. T., Carbone, S., and Dawidowicz, E. A., *Nature* **212**, 718 (1966).
95. Jendrasiak, G. L., Doctoral Dissertation, Michigan State University (1967).
96. Rosenberg, B., and Jendrasiak, G. L., *Chem. Phys. Lipids* **2**, 47 (1968).
97. Mueller, P., and Rudin, D. O., *Nature* **217**, 713 (1968).
98. Jendrasiak, G. L., *Chem. Phys. Lipids* **3**, 98 (1969).
99. Jain, M. K., Strickholm, A., and Cordes, E. H., *Nature* **222**, 871 (1969).
100. White, S., Ph.D. Thesis, University of Washington (1969).
101. Bergstrom, S., *Ark. Kemi Mineral. Geol.* **16A**, 1 (1943).
102. Henn, F. A., and Thompson, T. E., *J. Mol. Biol.* **31**, 227 (1968).
103. Lauger, P., Lesslauer, W., Marti, E., and Richter, J., *Biochim. Biophys. Acta* **135**, 20 (1967).
104. Lauger, P., Richter, J., and Lesslauer, W., *Ber. Bun. Ges.* **71**, 906 (1967).
105. Kallmann, H., and Pope, M., *J. Chem. Phys.* **32**, 300 (1960).
106. Rosenberg, B., and Bhowmik, B. B., *Chem. Phys. Lipids* **3**, 109 (1969).
107. Finkelstein, A., and Cass, A., *J. Gen. Physiol.* **52**, 145s (1968).
108. Tien, H. T., in *Advances Experimental Medicine and Biology*, Vol. 7 (M. Blank, ed.), Plenum Press, New York, 1970, pp. 135–154.
109. Bjerrum, N., *Selected Paper*, Emar Munksgaard, Copenhagen, 1949, pp. 108–119.
110. Kraus, R. M., and Fuoss, C. A., *J. Amer. Chem. Soc.* **55**, 1010 (1933).
111. Bass, L., and Moore, W. J., *Nature* **214**, 393 (1967).
112. Ohki, S., and Goldup, A., *Nature* **217**, 458 (1968).
113. Noguchi, S., and Koga, S., *J. Gen. Appl. Microbiol.* **15**, 41 (1969).
114. Bradley, J., and Dighe, A. M., *J. Colloid Interface Sci.* **29**, 157 (1969).
115. Van Zutphen, H., and van Deenen, L. L. M., *Chem. Phys. Lipids* **1**, 389 (1967).
116. Leslie, R. B., and Chapman, D., *Chem. Phys. Lipids* **1**, 143 (1967).
117. Hashimoto, M., *Bull. Chem. Soc. Japan* **41**, 2823 (1968).
118. Liberman, Y. A., and Babakov, A. V., *Biofizika* **13**, 362 (1968).
119. Liberman, E. A., and Topaly, V. P., *Biochim. Biophys. Acta* **163**, 125 (1968).
120. Le Blanc, O. H., *Biochim. Biophys. Acta* **193**, 350 (1969).
121. Ochs, A. L., and Burton, R. M., *Biophys. J.* A-27 (1968).
122. Haydon, D. A., in *Membrane Models and the Formation of Biological Membranes* (L. Bolis and B. A. Pethica, eds.), North-Holland Publishing Co., Amsterdam, 1968, pp. 91–97.
123. Dervichian, D. G., in *Molecular Association in Biological and Related Systems*, AICS, No. 84, American Chemical Society, Washington, D.C., 1968, pp. 78–87.
124. Van Zutphen, H., van Deenen, L. L. M., and Kinsky, S. C., *Biochim. Biophys. Res. Commun.* **22**, 393 (1966).
125. Papahadjopoulos, D., and Ohki, S., *Science* **164**, 1075 (1969).
126. Maddy, A. H., Huang, C., and Thompson, T. E., *Fed. Proc.* **25**, 933 (1966).
127. Smekal, E. and Tien, H. T., unpublished results (1969).
128. Mueller, P., and Rudin, D. O., *Biochem. Biophys. Res. Commun.* **26**, 398 (1967).
129. Gotlib, V. A., Buzhinsky, E. P., and Lev, A. A., *Biophys. Res.* **13**, 675 (1968).
130. Eisenman, G., Rudin, D. O., and Casby, J. C., *Science* **126**, 831 (1957).
131. Chappell, J. B., and Crofts, A. R., *Biochimica Biophysica Acta Library* (J. M. Tager, S. Papa, E. Quagliariello, and E. C. Slater, eds.), Elsevier, Amsterdam, 1966, Vol. 7.
132. Henderson, P. J., McGivan, J. D., and Chappell, J. B., *Biochem. J.* **111**, 521 (1969).

133. Lev, A. A., and Buzhinsky, E. P., *Z. Evol. Biokhim. Fiziol.* **9**, 102 (1967).
134. Andreoli, T. E., Tieffenberg, M., and Tosteson, D. C., *J. Gen. Physiol.* **50**, 2527 (1967).
135. Lardy, H. A., Graven, S. N., and Estrada, O. S., *Fed. Proc.* **26**, 1355 (1967).
136. Pressman, B. C., *Fed. Proc.* **27**, 1283 (1968).
137. Eisenman, G., Ciani, S. M., and Szabo, G., *Fed. Proc.* **27**, 1289 (1968).
138. Tien, H. T., *J. Phys. Chem.* **72**, 4512 (1968).
139. Mueller, P., Rudin, D. O., Tien, H. T., and Wescott, W. C., *Nature* **194**, 979 (1962).
140. Azzi, A., Chance, B., Radda, G. K., and Lee, C. P., *Proc. Nat. Acad. Sci. U.S.* **62**, 612 (1969).
141. Rubalcava, B., De Munoz, D. M., and Gilter, C., *Biochemistry* **8**, 2742 (1969).
142. Edelman, G. M., and McClure, W. O., *Accounts Chem. Res.* **1**, 65 (1968).
143. Stryer, L., *Science* **162**, 526 (1968).
144. Sjostrand, F. S., *Radiation Res. Suppl.*, **2**, 349 (1960).
145. Zahler, P., *Experientia* **25**, 449 (1969).
146. Del Castillo, J., Rodriguez, A., Remero, C. A., and Sanchez, V., *Science* **153**, 185 (1966).
147. Howard, R. E., and Burton, R. M., *J. Amer. Oil Chem. Soc.* **45**, 202 (1968).
148. Bean, R. C., Shepherd, W. C., and Chan, H., *J. Gen. Physiol.* **52**, 495 (1968).
149. Andreoli, T. E., Bangham, J. A., and Tosteson, D. C., *J. Gen. Physiol.* **50**, 1729 (1967).
150. Wood, R. E., Wirth, F. P., and Morgan, H. E., *Biochim. Biophys. Acta* **163**, 171 (1968).
151. Jung, C. Y., and Snell, F. M., *Fed. Proc.* **27**, 286 (1968).
152. Lippe, C., *J. Mol. Biol.* **39**, 669 (1969).
153. Lippe, C., *J. Mol. Biol.* **35**, 635 (1968).
154. Salton, M. R. J., *Ann. Rev. Microbiol.* **21**, 417 (1967).
155. Diana, A. L., and Tien, H. T., *Biophys. J.* **8**, A-25 (1968).
156. Hubbard, D., and Rynders, G. F., *J. Res. Nat. Bur. Stand.* **40**, 105 (1948).
157. Ling, G. N., and Kushnir, L. D., in *A Physical Theory of the Living State* (G. N. Ling, ed.), Blaisdell, New York, 1962, p. 275.
158. Colaccico, G., *Nature* **207**, 936 (1965).
159. Redwood, W. R., Mueldner, H., and Thompson, T. E. (Thompson, private communication, 1969).
160. Szent-Györgyi, A., *Introduction to a Submolecular Biology*, Academic Press, New York, 1960.
161. Bassham, J. A., and Calvin, M., *The Path of Carbon in Photosynthesis*, Prentice-Hall, Englewood Cliffs, New Jersey, 1957, Chapter 12.
162. Menke, W., *Brookhaven Symp. Biol.* **19**, 328 (1967).
163. Weier, T. E., and Benson, A. A., in *Biochemistry of Chloroplasts* (T. W. Goodwin, ed.), Academic Press, New York, 1966, p. 91.
164. Branton, D., *Ann. Rev. Plant Physiol.* **20**, 209 (1969).
165. Arnold, W., and Sherwood, H. K., *Proc. Nat. Acad. Sci. U.S.* **43**, 105 (1957).
166. Tien, H. T., Huemueller, W. A., and Ting, H. P., *Biochem. Biophys. Res. Commun.* **33**, 207 (1968).
167. Dilley, R. A., and Vernon, L. P., *Arch. Biochem. Biophys.* **111**, 365 (1965).
168. Packer, L., *Ann. N.Y. Acad. Sci.* **137**, 624 (1966).
169. Tien, H. T., *Nature* **219**, 272 (1968).
170. Mitchell, P., *Biol. Rev.* **41**, 445 (1966).

171. Bielawski, J., Thompson, T. E., and Lehninger, A. L., *Biochem. Biophys. Res. Commun.* **24,** 948 (1966).

172. Liberman, E. A., Mokhova, E. N., Skulachev, V. P., and Topaly, V. P., *Biofizika* **13,** 188 (1968).

173. Ting, H. P., and Tien, H. T., paper presented at First Cell Biology-Biophysics Midwest Meeting, Chicago, Illinois, March 14–15 (1969).

174. Van, N. T., and Tien, H. T., *J. Phys. Chem.* **74,** 3559 (1970).

175. Chance, B., and Williams, G. R., *Advan. Enzymol.* **17,** 65 (1956).

176. Robertson, R. N., *Biol. Rev.* **35,** 231 (1960).

177. Mitchell, P., and Moyle, J., in *Biochemistry of Mitochondria* (E. C. Slater, Z. Kaninga, and L. Wojtczak, eds.), Academic Press, New York, 1967.

178. Dilley, R. A., and Vernon, L. P., *Arch. Biochem. Biophys.* **111,** 365 (1965).

179. Cockrell, R. S., Harris, E. J., and Pressman, B. C., *Nature* **215,** 1487 (1967).

180. Chance, B., Lee, C. P., and Mela, L., *Fed. Proc.* **26,** 1341 (1967).

181. Thore, A., Keister, D. L., and San Pietro, A., *Biochemistry* **7,** 3499 (1968).

182. McCarty, R. E., *Biochem. Biophys. Res. Commun.* **32,** 37 (1968).

183. Liberman, E. A., Topaly, V. P., Tsofina, L. M., Jasaitis, A. A., and Skulachev, V. P., *Nature* **222,** 1076 (1969).

184. Sotnikov, P. S., and Melnik, E. I., *Biofizika* **13,** 185 (1968).

185. Ussing, H. H., and Zerahn, K., *Acta Physiol. Scand.* **23,** 110 (1951).

186. Skou, J. C., in *Progress in Biophysics and Molecular Biology* (J. A. V. Butler, ed.), Pergamon Press, London, 1964, Vol. 14, p. 131.

187. Racker, E., *Fed. Proc.* **26,** 1335 (1967).

188. Ling, G., and Gerard, R. W., *J. Cell. Comp. Physiol.* **34,** 382 (1949).

189. Hodgkin, A. L., and Huxley, A. F., *J. Physiol.* **117,** 500 (1952).

190. Kushnir, L. D., *Biochim. Biophys. Acta* **150,** 285 (1968).

191. Mueller, P., and Rudin, D. O., *J. Theor. Biol.* **4,** 268 (1963).

192. Bean, R. C., Aeronutronic Publication No. V-4184, Philco-Ford Corp. (1968).

193. Bean, R. C., Shepherd, W. C., Chan, H., and Eichner, J., *J. Gen. Physiol.* **53,** 741 (1969).

194. Mueller, P., and Rudin, D. O., *Nature* **213,** 603 (1967).

195. Monnier, A. M., *J. Gen. Physiol.* **51,** (pt. 2), 26s (1968).

196. Shashoua, V. E., *Nature* **215,** 846 (1967).

197. Del Castillo, A., Rodriguez, A., and Romero, C. Z., *Ann. N.Y. Acad. Sci.* **144,** 803 (1967).

198. Del Castillo, J., *J. Amer. Oil Chem. Soc.* **45,** 313 (1968).

199. Wald, G., *Science* **162,** 238 (1968).

200. Blasie, J. K., Dewey, M. M., Blaurock, A. E., and Worthington, C. R., *J. Mol. Biol.* **14,** 143 (1965).

201. Wolken, J. J., *J. Amer. Oil Chem. Soc.* **45,** 251 (1968).

202. Tien, H. T., and Kobamoto, N., *Nature* **224,** 1107 (1969).

203. Brown, K. T., *Vision Res.* **8,** 633 (1968).

204. Pak, W. L., and Cone, R. A., *Nature* **204,** 836 (1964).

205. Hagins, W. A., Zonada, H. V., and Adams, R. G., *Nature* **194,** 844 (1962).

206. Tien, H. T., in *Solid State Physics and Chemistry: An Introduction* (P. F. Weller, ed.), Marcel Dekker, New York, in press.

207. Teorell, T., in *Biophysics and Physiology of Biological Transport* (L. Bolis, V. Capraro, K. R. Porter, and J. D. Robertson, eds.), Springer-Verlag, New York, 1967, p. 340.

7

Cellular Narcosis and Hydrophobic Bonding

L. S. HERSH

Research and Development Laboratories
Corning Glass Works
Corning, New York

I. General Introduction

This chapter is an attempt to further develop Ferguson's (*1*, *2*) thermo-dynamic approach to cellular narcosis. Cellular narcosis is defined as the reversible inhibition of cellular function (e.g., contraction, luminescence, neural transmission) by any of a wide variety of nonpolar and weakly polar solutes. The structures of these solutes, which range from noble gases, simple paraffins, alcohols, ethers, halogenated alkanes to alkylamines and sterols, make a specific chemical interaction with the so-called active site or region seem very unlikely. It is this very lack of specificity that has both encouraged and prompted a thorough thermodynamic understanding of narcosis.

Of course, a thermodynamic analysis is completely independent of any particular molecular model. Thus, no specific attention has been paid to any of the current molecular theories of narcosis or anesthesia (3, 4). For a review of these approaches, there are several good articles available (5, 6, 7).

The present work is further restricted to the study of available biological data that involve a homologous series of primary alcohols. This is due primarily to the insufficiency of data concerning other homologs. The term "hydrophobic bonding" in the title is there primarily to bring attention to an extremely important unanswered question rather than to indicate that there is a serious attempt to investigate a correlation between narcosis and hydrophobic bonding.

II. Introduction

Our aim is to present an alternative approach to the thermodynamic analysis of the Meyer–Overton theory of narcosis proposed by Ferguson (1) and further developed by Mullins (8).

The key concept in the Meyer–Overton theory is the proposal made by H. H. Meyer (9) that narcosis results from the drug dissolving in a fatty phase within the cell. This hypothesis allows the action of the narcotic to be described in the general terms of thermodynamics: an equilibrium distribution that depends on the proper thermodynamic potentials. Overton (10) later published data showing a good correlation between narcotic potency and the same narcotic's olive oil/water partition coefficients, supporting Meyer's theory of narcosis. A later refinement was offered by K. H. Meyer and Hemmi (11) who suggested that narcosis was produced when a minimum value of narcotic mole fraction in the cellular phase was reached. An agent was inactive if the minimum mole fraction was unobtainable according to the Meyer–Hemmi contribution.

The composite model for narcosis derived from both the Meyer–Overton and Meyer–Hemmi approaches involves the following mechanism: narcosis is just produced when a minimum mole fraction of the narcotic in an oil-like cellular phase is reached. The minimum concentration of the narcotic in the aqueous phase necessary to produce this mole fraction in the cellular phase depends on the partition coefficient of the narcotic between the aqueous phase and the bipohase. The cell phase/aqueous solution distribution appears to be correlated with certain oil phase (e.g., olive oil, oleic acid)/aqueous phase partitioning of the same narcotic.

Ferguson's introduction (*1*) of the activity or relative fugacity concept caused the Meyer–Overton theory to be thermodynamically exact. However, Ferguson (*1*) was under the misconception that this use of the Raoult's law activity in the aqueous phase eliminated" . . . the disturbing effect of phase distribution . . ." We will attempt to show that Ferguson's choice of the Raoult's law activity or standard state for the narcotic introduced a different type of phase distribution for the narcotic. The use of a Raoult's law standard state shifted the phase distribution from that of biophase/aqueous solution to a biophase/pure liquid narcotic distribution. The use of the Henry's law activity or standard state, as we shall demonstrate, shifts the attention back to the original concern of the Meyer–Overton hypothesis: the change in molal or unitary free energy of the narcotic molecule going from an essentially water-like environment to the relatively nonpolar phase that we call the biophase or active site.

The importance of viewing narcosis from the standpoint of the Henry's law standard state is that there is a more obvious relation between narcotic activity and a large body of information derived by surface chemists working with what might be considered model systems. Also it emphasizes the possible connection between anesthesia and so-called "hydrophobic bonding."

III. Thermodynamics of Equilibrium

Our model system consists of a single substance that attains an equilibrium distribution between two contiguous phases. The important concepts to be understood are the escaping tendency, chemical potential, fugacity, activity, and activity coefficients. We will review these terms primarily in terms of our model—an equilibrium distribution of an organic polar molecule between an aqueous solution and some unknown phase.

If a substance (which we will refer to as the solute and indicate by the subscript 2) is distributed between two phases and the distribution has reached equilibrium, it may be said that the escaping tendency of the solute is the same in both phases. The term escaping tendency was introduced by G. N. Lewis in order to provide for the process of mass flow the same kind of intuitive picture we have for heat flow. In other words, temperature is to heat flow as escaping tendency is to mass flow.

The concept of escaping tendency is quite simple. The complication is caused by having two different quantitative escaping tendency scales: (1) The chemical potential or partial molal free energy and (2) fugacity or activity and activity coefficients. In the case of temperature we are much more reasonable and adopt one and only one temperature scale.

Considering again our model, under the condition of constant T and P, a direct measure of the escaping tendency is the partial molal free energy $\bar{F}_2$ or chemical potential of the solute μ_2. At equilibrium we have the identities

$$\bar{F}_2 = \left(\frac{F}{n_2}\right)_{T,P,N_{i,\,i\neq 2}} = \mu_2 \tag{1}$$

and the condition

$$(\bar{F}_2)^{\text{Phase I}} = (\bar{F}_2)^{\text{Phase II}} \tag{2}$$

or, equivalently,

$$(\mu_2)^{\text{Phase I}} = (\mu_2)^{\text{Phase II}} \tag{3}$$

In a physical sense the chemical potential that is equal to the partial molal free energy at constant temperature and pressure is an excellent measure of escaping tendency. It is completely analogous to the gravitational and electrostatic potential used in mechanical or electrical systems. However, because of mathematical difficulties, another measure of escaping tendency is more convenient for some purposes. In the case of solutions, the partial molal free energy of the solute approaches $-\infty$ as the solution becomes very dilute. So in order to avoid having to compare very large numbers, the term fugacity is introduced. The fugacity f_2 is equal to the vapor pressure when the vapor is a perfect gas and is defined by the equation

$$\bar{F}_2 = RT \ln f_2 + B_2 \tag{4}$$

where B_2 is that constant for a given substance at a given temperature which makes the fugacity equal to the vapor pressure in the ideal-gas state (12). Thus, as the vapor pressure of a substance approaches zero, the partial molal free energy goes towards the limit of $-\infty$. Now we can modify our definition of the equilibrium condition. A system is in equilibrium when, and only when, the fugacity of every substance is the same in all parts of the system.

The solution of one problem leads to another problem. There will be systems that involve substances that are practically nonvolatile and the calculation of the fugacities becomes very difficult. Therefore, the activity (a relative fugacity) defined by

$$a_2 = f_2/f°_2, \qquad (5)$$

where $f°_2$ is the fugacity in some standard state, is introduced. It is simpler to calculate relative fugacities or the activity than to determine individual fugacities. Now, however, we have introduced another complication. There are practical advantages in choosing different standard states in different situations and sometimes in the same situation. This leads to a solute having a single value of partial molal free energy or fugacity in any system or subsystem, but the activity will depend upon the choice of standard state. For instance, in the case of the distribution of a solute between two phases, we have at equilibrium

$$(f_2)^{\text{Phase I}} = (f_2)^{\text{Phase II}} \qquad (6)$$

If we choose different standard states for the solute in the two phases, we then have

$$(a_2)^{\text{I}} = \frac{(f°_2)^{\text{II}}}{(f°_2)^{\text{I}}} (a_2)^{\text{II}} \qquad (7)$$

The next section will describe the two most common choices of a standard state for a solute and their physical meanings.

IV. Activities and Standard States

In the case of solutions it is still more convenient to deal with concentrations rather than with relative fugacities or the activity. In order to achieve this we need a relation between the fugacity and some measure of concentration. There are two dependent empirical relations that describe the limiting behavior of a solution of two or more substances. For simplicity we will only consider a binary solution.

In the case of an extremely dilute solution, the fugacity of the solute is directly proportional to its mole fraction X_2. Thus, we have

$$f_2 = kX_2 \qquad (8)$$

which is called Henry's law. If we choose k as a standard state, then according to Eqs. (5) and (4) we obtain

$$\bar{F}_2 - F°_{2,\text{H}} = RT \ln (a_2)_{\text{H}} \qquad (9)$$

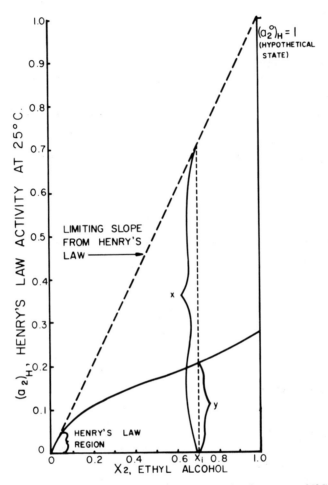

Fig. 1. $(a_2)_H$ versus X_2 for ethyl alcohol solution in water at 25°C.

where $F_{2,H}^{\circ} = RT \ln k$, and $F_{2,H}^{\circ}$ and $(a_2)_H$ are the Henry's law standard molal free energy and activity for the solute, respectively. At this point, it may help to refer to Fig. 1. It is important to realize that the Henry's law activity $(a_2^{\circ})_H$ refers to a hypothetical liquid consisting of pure solute whose fugacity is equal to the Henry's law constant k. Thus, $F_{2,H}^{\circ}$ is a measure of the escaping tendency of the solute in an imaginary pure liquid where the solute molecule behaves as if it were surrounded by solvent molecules.

The solid line in Fig. 1 describes the variation of $(a_2)_H$ with X_2 for ethyl alcohol at 25°C (*13*). Instead of dealing directly with $(a_2)_H$, it is preferred to tabulate values of the more slowly changing quantity $(\gamma_2)_H$, the Henry's law rational activity coefficient, defined by the relation

$$(\gamma_2)_H = \frac{(a_2)_H}{X_2} \tag{10}$$

In Fig. 1 the value of $(\gamma_2)_H$ may be calculated from the ratio y/x. It should be noted that the limiting value of $(\gamma_2)_H$ is 1 as $X_2 \to 0$.

The other common choice of standard state for the solute refers to the actual pure solute in the liquid state. (We are calling component 2 the solute even under the condition that $X_2 > X_1$.) In this case we have the relation

$$f_2 = f_2^\circ X_2 \tag{11}$$

which is obeyed only when $X_2 \to 1$ and is called Raoult's law. In this case the fugacity of the pure liquid solute f_2° is the standard state and Eqs. (5) and (4) yield

$$\bar{F}_2 - F_2^\circ = RT \ln (a_2)_R \tag{12}$$

where F_2° is the standard molal free energy of the pure liquid solute and $(a_2)_R$ is the Raoult's law activity. The expression for the activity coefficient is correspondingly

$$(\gamma_2)_R = \frac{(a_2)_R}{X_2} \tag{13}$$

which approaches the limit 1 as $X_2 \to 1$.

Figure 2 displays the relation between the Raoult's and Henry's law activities. Also, we see that again we may calculate the activity coefficient $(\gamma_2)_R$ from the ratio y/x.

The important difference between Eqs. (9) and (12) is that we are comparing the escaping tendency of the solute at any particular value of X_2 to a hypothetical pure liquid solute that behaves solvent-like in the former case and to the actual pure solute in the latter case. This difference in standard states may be formulated quantitatively by the following derivation. At any particular value of X_2 we have the identity

$$\bar{F}_2 = F_2^\circ + RT \ln (\gamma_2)_R X_2 = F_{2,H}^\circ + RT \ln (\gamma_2)_H X_2 \tag{14}$$

Rearranging gives us

$$\ln \frac{(\gamma_2)_H}{(\gamma_2)_R} = \frac{[F_2^\circ - F_{2,H}^\circ]}{RT} \tag{15}$$

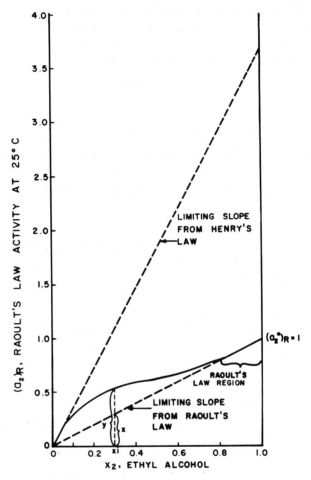

Fig. 2. $(a_2)_R$ versus X_2 for ethyl alcohol solution in water at 25°C.

The ratio of the Henry's law to the Raoult's law activity coefficient is an exponential function of the difference in standard states. One could use Eq. (15) to calculate one of the activity coefficients from a knowledge of the term on the right-hand side and one of the activity coefficients at some value of X_2. In the next section, however, we shall compile values of the Henry's law activity coefficient by the use of a standard technique. The solutions we are interested in are the aqueous solutions of the *n*-alkanols: methyl, ethyl, and *n*-propyl, and *n*-butyl.

V. Activities and Activity Coefficients for Some Alcohols in Water

The values for the Raoult's law activity coefficients $(\gamma_2)_R$ are taken from Butler et al. (*14*) and are given in Table 1 and Fig. 3. If we consider $(\gamma_2)_R$ as a measure of the relative fugacity or escaping tendency, we see that for any of the alcohols considered, the escaping tendency tends to increase as the average environment of any particular alcohol molecule

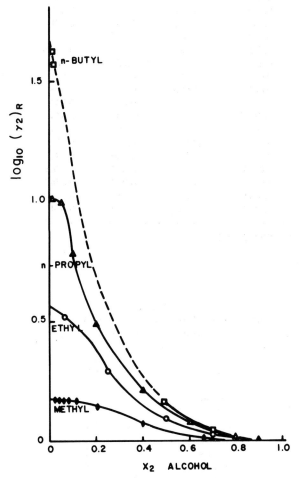

Fig. 3. $\text{Log}_{10}(\gamma_2)_R$ versus X_2, mole fraction of alcohol, for methyl, ethyl, *n*-propyl, and *n*-butyl alcohols in water at 25°C (*14*). (– – –) Nonmiscibility.

TABLE 1
ACTIVITY COEFFICIENTS IN ALCOHOL–WATER SOLUTIONS

X_2	$(\gamma_2)_H$	$(\gamma_2)_R$	X_2	$(\gamma_2)_H$	$(\gamma_2)_R{}^a$
	Methyl			Ethyl	
0.000	1.00	1.51	0.000	1.000	3.69
0.0202	0.998	1.505	0.064	0.896	3.310
0.0403	0.996	1.503	0.25	0.536	1.980
0.0620	0.993	1.498	0.50	0.338	1.247
0.0791	0.999	1 505	0.70	0.288	1.063
0.1145	0.983	1.482	1.00	0.271	1.00
0.2017	0.929	1.403			
0.3973	0.785	1.186		n-Propyl	
0.6579	0.682	1.029	0.000	1.000	14.4
0.8137	0.673	1.008	0.010	0.921	12.3
1.000	0.663	1.000	0.020	0.868	11.6
			0.050	0.742	9.92
	n-Butyl		0.100	0.454	6.05
0.000	1.000	52.9	0.200	0.234	3.12
0.010	0.827	42.7	0.400	0.122	1.63
0.0188	0.722	37.2	0.600	0.0887	1.19
0.4876	0.0278	1.44	0.800	0.0764	1.02
0.700	0.0213	1.10	0.900	0.0740	0.99
0.850	0.0195	1.01	0.950	0.0752	1.01
1.000	0.0193	1.000	1.000	0.0747	1.00

a $(\gamma_2)_R$ from Butler et al. (14).

TABLE 2
HENRY'S LAW CONSTANT IN WATER–ALCOHOL
SOLUTIONS AT 25°C (14, 15)

Alcohol	k (mm Hg)
Methyl	191
Ethyl	218
n-Propyl	291
n-Butyl	359
n-Amyl	532
n-Hexyl	649
n-Heptyl	798
n-Octyl	1020

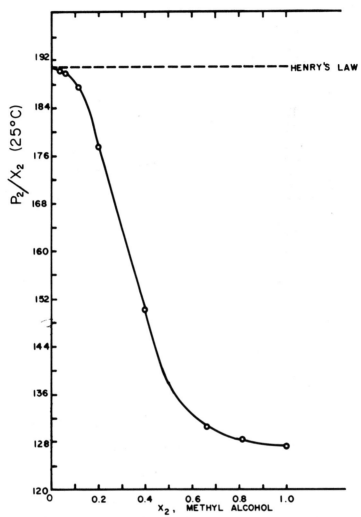

Fig. 4. P_2/X_2 versus X_2 for methyl alcohol in water at 25°C. Extrapolation to $X_2 = 0$ for the Henry's law constant k.

becomes more water-like. Thus $(\gamma_2)_R$ goes from a limiting value of 1 in pure alcohol to values greater than 1 as we increase the proportions of water. This effect becomes greater at any particular value of X_2 as we increase the number of methylene groups.

The Henry's law activity coefficients for the same alcohol solutions were calculated with the data of Butler et al. (*14, 15*). The value for k,

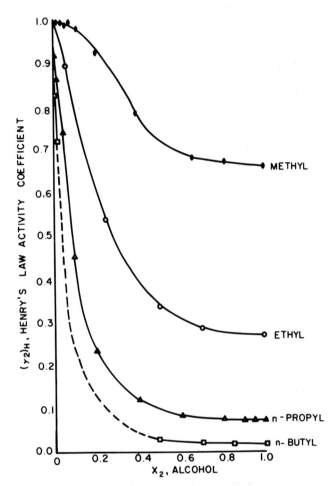

Fig. 5. $(\gamma_2)_H$ versus X_2 for methyl, ethyl, n-propyl, and n-butyl alcohols at 25°C. (– – –) Nonmiscibility.

the Henry's law constant, was obtained by extrapolating values of P_2/N_2 to $N_2 = 0$. An example of this procedure is shown in Fig. 4, where the data for methyl alcohol are shown. The values for k (or $F_{2,H}^{\circ}$) obtained for all the alcohols are presented in Table 2. The calculated values of $(\gamma_2)_H$ are tabulated in Table 2 and shown in Fig. 5. Here we observe the same relative variation of the chemical potential of the alcohols expressed in an opposite manner. The relative escaping tendency decreases as the average nature of an alcohol molecule's neighbors go from water-like to alcohol-like.

In the next section we will use the Henry's law activity coefficients to characterize the narcotic behavior of some n-alkanols.

VI. Thermodynamic Analysis of Biological Data

The thermodynamic condition for the equilibrium distribution of a biologically active species between two phases at constant temperature and pressure is

$$\bar{F}_{2,w} = \bar{F}_{2,x} \tag{16}$$

where the subscript w denotes the aqueous phase, x the unknown biophase, and 2 refers to the narcotic. If there were several intermediate phases between the aqueous phase and the active biophase we would have

$$\bar{F}_{2,w} = \bar{F}_{2,x''} = \bar{F}_{2,x'''} = \bar{F}_{2,x} \tag{17}$$

in which case Eq. (16) is still true.

If we express the partial molal free energy of the alcohol in terms of the relative fugacity or activity, we have a choice for the standard state in the aqueous phase. First, however, let us examine the situation in the biophase. We will make the assumption that when we have reached the isonarcotic activity in the aqueous phase, the mole fraction of the narcotic in the biophase is very small (or $X_2 \ll 1$). This leads to the fugacity of the narcotic being approximately proportional to its mole fraction. In other words, the solute approximately obeys Henry's law in the biophase. Then for some value of X_2 in the biophase we have

$$f_2 = k_2 X_2 = (F_2^\circ)_{\mathrm{H},x} X_2 \tag{18}$$

or equivalently

$$\bar{F}_{2,x} = F_{2,\ \mathrm{H},x}^\circ + RT \ln X_2 \tag{19}$$

If there are slight departures from Henry's law behavior, we may more generally use the expression

$$\bar{F}_{2,x} = F_{2,\ \mathrm{H},x}^\circ + RT \ln (a_2)_{\mathrm{H},x} \tag{20}$$

where $(a_2)_{\mathrm{H},x}$ is the Henry's law activity of the narcotic in the biophase. The nature of the thermodynamic proof of the validity of Henry's law insures that Eq. (20) is obeyed whether the solvent is a pure substance or is itself a solution (12).

Getting back to the aqueous phase, we will initially formulate the partial molal free energy of the narcotic in terms of some unknown standard state as

$$\bar{F}_{2,w} = F_{2,y,w}^{\circ} + RT \ln (a_2)_{y,w} \qquad (21)$$

Then equating Eqs. (21) and (20) according to Eq. (16) and rearranging, we have

$$\log_{10}(a_2)_{y,w} = \frac{[F_{2,\,\mathrm{H},x}^{\circ} - F_{2,y,w}^{\circ}]}{2.3RT} + \log_{10}(a_2)_{\mathrm{H},x} \qquad (22)$$

Ferguson's contribution to the Meyer–Overton theory of anesthesia was to use the activity concept for the narcotic (1). He chose the Raoult's law standard state for the narcotic in an aqueous solution. If we apply this choice to Eq. (20), we have at the minimum level of narcotic

$$\log_{10}(a_2)_{\mathrm{R},w} = \frac{[F_{2,\,\mathrm{H},x}^{\circ} - F_2^{\circ}]}{2.3RT} + \log_{10}(a_2)_{\mathrm{H},x} \qquad (23)$$

Ferguson noted that for the homologous series of n-alkyl alcohols $(a_2)_{\mathrm{R},w}$ was fairly constant with a tendency to increase slightly for the higher members. In terms of Eq. (23) we see that the relative constancy of $(a_2)_{\mathrm{R},w}$ may be due to one of two possibilities: (1) that both $[F_{2,\,\mathrm{H},x}^{\circ} - F_2^{\circ}]$ and $(a_2)_{\mathrm{H},x}$ remain constant or (2) that whatever change occurs for these terms, they nearly compensate each other. We will return to this question at a later point. For now, let us examine a different approach to the thermodynamic analysis, an approach that surface chemists use very frequently in the study of colloidal solutions.

First, we will use the Henry's law standard state for the narcotic in the aqueous solution. Rewriting Eq. (22) we have

$$\log_{10}(a_2)_{\mathrm{H},w} = \frac{[F_{2,\,\mathrm{H},x}^{\circ} - F_{2,\mathrm{H},w}^{\circ}]}{2.3RT} + \log_{10}(a_2)_{\mathrm{H},y} \qquad (24)$$

Second, we know from other studies of the distribution of polar organic molecules between an aqueous phase and some oil phase that the difference in unitary free energy may be approximately considered to be an additive function of the methylene group and the head polar group. For the n-alkyl alcohols we have the relation

$$[F_{2,\mathrm{H},x}^{\circ} - F_{2,\,\mathrm{H},w}^{\circ}] = \Delta F^{\circ} = N\Delta F(\text{—CH}_2\text{—}) + \Delta F(\text{—OH—}) \qquad (25)$$

where N is the number of methylene groups.* Combining Eqs. (24) and

* Strictly speaking N is the number of carbon atoms since we are ignoring the fact that there is always one methyl group present.

(25) we obtain

$$\log_{10}(a_2)_{\mathrm{H},w} = \frac{N\Delta F(\mathrm{-CH_2-})}{2.3RT} + K \qquad (26)$$

where

$$K = \frac{\Delta F(\mathrm{-OH-})}{2.3RT} + \log_{10}(a_2)_{\mathrm{H},x}$$

Then, if we examine the narcotic behavior of a homologous series of some narcotic, such as the alcohols, in terms of the Henry's law activity in the aqueous phase, a linear relation between $\log_{10}(a_2)_{\mathrm{H},w}$ and N will occur

TABLE 3
DESCRIPTION OF BIOLOGICAL STUDIES

Study	Description	Ref.
1	Luminescence in bacteria	16
2	Reflex response in tadpoles	16
3	Spontaneous movement of tadpoles	16
4	Nerve impulse block of stellate ganglion of cat synaptic pathway	16
5	Spontaneous contraction of frog heart	16
6	Enzymatic inhibition in goose erythrocytes	16
7	Response of frog gastrocnemius muscle	16
8	Nerve impulse block of stellate ganglion of cat nonsynaptic	16
9	Nerve impulse block of frog sciatic nerve	48
10	Same as 9	29

only if the mole fraction or Henry's law activity in the biophase is constant and the additivity of the unitary free energy of methylene groups is a reasonable approximation.

Values of $X_{2,w}$ were obtained primarily from the data collected by Brink and Posternak (16) and the isonarcotic activity $(a_2)_{\mathrm{H},w}$ was calculated by the use of the Henry's law activity coefficients. The value of $(\gamma_2)_{\mathrm{H}}$, for alcohols $n\text{-}C_5$ through $n\text{-}C_8$ with values of $X_2 < 10^{-3}$ was estimated to be approximately 1. $X_{2,w}$ is defined as the minimum concentration, expressed as the mole fraction, of an aqueous solution of alcohol molecules which produce some reversible depression of various biological activities. The description of biological effect measured in each study is given in Table 3.

The plot of $\log (a_2)_{\mathrm{H}}$ versus N for all ten studies is presented in Figs. 6–8. The linear relation between $\log_{10}(a_2)_{\mathrm{H},w}$ and N allows for the

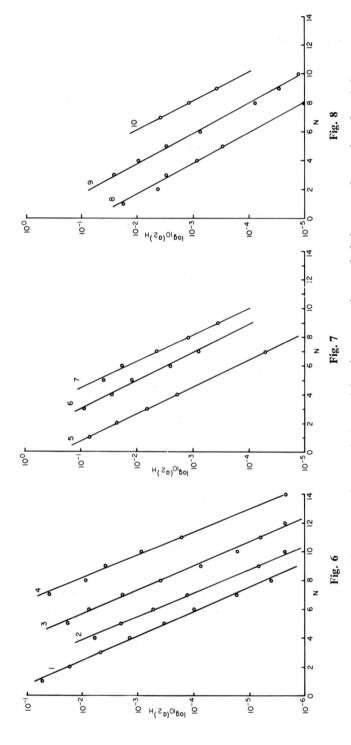

Figs. 6, 7, and 8. $Log_{10}(a_2)_H$, minimum narcotic Henry's law activity, for some primary alcohols versus the number of methylene groups N (see footnote, p. 362). Data from systems 1–10 (see Table 3).

calculation of $\Delta F(-CH_2-)$ and K from the slope and intercept, respectively, according to Eq. (26). These values are presented in Table 4. The linearity also supports the Meyer–Hemmi hypothesis, i.e., that equal narcotic effects are produced by equal concentrations in the biophase, except that we would modify equal concentrations to equal Henry's law activities.

The constancy of the Henry's law activity in the biophase for any given study with an homologous series of alcohols helps rationalize the

TABLE 4

VALUES OF $\Delta F(-CH_2-)$ AND K FROM NARCOTIC STUDIES

Study	$-\Delta F(-CH_2-)$ cal/mole	$-K$	N
1	770	0.7	1–8
2	790	1.2	2–5, 7–10
3	800	1.1	1–8
4	830	0.8	1–5, 8
5	700	0.6	1–4, 7
6	720	0.4	1–5
7	740	0.7	1–5
8	650	1.2	1–5, 8
9	650	1.2	1–4, 6–8
10	650	2.0	3–5

variation of the Raoult's law activity of the narcotic in the aqueous phase noted by Ferguson (1). If we refer back to Eq. (23) we note that a variation in $(a_2)_{R,w}$ with the number of methylene groups must be due to a corresponding variation of $[F^\circ_{2,H,x} - F^\circ_2]$. In fact, one would expect this to be a slowly varying additive function of N. The convenience of employing the Henry's law activity instead of the Raoult's law activity is that values of $\Delta F(-CH_2-)$ based on the Henry's law standard state in the aqueous phase have been compiled for various physical systems. Thus it may be possible to gain some insight into the nature of the biophase or active site by the comparison of values of $\Delta F(-CH_2-)$ obtained from the narcotic studies with those values obtained from what may be considered as model systems. This will be done in the next section with the full awareness of the limitations of thermodynamic analogies.

VII. Values of $\Delta F(-CH_2-)$ from Physical Systems

The best systems to examine for comparative values of $\Delta F(-CH_2-)$ are those involving a single phase; then there is no ambiguity associated with the standard state. In other words, the problem of how to divide the free energy into a unitary and cratic (17) part does not exist. Two cases satisfying these criteria are the solubility of a liquid organic molecule in water and micelle phenomena.

The simplest system concerning solubility is the solution of the liquid alkanes. The free energy of the process

$$C_N H_{2N+2}(aq) \rightleftarrows C_N H_{2N+2}(l) \tag{27}$$

can be calculated from the available data (18) for the solubility of liquid n-alkanes at 25°C. The total ΔF for Eq. (27) can be represented to within ± 0.04 kcal/mole for $N = 5, 6, 7$, and 8 by the following contributions (18): $\Delta F(-CH_2-) = -850$ cal/mole and $\Delta F(-CH_3-) = -2.18$ kcal/mole.

This value for $\Delta F(-CH_2-)$ for the n-alkanes agrees quite well with the value obtained from the solubility of the n-alkyl alcohols at 25°C. Table 5 shows the solubility of the n-alkanols in water at 25°C. The solubility

TABLE 5
SOLUBILITY OF n-ALKYL ALCOHOLS AT 25°C

Alcohol	X_2	$(\gamma_2)_H$	$(a_2)_{H,sat}$	Ref.
n-Butyl	1.8×10^{-2}	0.72	1.3×10^{-2}	15, 19
n-Pentyl	4.5×10^{-3}	1	4.5×10^{-3}	15, 19
n-Hexyl	1.1×10^{-3}	1	1.1×10^{-3}	15, 19
n-Heptyl	2.7×10^{-4}	1	2.7×10^{-3}	15, 19
n-Octyl	7.4×10^{-4}	1	7.4×10^{-4}	15, 19
n-Nonyl	1.74×10^{-5}	1	1.7×10^{-5}	19
n-Decyl	4.2×10^{-6}	1	4.2×10^{-6}	19

of the alcohols from n-butyl to n-octyl was determined by taking the average of the values obtained by Butler et al. (15) and Shinoda (19). The values of the solubility for alcohols n-nonyl and n-decyl were obtained by Shinoda (19). Figure 9 shows the plot of $\log (a_2)_{H,sat}$ against N. The

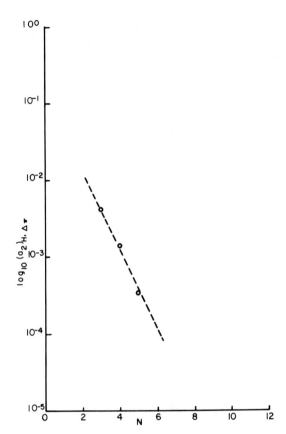

Fig. 9. $Log_{10}(a_2)_{H,sat}$, Henry's law activity, for saturated solutions of *n*-butyl, *n*-pentyl, *n*-hexyl, *n*-heptyl, *n*-octyl, *n*-nonyl, and *n*-decyl alcohols in water at 25°C versus N.

slope and intercept of this plot yielded $\Delta F(—CH_2—) = -810$ cal/mole and $\Delta F(—OH—) = 740$ cal/mole, respectively.

The calculation of $\Delta F(—CH_2—)$ from micelle phenomena requires that we assume that the micelle can be treated as a thermodynamic phase (*20*). Molyneux et al. (*18*) have recently reviewed values of the free energy increment per methylene group for various colloidal electrolytes. They found an average value of -650 ± 20 cal/mole for nine different types of amphiphile.

Another phenomenon of comparative interest is the effect of alcohols on the critical micelle concentration (CMC) of various colloidal electrolytes. Preliminary x-ray studies indicate that the alcohol molecules penetrate between the colloidal electrolyte in the micelle, producing a larger mixed micelle (21). Also there is tentative evidence that saturated aqueous solutions yield equal numbers of polar organic molecules in the micelle (21).

Studies of the effect of the primary alcohols with chain length of 2–10 carbon atoms on the CMC of fatty acid salts and dodecylammonium chloride have yielded two important empirical relations (22). The first is given by

$$\log_{10} \frac{dc}{dC_A} = 0.5N - I \tag{28}$$

where dc/dC_A is the rate of change of the CMC with alcohol concentration, N is the number of carbon atoms in the alcohol, and I is some constant. The second relation shows that the CMC is a linear function of alcohol concentration, which may be formulated as

$$\frac{C° - C}{C_A} = \frac{\Delta C}{C_A} = \frac{dC}{dA} \tag{29}$$

where $C°$, C, C_A are the CMC without alcohol, the CMC with alcohol, and the alcohol concentration, respectively. If we combine Eqs. (28) and (29) and solve for $\log C_A$, we have

$$\log_{10} C_A = -0.5N + I + \Delta C \tag{30}$$

where ΔC is some constant decrease in the CMC. If we arbitrarily choose a value of ΔC that is sufficiently small, C_A effectively may be set equal to the Henry's law activity $(a_2)_H$, giving us

$$\log_{10}(a_2)_H = -0.5N + K' \tag{31}$$

where $K' = I + \Delta C$.

Equation (31) is derivable in terms of $\Delta F(-CH_2-)$ by assuming that equal activities of the alcohol molecule in the micelle phase cause an equal lowering of the CMC. Following the same procedure used for the derivation of Eq. (26) we have

$$\log (a_2)_{H,\Delta C} = \frac{\Delta F(-CH_2-)N}{2.3RT} + K'' \tag{32}$$

where K'' is a new constant and $(a_2)_{H,\Delta C}$ is the Henry's law activity of the alcohol in the aqueous phase producing some small constant decrease in the CMC for members of a homologous series of straight-chain alcohols. Evaluating $\Delta F(-CH_2-)$ from Eqs. (31) and (32) yields -680 cal/mole (20°C).

TABLE 6

VALUES OF $\Delta F(\text{—CH}_2\text{—})$ AND $\Delta F(\text{—OH—})$ FOR VARIOUS SYSTEMS (25°C)

	$\Delta F(\text{—CH}_2\text{—})$ (cal/mole)	$\Delta F(\text{—OH—})$ (cal/mole)	Ref.
Oil/water phases			
n-Alkanes	−850	—	18
n-Alkanols	−810	770	15, 19
Micelle/water			
n-Alkyl amphiphiles	−650	—	18
ΔCMC by n-alkanols	−680	—	22
n-Alkanols at interfaces			
Petroleum ether/water	−820	−700	23
Air/water	−750	−580	24

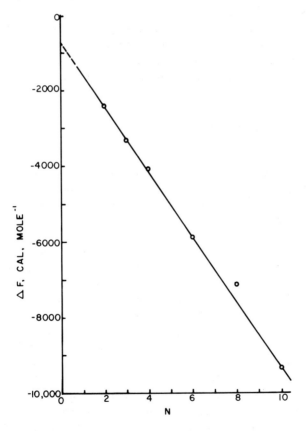

Fig. 10. Free energies of adsorption ΔF at a petroleum ether/water interface versus N for ethanol, 1-propanol, 1-butanol, 1-hexanol, 1-octanol, and 1-decanol at 20°C (23).

369

Finally, we record in Table 6 values of $\Delta F(\text{—CH}_2\text{—})$ calculated from the transfer of the alcoholic methylene groups from extremely dilute aqueous solutions to two different interfaces—the petroleum ether/water and air/water. It is assumed here that the effect of concentration at the interface is negligible. Figure 10 shows the values of the total unitary free energy of adsorption at a petroleum ether/water interface for the n-alkyl alcohols (ethyl, butyl, hexyl, octyl) based on the Henry's law standard state in the aqueous phase, computed by Haydon and Taylor (23). The slope and intercept yield the values $\Delta F(\text{—CH}_2\text{—}) = -820$ cal/mole and $\Delta F(\text{—OH—}) = -700$ cal/mole, respectively. Note that the value for $\Delta F(\text{—CH}_2\text{—})$ at the petroleum ether/water interface is quite similar to the value found for the n-alkyl alcohol/water phase distribution, but the value for $\Delta F(\text{—OH—})$ changes from -700 cal/mole for the former case to 770 cal/mole for the latter case.

Similar values are obtained for the adsorption at the aqueous solution/air interface. Here we have used values determined by the study of surface tension of the n-alkyl alcohols ($N = 2$–8) at 25°C by Posner, Anderson, and Alexander (24). These are also presented in Table 6.

VIII. Comparison of $\Delta F(\text{—CH}_2\text{—})$

The primary conclusion one can draw from Table 6 is that the transfer to the micelle phase involves a substantially less negative value of $\Delta F(\text{—CH}_2\text{—})$ than for both the bulk oil phase or oil interface (18). Also, it appears that $\Delta F(\text{—OH—})$ is a much more sensitive probe for distinguishing between an interfacial or bulk phase distribution. If we compare these $\Delta F(\text{—CH}_2\text{—})$ with the calculated $\Delta F(\text{—CH}_2\text{—})$ obtained from the narcotic studies, we see that there is a suggestion of additional information concerning the active site in the biophase. The narcotic $\Delta F(\text{—CH}_2\text{—})$ listed in Table 4 appears to fall into two groups: one that ranges about -750 cal/mole and the other at -650 cal/mole. The latter value was consistently obtained with systems involving the nerve impulse block of isolated nerve or muscle fiber, while the former and higher $\Delta F(\text{—CH}_2\text{—})$ were obtained from studies that dealt with either a complete organism or organ. The problem of sorting out values of $\Delta F(\text{—OH—})$ for the narcotics is twofold: (1) the term $RT \ln (a_2)_{\text{H},x}$ must be subtracted from K, the constant in Eq. (26), in order to obtain $\Delta F(\text{—OH—})$; (2) the use of polyhydric compounds as an alternative approach has only been tried with the lower molecular weight substances, such as glycerol or glycol, which are completely inactive (*25*).

The value of $\Delta F(-CH_2-)$ obtained for the isolated nerve or muscle fiber under reversible block is quite similar to that obtained in the micelle studies. While a micelle is at best a crude model of an excitable cell membrane, this correspondence of $\Delta F(-CH_2-)$ is of special interest because of other physical evidence that provides additional support. Very pertinent spectroscopic information is obtained from the spin label experiments of Hubbell and McConnell (*26*). They found that compounds such as butanethiol, octylamine, and tetracaine increased the solubility of their organic spin-labeled compound to about the same extent in *both* phospholipid vesicles and excitable membranes. They interpreted this as an indication that these molecules, e.g., butanethiol, caused a local disordering or "melting" of the hydrophobic regions of the membranes which increased the volume available to the spin labeled compound. Another way of summarizing their findings is that they found evidence that the vagus nerve of the rabbit and the excitable membrane of muscle contained liquid-like hydrophobic regions of low viscosity very similar to that found in phospholipid vesicles. Hubbell and McConnell also performed similar experiments with a limited number of micelle solutions (*26*). Their results were only qualitative but there was indication that the spectra obtained with micelle solutions were similar to those found for both the aqueous phospholipid dispersions and isolated nerve tissue.

Another important model system used in narcotic studies is the water-insoluble monlayer at the air/liquid interface. Studies by Skou (*27–30*), Gershfeld (*31, 32*), Clements and Wilson (*33, 34*), and Hersh (*35*) all indicate a definite correlation of biological activity with a change in surface tension at various lipid monolayer/aqueous solution interfaces. Unfortunately, there is only one limited study by Skou (*29*) involving a homologous series. Skou used a monolayer of lipoids extracted quantitatively from sciatic frog nerves and several local anesthetics including

TABLE 7

HENRY'S LAW ACTIVITY OF SOME ALCOHOLS CAUSING A SURFACE PRESSURE INCREASE OF 9.3 dynes/cm AT A LIPOID MONOLAYER/AQUEOUS SOLUTION INTERFACE (*20*)

Alcohol	Mmoles/ liter	X_2	$(\gamma_2)_H$	$(a_2)_{H,\Delta\pi}$
1-Propanol	0.240	4.30×10^{-3}	0.973	4.2×10^{-3}
1-Butanol	0.080	1.44×10^{-3}	0.986	1.4×10^{-3}
1-Pentanol	0.019	0.34×10^{-3}	1.0	3.4×10^{-4}

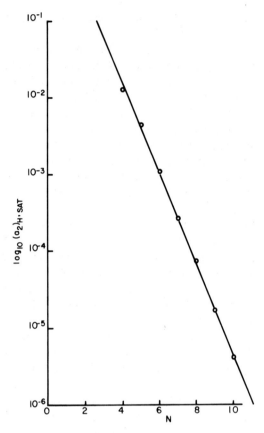

Fig. 11. $\text{Log}_{10}(a_2)_{\text{H},\Delta\pi}$, the Henry'a law activity at $\Delta\pi = 9.3$ dynes/cm versus N for n-propyl, n-butyl, and n-pentyl alcohols at 22°C (8, 29).

3 primary alcohols, 3-propyl, n-butyl, and n-amyl. By using concentrations of these alcohols yielding a surface pressure increase $\Delta\pi$ of 9.3 dynes/cm, the Henry's law activity may be calculated as shown in Table 7 and the relation

$$(a_2)_{\text{H},\Delta\pi} = \frac{\Delta F(-\text{CH}_2-)}{2.3RT} + K''' \tag{33}$$

may be derived in a manner similar to the derivation of Eq. (32). Here, $(a_2)_{\text{H},\Delta\pi}$ is the Henry's law activity of the alcohol at the point where $\Delta\pi = 9.3$ dynes/cm. The plot of $\text{log}_{10}(a_2)_{\text{H},\Delta\pi}$ versus N is shown in Fig. 11.

The value of the slope yields an approximate value for $\Delta F(\text{—}CH_2\text{—})$ of -700 cal/mole which is fairly close to the $\Delta F(\text{—}CH_2\text{—})$ obtained with the narcotic studies involving isolated nerve and muscle.

IX. Anesthesia and Hydrophobic Bonding

The listing of $\Delta F(\text{—}CH_2\text{—})$ for the anesthetic studies points to the connection between the narcotic phenomena and so-called "hydrophobic bonding." The tendency of nonpolar groups of proteins to cluster together in aqueous environments is loosely characterized as hydrophobic bonding (36). There is no bond in the true sense of the word; the main purpose of the term is to emphasize that the source of immiscibility is the change in standard entropy rather than enthalpy (37).

The model systems used by Kauzmann (36) to develop the concept of hydrophobic bonding in proteins are quite similar to those considered as model systems for narcosis: (1) the solubility of an organic molecule such as an alcohol in water and (2) the distribution of a small organic molecule (about 4 carbons) between an organic solvent and water. The standard molal or unitary free energy change for these systems, considering the direction as water to oil phase, is highly negative but the *enthalpy* and *entropy* are positive. Therefore, according to

$$\Delta F^\circ = \Delta H - T\Delta S^\circ \tag{34}$$

the negative value of ΔF° is "determined" by the positive ΔS°. Values of ΔS° for the vaporization of certain small organic molecules from aqueous solutions were also used by Frank and Wen (38) to support their picture of "icebergs" of water surrounding the nonpolar segments of organic molecules in water. These ideas were subsequently theoretically developed by several authors (39, 40). It is not our intention to review the subject of hydrophobic bonding; that is presented in another chapter. The question we want to raise is whether the negative values of $\Delta F(\text{—}CH_2\text{—})$ derived from the narcotic or anesthetic studies are also primarily due to entropy effects.

Unfortunately, there are no experimental studies to which we can refer. There have been a few investigations concerned with the temperature coefficient of the activity or concentration of a particular narcotic (41–44). The problem with this approach is that one has three differential terms to consider. This is shown by taking the partial derivative of Eq. (26) with respect to temperature. Thus if we consider the alcohols as an example we obtain

$$\frac{\partial \log_{10}(a_2)_{\mathrm{H},w}}{\partial T} = \frac{1}{2.3RT}\frac{\partial}{\partial T}(N\Delta F(-\mathrm{CH_2}-) + \Delta F(-\mathrm{OH}-)$$

$$+ \frac{\partial \log(a_2)_{\mathrm{H},x}}{\partial T} - \frac{\Delta F^\circ}{2.3RT^2} \quad (35)$$

A much simpler approach is to examine some graded, quantitative response (44) for a homologous series at several different temperatures. Then it is a simple matter to calculate $\Delta S(-\mathrm{CH_2}-)$ and $\Delta H(-\mathrm{CH_2}-)$ from the relations

$$\Delta S(-\mathrm{CH_2}-) = -\partial \frac{\Delta F(\mathrm{CH_2})}{\partial T} \quad (36)$$

and

$$\Delta H(-\mathrm{CH_2}-) = \Delta F(-\mathrm{CH_2}-) + T\Delta S(-\mathrm{CH_2}) \quad (37)$$

At the present time the only information available for establishing whether narcosis or anesthesia is related to hydrophobic bonding comes from studies concerning model systems. There have been several recent studies of the thermodynamic behavior of the n-alkanols in water, which tend to show a decrease in the importance of ΔS° and the corresponding increase of the ΔH term as the carbon chain increases (45, 46). Corkill et al. (46) have concluded from their calorimetric study of heats of solution of some 1-alkanols in water that the amount of structure produced in water by the alkyl chain is greater for that part of the chain within the "sphere of action" of the polar group than for that outside. If this increase in importance of ΔH in determining ΔF according to Eq. (34) is also true for narcosis, then approaches to the anesthetic action, which employs molecular interaction parameters such as the work of Agin et al. (47), would have thermodynamic support. The question of whether hydrophobic bonding is pertinent to anesthesia is extremely important and should be attacked directly.

X. Summary

The Henry's law constant was introduced as the standard state for a homologous series of alcohols. Values of $\Delta F(-\mathrm{CH_2}-)$, based on a Henry's law standard state, were calculated from the narcotic behavior of these same alcohol homologs and were compared with $\Delta F(-CH_2-)$ obtained in model systems. A similarity in $\Delta F(-CH_2-)$ was noted for the narcotic studies involving isolated muscle and nerve, and those concerned with micellar solutions. Also, attention was directed to a possible connection between hydrophobic bonding and cellular narcosis.

REFERENCES

1. Ferguson, J., *Proc. Roy. Soc. (London)* **B127**, 387 (1939).
2. Ferguson, J., Symposium: *Mecanisme de la Narcose*, CNRS Paris, 1951, p. 25.
3. Miller, S. L., *Proc. Nat. Acad. Sci. U.S.* **47**, 1515 (1961).
4. Pauling, L., *Science* **134**, 15 (1961).
5. Seeman, P., *Int. Rev. Neurobiol.* **9**, 145 (1966).
6. Ritchie, J. M., and Greengard, P., *Ann. Rev. Pharmacol.* **6**, 405 (1966).
7. Butler, T. C., *Pharmacol. Rev.* **2**, 121 (1950).
8. Mullins, L. J., *Chem. Rev.* **54**, 289 (1954).
9. Meyer, H. H., *Arch. Exp. Path. Pharmak.* **42**, 119 (1899).
10. Overton, E., *Studien uber die Narkose*, Fischer, Jena, 1901.
11. Meyer, K. H., and Hemmi, H., *Biochem. Z.* **277**, 39 (1935).
12. Lewis, G. N., and Randall, M. (revised by K. S. Pitzer and L. Brewer), *Thermodynamics*, McGraw-Hill, New York, 1961, pp. 204, 232.
13. Shaw, R., and Butler, J. A. V., *Proc. Roy. Soc. (London)* **A129**, 519 (1930).
14. Butler, J. A. V., Thomson, D. W., and MacLennan, W. H., *J. Chem. Soc.* **1933**, 674 (1933).
15. Butler, J. A. V., Ramchandani, C. N., and Thomson, D. W., *J. Chem. Soc.* **1935**, 380 (1935).
16. Brink, F., and Posternak, J. M., *J. Cell. Comp. Physiol.* **32**, 211 (1948).
17. Gurney, R. W., *Ionic Processes in Solution*, Dover, New York, p. 88.
18. Molyneux, P., Rhodes, C. T., and Swarbrick, J., *Trans. Faraday Soc.* **61**, 1043 (1965).
19. Kinoshita, K., Ishikawa, H., and Shinoda, K., *Bull. Chem. Soc. (Japan)* **31**, 1081 (1958).
20. Hutchinson, E., Inaba, A., and Bailey, L. G., *Z. Phys. Chem. (Frankfurt)* **5**, 344 (1955).
21. Harkins, W. D., Mattoon, R. W., and Mittelman, R., *J. Chem. Phys.* **15**, 763 (1947).
22. Herzfeld, S. H., Corrin, M. L., and Harkins, W. D., *J. Phys. Chem.* **54**, 271 (1950).
23. Haydon, D. A., and Taylor, F. H., *Phil. Trans.* **252**, 225 (1960).
24. Posner, A. M., Anderson, J. R., and Alexander, A. E., *J. Colloid Sci.* **7**, 625 (1952).
25. Agin, D., Hersh, L. S., and Holtzman, D., unpublished data.
26. Hubbell, W. L., and McConnell, H. M., *Proc. Nat. Acad. Sci. U.S.* **61**, 12 (1968).
27. Skou, J. C., *Acta Pharmacol. Toxicol.* **10**, 317 (1954).
28. Skou, J. C., *Acta Pharmacol. Toxicol.* **10**, 325 (1954).
29. Skou, J. C., *Biochim. Biophys. Acta* **30**, 625 (1958).
30. Skou, J. C., *J. Pharm. Pharmacol.* **13**, 204 (1961).
31. Gershfeld, N. L., *J. Phys. Chem.* **66**, 1923 (1962).
32. Shanes, A. M., and Gershfeld, N. L., *J. Gen. Physiol.* **44**, 345 (1960).
33. Clements, J. A., and Wilson, K. M., *Proc. Nat. Acad. Sci. U.S.* **48**, 1008 (1962).
34. Clements, J. A., and Wilson, K. M., *Int. Anestiol. Clin.* **1**, 969 (1963).

35. Hersh, L. S., *Mol. Pharmacol.* **3** (6) 581 (1967).

36. Kauzmann, W., *Advan. Prot. Chem.* **14,** 1 (1959).

37. Nemethy, G., Sheraga, H. A., and Kauzmann, W., *J. Phys. Chem.* **72,** 1842 (1968).

38. Frank, H. S., and Evans, M. W., *J. Chem. Phys.* **13,** 507 (1945).

39. Nemethy, G., *Angew. Chem.* **6** 195 (1967).

40. Stillinger, F. H. Jr., and Ben-Naim, A., *J. Phys. Chem.* **73,** 900 (1969).

41. Cherkin, A., and Catchpool, J. F., *Science* **144,** 1460 (1964).

42. Horowitz, S. B., and Fenichel, I. R., *Ann. N.Y. Acad. Sci.* **125,** 572 (1965).

43. Eger, E. I., II, Saidman, L. J., and Brandstater, B., *Anesthesiology* **26,** 764 (1965).

44. Fenichel, I. R., and Horowitz, S. B., *Science* **152,** 1110 (1966).

45. Benjamin, L., *J. Phys. Chem.* **68,** 3575 (1964).

46. Corkill, J. M., Goodman, J. F., and Tate, J. R., *Trans. Faraday Soc.* **65,** 1742 (1969).

47. Agin, D., Hersh, L. S., and Holtzman, D., *Proc. Nat. Acad. Sci. U.S.* **53,** 952 (1962).

48. Laget, P., Posternak, J. M., and Mangold, R., Symposium: *Mecanisme de la Narcose*, CNRS, Paris, 1951.